MATHEMATICS FOR ENGINEERS

Volume 2

Integral Calculus, Taylor and Fourier Series,
Calculus for Multivariable Functions, 1st Order
Differential Equations, Laplace Transform

MATHEMATICS FOR ENGINEERS

Volume 2

Integral Calculus, Taylor and Fourier Series,
Calculus for Multivariable Functions, 1st Order
Differential Equations, Laplace Transform

Thomas Westermann

University of Applied Sciences Karlsruhe, Germany

NEW JERSEY · LONDON · SINGAPORE · BEIJING · SHANGHAI · HONG KONG · TAIPEI · CHENNAI

Published by

World Scientific Publishing Co. Pte. Ltd.

5 Toh Tuck Link, Singapore 596224

USA office: 27 Warren Street, Suite 401-402, Hackensack, NJ 07601

UK office: 57 Shelton Street, Covent Garden, London WC2H 9HE

Library of Congress Control Number: 2024052580

British Library Cataloguing-in-Publication Data
A catalogue record for this book is available from the British Library.

Translation from the German language edition:
Mathematik für Ingenieure by Prof Dr Thomas Westermann
Copyright © Springer-Verlag Berlin Heidelberg 2020

MATHEMATICS FOR ENGINEERS
Volume 2: Integral Calculus, Taylor and Fourier Series, Calculus for Multivariable Functions,
1st Order Differential Equations, Laplace Transform

ISBN 978-981-12-9921-6 (hardcover)
ISBN 978-981-98-0078-0 (paperback)
ISBN 978-981-12-9922-3 (ebook for institutions)
ISBN 978-981-12-9923-0 (ebook for individuals)

For any available supplementary material, please visit
https://www.worldscientific.com/worldscibooks/10.1142/14018#t=suppl

Desk Editor: Tan Rok Ting

Preface

This second volume in our series is intended primarily as a companion text for the second semester mathematics preliminaries for students and lecturers of electrical engineering and other engineering disciplines.

In a clear and concise manner, and without too much abstraction, it introduces students to the topics covered in the basic mathematics lectures. Volume 2 also provides students at universities and applied universities with a largely accurate, but always illustrative, presentation as a practical aid to entry into higher mathematics.

On the one hand, integral calculus plays a central role in undergraduate and graduate engineering courses. This topic is the focus of the second volume. On the other hand, functions of more than one variable are needed to describe real world behavior. Partial derivatives and related constructions are central to this topic. The transition to the third semester is the introduction to first-order differential equations and the Laplace transform.

Mathematical concepts are clearly motivated, systematically equated and visualized in many animations. Mathematical proofs are almost completely avoided. Instead, many applications not only support the application of mathematics, but also contribute to a better understanding of mathematics.

The successful acceptance of the German book and the positive feedback have led us to transfer the presentation and concept of the 8th edition as far as possible to the English lecture notes.

Important formulas and statements are clearly highlighted in order to increase the readability of the books. More than 300 images and sketches support the character of modern textbooks. The color-coded layout provides a clear overview of the presentation of the content, e.g. by adding new terms and definitions in light grey, important statements and sentences in grey.

There are additional styles to make the book easier to read:

- The symbol ⚠ **Caution:** draws your attention to passages that are often misinterpreted, overlooked or ignored.
- Tips and rules help you work through the examples and exercises.
- Definitions and important phrases are highlighted in grey boxes.
- Numerous summaries are highlighted in color.
- Important formulas and results are marked.
- Examples and applications are clearly arranged throughout the text.
- 380 fully worked examples,
- over 360 problems with solutions,
- and more than 200 illustrations and sketches will help you very well for **study and prepare for exams.**

Alongside the topics covered in the book, additional material is available on the website, as well as MAPLE worksheets that can be downloaded for the current version of MAPLE. The description can be found under the MAPLE tab on the book's website:

$$\textit{http://www.imathhome.de/books/mathe/start.htm}$$

In the book, the following two symbols explicitly refer to additional information information that can be found on the home page:

① Animations, in gif format are available: By clicking on the appropriate location on the web the animations are played through the Internet browser.

② References indicate the MAPLE descriptions. All MAPLE worksheets are available on the website. An overview of all worksheets can be found in *index.mws.*

I would especially like to thank Mayur Shelke for his valuable and intensive help in translating the German book into an adequate English textbook. I would also like to thank the publisher for the careful proofreading and excellent fine-tuning of the English version. Special thanks go to Ms. Andrea Wolf and Ms. Rok Ting of World Scientific Publishing, who made it possible for me to publish these lecture notes in this important and renowned publishing house.

Karlsruhe, July 2024   *Thomas Westermann*

Table of Contents

Chapter 8
Integral Calculus

Historically, the integral operation, like the derivative of a function, is motivated by the physical description of a motion (path, velocity and acceleration). But many advanced topics in mathematics, such as the Laplace and Fourier transforms and the formulas for solving first-order differential equations, are also based on the integral concept. Integration is also important for calculating areas, volumes of bodies, center of gravity, etc.

8

8 Integral Calculus

Historically, the integral operation, like the derivative of a function, is motivated by the physical description of a motion (path, velocity and acceleration). But many advanced topics in mathematics, such as the Laplace and the Fourier transforms as well as solving first-order differential equations, are also based on the integral concept. Integration is also important for calculating areas, volumes of bodies, center of gravity, etc.

8.1 Integration

Let us start with a geometric problem: Given is a non-negative function $f(x)$. What is the area under the curve with the x-axis within the interval $[a, b]$?

Example 8.1. To find the area A_0^b under the graph of the function $f(x) = x^2$ in the range $[0, b]$, we divide the interval $[0, b]$ into n sub-intervals by a division Z_n

$$x_0 = 0, \; x_1 = \frac{b}{n}, \; x_2 = 2\frac{b}{n}, \ldots, \; x_{n-1} = (n-1)\frac{b}{n}, \; x_n = b.$$

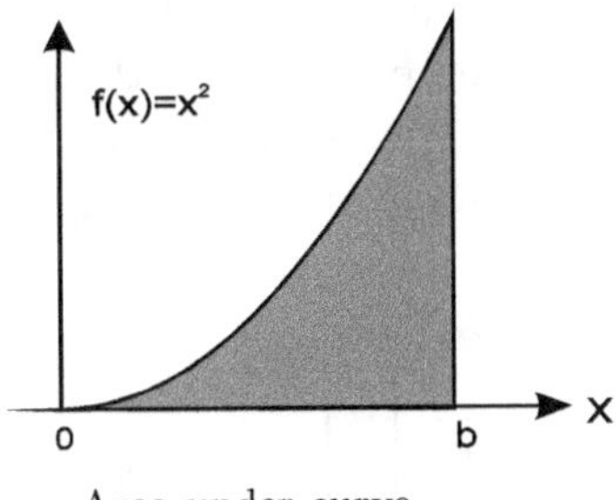

Area under curve

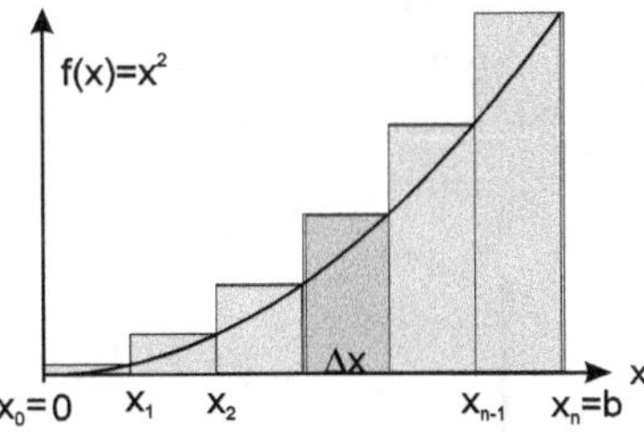

Approximation by rectangles

For each sub-interval of interval length $\Delta x = x_k - x_{k-1} = \frac{b}{n}$ we select the right point $x_k = k \cdot \Delta x$ and evaluate the function

$$f(x_k) = x_k^2 = (k \cdot \Delta x)^2.$$

The area of the corresponding rectangle is

$$\Delta x \cdot f(x_k) = \Delta x \cdot k^2 \cdot \Delta x^2.$$

Then we add all the rectangle areas to

$$S_n = \Delta x\, f(x_1) + \Delta x\, f(x_2) + \cdots + \Delta x\, f(x_n)$$

$$= \sum_{k=1}^{n} \Delta x\, f(x_k) = (\Delta x)^3 \sum_{k=1}^{n} k^2.$$

According to Volume 1 (Section 1.2.2: Mathematical Induction, Problem 1.5), this sum can be simplified to

$$\sum_{k=1}^{n} k^2 = \frac{1}{6}\, n\,(n+1)\,(2n+1)$$

$$\Rightarrow S_n = (\Delta x)^3 \frac{1}{6}\, n\,(n+1)\,(2n+1) = \frac{b^3}{n^3}\frac{1}{6}\, n\,(n+1)\,(2n+1) \overset{n\to\infty}{\longrightarrow} \frac{b^3}{3}.$$

By refining the decomposition Z_n of the interval, the area below $f(x) = x^2$ is approximated with arbitrary precision. For $n \to \infty$, the subtotal S_n converges to the area below the graph of x^2: $A_0^b = \frac{b^3}{3}$. $\square$

The procedure from this tutorial is generalized by using the following construction to calculate the area under any $f(x)$ curve with the x-axis: First, the curve is approximated by a piecewise constant function (step function). All the rectangles are added to give an approximation of the area under the curve. Finally, by increasing the number of subdivisions the function is approximated more and more precisely by the step function. So the area under the curve is approximated by the sum of the rectangle areas. The following definition for the definite integral is more precise:

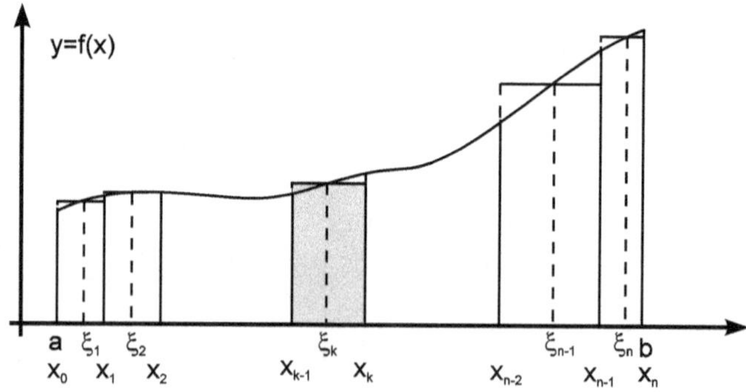

Figure 8.1. Subtotal

Definition: (Definite integral; Riemann Integral)
Given is a continuous function $f : [a,\ b] \to \mathbb{R}$ with $y = f(x)$.

(1) Let Z_n be an equidistant division of the interval $a \le x \le b$ into n sub-intervals

$$a = x_0 < x_1 < x_2 < \cdots < x_{n-1} < x_n = b.$$

The lengths are $\Delta x_k = x_k - x_{k-1} = \frac{b-a}{n}$. Let $\xi_k \in [x_{k-1},\ x_k]$ be any intermediate point in the interval. Then

$$S_n = \sum_{k=1}^{n} \Delta x_k\, f(\xi_k)$$

is the Riemann subtotal with respect to the decomposition Z_n.

(2) The definite integral (Riemann Integral) of the continuous function f in the range from $x = a$ to $x = b$ is the limit of the Riemann subtotal S_n for $n \to \infty$:

$$\int_a^b f(x)\, dx := \lim_{n \to \infty} \sum_{k=1}^{n} \Delta x_k\, f(\xi_k).$$

 **Visualization:** More illustrative than any precise mathematical definition is a descriptive interpretation. The animation shows the transition from the subtotal to the integral by increasing the number of subdivisions of the $[a,\ b]$ interval. This animation shows the convergence from the discrete subtotal to the integral. The two figures below show the approximation of the integral $\int_0^1 \left(x^2 + 1\right) dx$ for a subdivision of $N = 50$ (left) and $N = 100$ (right).

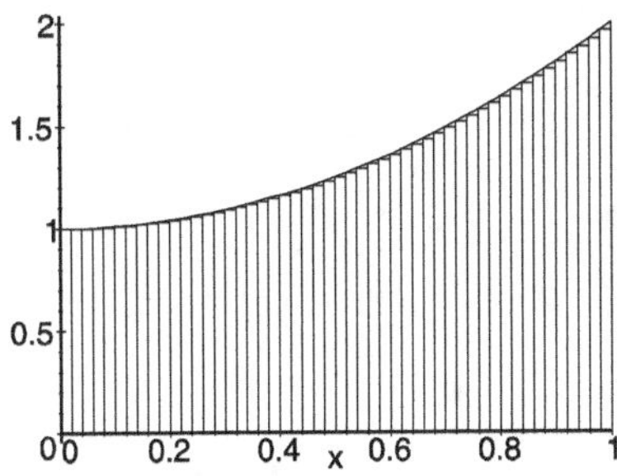
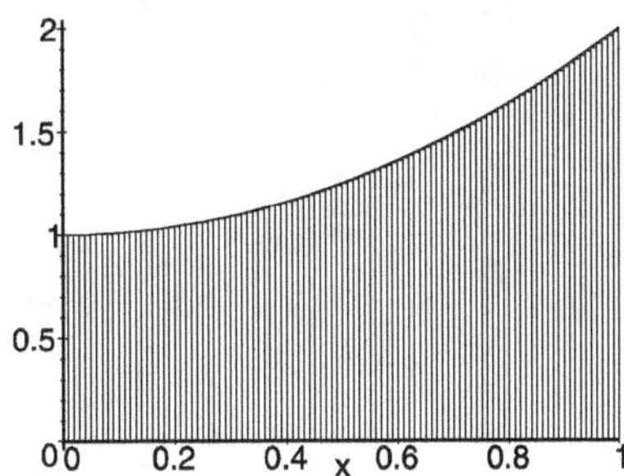

Remarks:

(1) This integral concept was introduced by the mathematician *Riemann* $(1826 - 1866)$ and is therefore called the Riemann Integral. Since this is the only integral we are concerned with, we will refer to it simply as the integral.

(2) If f is a continuous function, then for any subdivision Z_n with $\Delta x_k = x_k - x_{k-1} \xrightarrow{n \to \infty} 0$ and any choice of $\xi_k \in [x_{k-1}, x_k]$, the subtotal converges to the same value. We say that the integral is *well-defined*.

(3) More generally, a function can be integrated if, for any given subdivision Z_n and any $\xi_k \in [x_{k-1}, x_k]$, the subtotal converges to the same value.

(4) This algebraic definition of the integral corresponds exactly to the procedure for calculating the area of the first example for a non-negative function f. However, the algebraic definition is more general and therefore goes beyond the area calculation.

(5) There are common names for the symbols occurring in the integral $\int_a^b f(x)\,dx$: x: *integral variable;* $f(x)$: *integrand;* a: *lower limit;* b: *upper limit.*

Example 8.2. A mass moves with the velocity $v(t)$ along the x-axis. At time $t = 0$ it is at position $x = 0$. Find the distance $x(T)$ at time $t = T$.

If the velocity is constant, $v(t) = v_0$, then the distance is $x(T) = v_0\,T$. For a non-constant velocity $v(t)$, we divide the time interval $[0, T]$ into sub-intervals $0 = t_0 < t_1 < t_2 < \cdots < t_n = T$. In each sub-interval $v(t)$ is assumed to be constant:

$$v(t) \approx v(t_{k-1}) \qquad \text{for } t \in [t_{k-1}, t_k] \quad k = 1, \ldots, n.$$

Then the distance $x(t_k)$ at the time $t = t_k$ $(k = 1, 2, \ldots, n)$ is approximated by

$$x(t_1) \approx (t_1 - t_0) \cdot v(t_0) = \Delta t_1\, v(t_0)$$
$$x(t_2) \approx x(t_1) + (t_2 - t_1) \cdot v(t_1) = x(t_1) + \Delta t_2\, v(t_1)$$
$$x(t_3) \approx x(t_2) + (t_3 - t_2) \cdot v(t_2) = \Delta t_1\, v(t_0) + \Delta t_2\, v(t_1) + \Delta t_3\, v(t_2)$$
$$\vdots$$
$$x(t_n) = x(T) \approx \sum_{k=1}^{n} \Delta t_k\, v(t_{k-1}).$$

The approximate value obtained for $x(T)$ is therefore the Riemann sub-sum S_n. The exact value of the distance is obtained by taking $n \to \infty$, which gives the definite integral

$$x(T) = \int_0^T v(t)\, dt.$$

$\square$

⊛ The Indefinite Integral

For a positive function the definite integral $\int_a^b f(t)\, dt$ represents the area between the curve $f(t)$ and the time axis. If we consider the lower limit as fixed, the upper limit as flexible, the value of the integral depends on this upper limit.

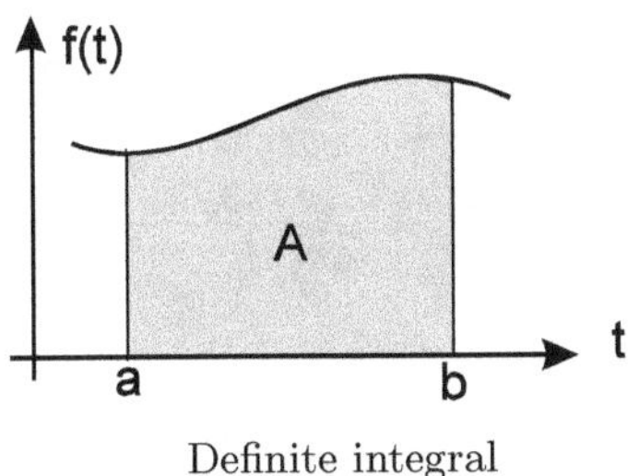

Definite integral

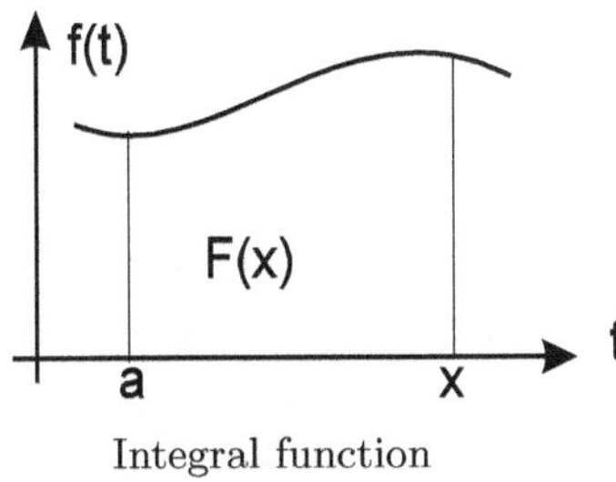

Integral function

To symbolize the dependence on the upper limit, we replace b by x and define a function $F(x) := \int_a^x f(t)\, dt$:

Definition: (Indefinite Integral, Integral Function)
The indefinite integral

$$F(x) := \int_a^x f(t)\, dt$$

is the integral function $F(x)$, for which the upper limit of the integral is variable.

For the indefinite integral $F(x) = \int_a^x f(t)\, dt$, the area between the function $f(t)$ and the t-axis depends on the upper limit.

Example 8.3. If we choose $f(t) = t^2$ and $a = 0$, then the corresponding integral function according to Example 8.1 is the function $F(x) = \frac{1}{3}x^3$. There is a relationship between the integral function and its integrand: $F'(x) = f(x)$, which is generally true, as shown in Section 8.2. □

⊗ Numerical Integration

Unfortunately, we have to admit that sometimes even relatively simple functions cannot be integrated elementarily. Examples are e^{-x^2} or $\frac{\sin x}{x}$. In these cases it is necessary to use numerical methods: According to the definition of the definite integral, we divide the interval $[a, b]$ into n sub-intervals $[x_i, x_{i+1}]$ with the interval length $h := \frac{b-a}{n}$ and set

$$x_0 = a; \quad x_{i+1} = x_i + h \quad (i = 0, \ldots, n-1); \quad x_n = b.$$

If the function $f(x)$ to be integrated is replaced in each interval $[x_i, x_{i+1}]$ by a constant $f(\xi_{i+1})$, $\xi_{i+1} \in [x_i, x_{i+1}]$, the integral is approximated by the subtotal

$$S_n \approx \sum_{i=0}^{n-1} A_i = \sum_{i=0}^{n-1} f(\xi_{i+1})(x_{i+1} - x_i) = h \sum_{i=0}^{n-1} f(\xi_{i+1}).$$

So the simplest approximation is

$$\int_a^b f(x)\, dx \approx h\left(f(\xi_0) + f(\xi_1) + \cdots + f(\xi_{n-1})\right).$$

8.2 Fundamental Theorem of Calculus

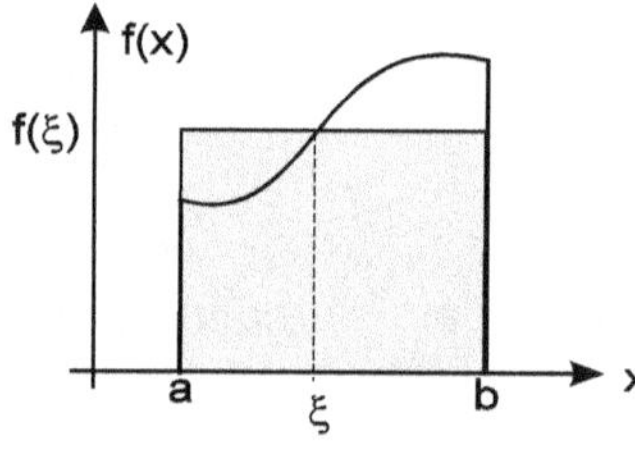

The construction of the definite integral may seem complicated, but it turns out that in many cases the calculation becomes simple. This fact is due to the relationship between the derivative of the integral function and the integrand, which we will investigate now. First, we present a generalization of the mean value theorem of the differential calculus: The area under a curve $f(x)$ can be replaced by an area-equivalent rectangle with the same base side and adapted height $f(\xi)$, where $f(\xi)$ is the *integral mean value* of the function f in the interval $[a, b]$:

Mean Value Theorem of Integral Calculus

If $f : [a, b] \to \mathbb{R}$ is continuous, then there exists a $\xi \in (a, b)$ with the property that

$$f(\xi) \cdot (b - a) = \int_a^b f(x)\, dx.$$

Proof: Due to its definition, the integral over a constant function $f(x) = c$ is

$$\int_a^b c\, dx = c \cdot (b - a).$$

If $f(x)$ is not constant, we define the minimum value m and the maximum value M of the function in the interval $[a, b]$:

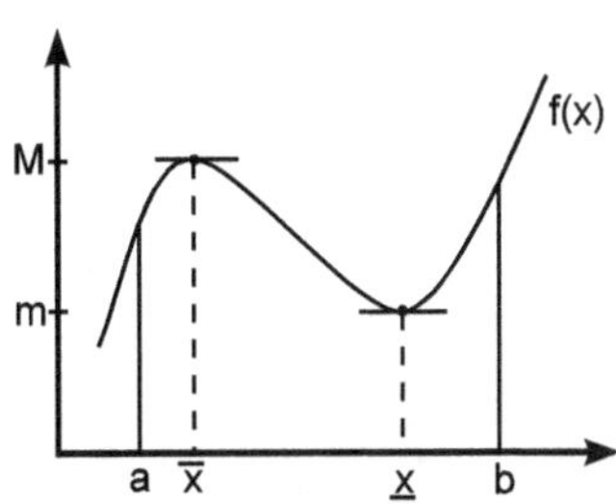

$$m := \min_{x \in [a,\, b]} f(x), \qquad M := \max_{x \in [a,\, b]} f(x).$$

Then there is a $\underline{x}$ with $f(\underline{x}) = m$ and a $\bar{x}$ with $f(\bar{x}) = M$ such that

$$f(\underline{x}) \le f(x) \le f(\bar{x}).$$

Since $f(\underline{x})$ and $f(\bar{x})$ are constant numbers, the following applies

$$f(\underline{x})\,(b - a) = \int_a^b f(\underline{x})\, dx \le \int_a^b f(x)\, dx \le \int_a^b f(\bar{x})\, dx = f(\bar{x})\,(b - a)$$

$$\Rightarrow \quad f(\underline{x}) \le \frac{1}{b - a} \int_a^b f(x)\, dx \le f(\bar{x}).$$

Since f is a continuous function, there exists a $\xi \in (a, b)$ such that

$$f(\xi) = \frac{1}{b - a} \int_a^b f(x)\, dx.$$

In the above consideration, we have used the monotonic property of the definite integral. It tells us that with $g(x) \le f(x) \le h(x)$ we have

$$\int_a^b g(x)\, dx \le \int_a^b f(x)\, dx \le \int_a^b h(x)\, dx.$$

This monotonic property is checked directly with the algebraic definition of the integral. $\qquad\square$

For the sake of completeness, we give (without proof) a more general formulation of the mean value theorem:

General Mean Value Theorem of Integral Calculus

Let $f, \varphi : [a, b] \to \mathbb{R}$ be continuous functions and $\varphi \geq 0$. Then there is a $\xi \in (a, b)$ with the property that

$$\int_a^b f(x)\, \varphi(x)\, dx = f(\xi) \int_a^b \varphi(x)\, dx.$$

We now establish the relationship between differential and integral calculus. This relationship is not only of theoretical importance, it also provides a practical method for calculating definite integrals.

Theorem of Integral Functions

Let $f : [a, b] \to \mathbb{R}$ be continuous and $F(x) := \int_a^x f(t)\, dt$ be an integral function with respect to f. Then F is differentiable with

$$F'(x) = f(x).$$

Important: The theorem of the integral function says that the derivative of the integral function $F(x)$ gives the integrand $f(x)$!

Proof: We look at the difference of the areas

$$\Delta F = F(x + h) - F(x) = \int_a^{x+h} f(t)\, dt - \int_a^x f(t)\, dt = \int_x^{x+h} f(t)\, dt$$

and apply the mean value theorem of integral calculus to the integral on the right: $\int_x^{x+h} f(t)\, dt = h\, f(\xi_h)$ with $\xi_h \in (x, x + h)$. Then we form the difference quotient

$$\frac{1}{h} \Delta F = \frac{F(x + h) - F(x)}{h} = f(\xi_h).$$

For $h \to 0$ the left side converges to $F'(x)$. The right side converges to $f(x)$. Since $\xi_h \xrightarrow{h \to 0} x$ and f is continuous, we have: $f(\xi_h) \xrightarrow{h \to 0} f(x)$. $\square$

Examples 8.4:

① For $f(x) = 1$, $F(x) = \int_0^x 1\, dt = x$, because $F'(x) = 1$.

② For $g(x) = x$, $G(x) = \int_0^x t\, dt = \frac{x^2}{2}$, because $G'(x) = x$.

③ For $h(x) = x^2$, $H(x) = \int_0^x t^2\, dt = \frac{x^3}{3}$, because $H'(x) = x^2$.   □

⊘ Antiderivative functions

In general, we have shown that the derivative of the integral function $F(x) = \int_a^x f(t)\, dt$ returns the integrand $f(x)$. We call such functions $F(x)$ with $F'(x) = f(x)$ *antiderivative functions* or antiderivatives for short:

> **Definition:** *Each function* $F(x)$ *with* $F'(x) = f(x)$ *is called* **antiderivative function** *of* $f(x)$.

Using this notation, it is possible to reformulate the theorem about the integral functions: **Any indefinite integral**

$$I(x) = \int_a^x f(t)\, dt$$

is an antiderivative function of $f(x)$, i.e. $I'(x) = f(x)$.

Example 8.5. The following table gives antiderivative functions $F(x)$ for some important functions. The property $F' = f$ can be calculated directly.

Table 8.1: Elementary antiderivative functions

$f(x)$	$x^n\ (n \neq -1)$	x^{-1}	$\sqrt[n]{x}$	e^x	$\sin x$	$\cos x$
$F(x)$	$\frac{1}{n+1}x^{n+1}$	$\ln x$	$\frac{1}{\frac{1}{n}+1}x^{\frac{1}{n}+1}$	e^x	$-\cos x$	$\sin x$

Remark: For every continuous function there are an infinite number of antiderivative functions, because e.g. for x^n both $\frac{1}{n+1}x^{n+1}$ and $\frac{1}{n+1}x^{n+1}+2$ as well as $\frac{1}{n+1}x^{n+1}+C$ are antiderivative functions. However, two antiderivatives of a function differ by at most one constant:

Theorem: If F_1 and F_2 are two antiderivative functions of f, then they differ by at most one additive constant $C \in \mathbb{R}$:

$$F_1(x) = F_2(x) + C.$$

Proof: Since F_1 and F_2 are antiderivative functions of the same f, we know that $F_1'(x) = f(x) = F_2'(x)$. Therefore,

$$(F_1(x) - F_2(x))' = 0 \quad \Rightarrow \quad F_1(x) - F_2(x) = const. \qquad \square$$

Consequently, any indefinite integral can be written in the form of

$$\int_a^x f(t)\, dt = F(x) + C,$$

where $F(x)$ is an arbitrary antiderivative function and C is a constant. For every continuous function $f(x)$ there are infinitely many indefinite integrals. Therefore, this set of functions is characterized by the omission of the integration limits

$$\int f(x)\, dx = \{\text{Set of all indefinite integrals from } f(x)\}.$$

Set of Antiderivatives

Since all antiderivative functions differ by only one constant, we write

$$\int f(x)\, dx = F(x) + C,$$

and call C the *integration constant*.

Examples 8.6:

① $\displaystyle\int e^x\, dx = e^x + C.$

② $\displaystyle\int x^k\, dx = \frac{1}{k+1} x^{k+1} + C \qquad (k \neq -1).$

③ $\displaystyle\int \frac{1}{x}\, dx = \ln(x) + C \qquad (x > 0).$

④ $\displaystyle\int \cos(x)\, dx = \sin(x) + C. \qquad \square$

Example 8.7: $\int \frac{1}{x}\, dx = \ln(|x|) + C.$

For $x > 0$, $(\ln(x))' = \frac{1}{x}$, so we get the indefinite integral $\int \frac{1}{x}\, dx = \ln(x) + C$. For $x < 0$, $(\ln(-x))' = \frac{1}{x}$, so we get the indefinite integral $\int \frac{1}{x}\, dx = \ln(-x) + C$. In summary, the given compact formula applies.

⚠ **Remark:** One reason why integral calculus is more difficult than differential calculus is that not every antiderivative function can be represented by an elementary function. The functions

$$f(x) = e^{-x^2}, \quad f(x) = \frac{\sin x}{x}$$

have no elementary representable antiderivative functions! However, antiderivative functions are neither geometrically nor physically as important as the definite integrals. So far we have only explicitly calculated $\int_0^b x^2\, dx = \frac{b^3}{3}$. The following fundamental theorem of calculus subordinates the calculation of definite integrals to a simpler task, namely the search for antiderivative functions:

Fundamental Theorem of Calculus

Let $f : [a, b] \to \mathbb{R}$ be a continuous function and F an antiderivative of f. Then

$$\int_a^b f(x)\, dx = F(b) - F(a).$$

Proof: Let $x \in [a, b]$ and $F_0(x) := \int_a^x f(t)\, dt$. Then $F_0(x)$ an antiderivative function of f with $F_0(a) = 0$ and $F_0(b) = \int_a^b f(t)\, dt$. If $F(x)$ is an arbitrary antiderivative function of f: $F - F_0 = const = c$, then

$$F(b) - F(a) = F_0(b) + c - (F_0(a) + c)$$

$$= F_0(b) - F_0(a) = F_0(b) = \int_a^b f(t)\, dt. \qquad \square$$

Table of Antiderivative Functions. The following table compiles the set of antiderivative functions $\int f(x)\, dx = F(x)$ for many elementary functions. The validity of most of them can be checked very easily with the relation $F'(x) = f(x)$.

Table 8.2: Antiderivative Functions

$f(x) = F'(x)$	Antiderivative F: $F(x) = \int f(x)\, dx + C$	Domain $\mathbb{D}_f$
$k \ (k \in \mathbb{R})$	$kx + C$	$\mathbb{R}$
$x^\alpha \ (\alpha \neq -1)$	$\frac{1}{\alpha+1}\, x^{\alpha+1} + C$	$\mathbb{R}_{>0}$
x^{-1}	$\ln|x| + C$	$\mathbb{R} \setminus \{0\}$
$\sin x$	$-\cos x + C$	$\mathbb{R}$
$\cos x$	$\sin x + C$	$\mathbb{R}$
$\tan x$	$-\ln|\cos x| + C$	$\mathbb{R} \setminus \left\{ x = \frac{\pi}{2} + k\pi,\, k \in \mathbb{Z} \right\}$
$\cot x$	$\ln|\sin x| + C$	$\mathbb{R} \setminus \{ x = k\pi,\, k \in \mathbb{Z} \}$
$a^x \ (a > 0, \neq 1)$	$\frac{a^x}{\ln a} + C$	$\mathbb{R}$
e^x	$e^x + C$	$\mathbb{R}$
$e^{ax} \ (a \neq 0)$	$\frac{1}{a}\, e^{ax} + C$	$\mathbb{R}$
$\ln x$	$x \cdot \ln x - x + C$	$\mathbb{R}_{>0}$
$\frac{1}{\cos^2 x}$	$\tan x + C$	$\mathbb{R} \setminus \left\{ x = \frac{\pi}{2} + k\pi,\, k \in \mathbb{Z} \right\}$
$\frac{1}{\sin^2 x}$	$-\cot x + C$	$\mathbb{R} \setminus \{ x = k\pi,\, k \in \mathbb{Z} \}$
$\sin^2 x$	$\frac{1}{2}(x - \sin x \cdot \cos x) + C$	$\mathbb{R}$
$\cos^2 x$	$\frac{1}{2}(x + \sin x \cdot \cos x) + C$	$\mathbb{R}$
$\tan^2 x$	$\tan x - x + C$	$\mathbb{R} \setminus \left\{ x = \frac{\pi}{2} + k\pi,\, k \in \mathbb{Z} \right\}$
$\cot^2 x$	$-\cot x - x + C$	$\mathbb{R} \setminus \{ x = k\pi,\, k \in \mathbb{Z} \}$

$f(x) = F'(x)$	Antiderivative function F: $F(x) = \int f(x)\, dx + C$	Domain $\mathbb{D}_f$
$\arcsin x$	$x \cdot \arcsin x + \sqrt{1-x^2} + C$	$(-1, 1)$
$\arccos x$	$x \cdot \arccos x - \sqrt{1-x^2} + C$	$(-1, 1)$
$\arctan x$	$x \cdot \arctan x - \frac{1}{2} \ln\left(x^2 + 1\right) + C$	$\mathbb{R}$
$\operatorname{arccot} x$	$x \cdot \operatorname{arccot} x + \frac{1}{2} \ln\left(x^2 + 1\right) + C$	$\mathbb{R}$
$\dfrac{1}{\sqrt{1-x^2}}$	$\arcsin x + C$	$(-1, 1)$
$\dfrac{-1}{\sqrt{1-x^2}}$	$\arccos x + C$	$(-1, 1)$
$\dfrac{1}{1+x^2}$	$\arctan x + C$	$\mathbb{R}$
$\dfrac{-1}{1+x^2}$	$\operatorname{arccot} x + C$	$\mathbb{R}$

$f(x) = F'(x)$	Antiderivative function F: $F(x) = \int f(x)\, dx + C$	Domain $\mathbb{D}_f$
$\sinh x$	$\cosh x + C$	$\mathbb{R}$
$\cosh x$	$\sinh x + C$	$\mathbb{R}$
$\tanh x$	$\ln(\cosh x) + C$	$\mathbb{R}$
$\coth x$	$\ln|\sinh x| + C$	$\mathbb{R} \setminus \{0\}$
$\dfrac{1}{\cosh^2 x}$	$\tanh x + C$	$\mathbb{R}$
$\dfrac{1}{\sinh^2 x}$	$-\coth x + C$	$\mathbb{R} \setminus \{0\}$
$\dfrac{1}{\sqrt{1+x^2}}$	$\operatorname{ar} \sinh x + C$	$\mathbb{R}$
$\dfrac{1}{\sqrt{x^2-1}}$	$\operatorname{ar} \cosh x + C$	$(1, \infty)$
$\dfrac{1}{1-x^2}$	$\operatorname{ar} \tanh x + C$	$(-1, 1)$
$\dfrac{1}{1-x^2}$	$\operatorname{ar} \coth x + C$	$\mathbb{R} \setminus [-1, 1]$

The calculation of definite integrals is done in two steps:

> **Calculation of Definite Integrals**
>
> (1) Determine an antiderivative $F(x)$ of the integrand $f(x)$.
>
> (2) With this antiderivative function, the difference $F(b) - F(a)$ is calculated:
>
> $$\int_a^b f(x)\, dx = \left[F(x)\right]_a^b = F(x)\Big|_a^b = F(b) - F(a).$$
>
> Then $\left[F(x)\right]_a^b = F(x)\Big|_a^b$ is used as a shorthand for the difference $F(b) - F(a)$.

Examples 8.8 (Calculation of Definite Integrals):

① $\displaystyle\int_a^b x^3\, dx = ?$:

According to Table 8.2 an antiderivative function is $F(x) = \frac{1}{4}x^4$, so

$$\int_a^b x^3\, dx = \frac{1}{4}x^4\Big|_a^b = \frac{1}{4}\left(b^4 - a^4\right).$$

For $0 \le a < b$ this is the area of the curve $y = x^3$ with the x-axis in the range of $a \le x \le b$.

② $\displaystyle\int_0^\pi \sin x\, dx = ?$:

An antiderivative function of $\sin(x)$ is $F(x) = -\cos(x)$, so

$$\int_0^\pi \sin x\, dx = -\cos x\Big|_0^\pi = -\cos\pi - (-\cos(0)) = 1 - (-1) = 2.$$

This is the area under the sine function in the first half period.

③ $\displaystyle\int_0^{2\pi} \sin x\, dx = ?$: We have the same antiderivative as in the previous example, but now the range is from 0 to 2π, giving

$$\int_0^{2\pi} \sin x\, dx = -\cos x\Big|_0^{2\pi} = -\cos(2\pi) - (-\cos(0)) = -1 - (-1) = 0.$$

Between 0 and π the sine function is positive, while between π and 2π it is negative. And so the areas contribute so that the overall result is zero.

Application Example 8.9 (**Expansion Work of Gases**).

In a cylinder of the base area $A\ [cm^2]$ a gas is compressed by a movable piston. The height of the piston from the bottom of the cylinder is $x\ [cm]$. The pressure of the gas in the cylinder is $p\,(x)\ \left[\frac{g}{cm\,s^2}\right]$. If the piston is moved from $x = a$ to $x = b$, the *work* of the gas is given by

$$W = \int_a^b A \cdot p\,(x)\ dx.$$

As a special case, we consider the *isothermal expansion* of an ideal gas with the equation of state

$$p\,(x) \cdot V\,(x) = p\,(a) \cdot V\,(a) = const$$
$$(\textit{Boyle-Mariotte's law})$$

With the volume $V\,(x) = A \cdot x$, we get

$$p\,(x) = \frac{p\,(a) \cdot V\,(a)}{V\,(x)} = \frac{p\,(a) \cdot V\,(a)}{A \cdot x}.$$

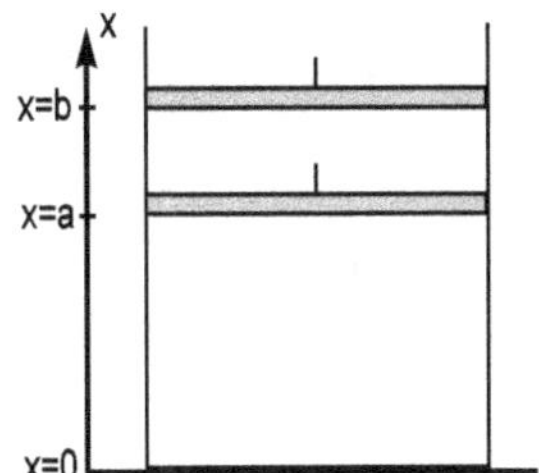

$$W = A \int_a^b \frac{p\,(a) \cdot V\,(a)}{A \cdot x}\ dx = p\,(a) \cdot V\,(a) \int_a^b \frac{1}{x}\ dx.$$

From Table 8.2 we obtain

$$W = p\,(a) \cdot V\,(a) \cdot \Big[\ln x\Big]_a^b = p\,(a) \cdot V\,(a) \cdot [\ln\,(b) - \ln\,(a)]$$
$$= p\,(a) \cdot V\,(a) \cdot \ln\left(\frac{b}{a}\right).\qquad\qquad\square$$

8.3 Rules of Integral Calculus

The calculation of definite integrals is simplified with the integration rules. They follow directly from the definition of the definite integral as the limit value of the subtotal. All functions occurring are assumed to be continuous.

Constant Factor Rule

A constant factor c can be placed in front of the integral:

$$\int_a^b c \cdot f\,(x)\ dx = c \cdot \int_a^b f\,(x)\ dx.$$

Example 8.10. $\displaystyle \int_0^{\pi/2} 4\cos x\, dx = 4 \int_0^{\pi/2} \cos x\, dx$

$$= 4 \left[\sin x\right]_0^{\pi/2} = 4 \left(\sin(\frac{\pi}{2}) - \sin(0)\right) = 4. \qquad \square$$

Sum Rule

A sum of functions is integrated term by term:

$$\int_a^b \left(f_1(x) + f_2(x)\right) dx = \int_a^b f_1(x)\, dx + \int_a^b f_2(x)\, dx.$$

Example 8.11. $\displaystyle \int_0^1 \left(-3x^2 + x\right) dx = -3 \int_0^1 x^2\, dx + \int_0^1 x\, dx$

$$= -3 \left[\frac{x^3}{3}\right]_0^1 + \left[\frac{x^2}{2}\right]_0^1 = -\frac{1}{2}. \qquad \square$$

Remarks:

(1) The factor and sum rules also apply analogously to indefinite integrals.

(2) Previously, $a < b$ was always assumed. The factor and sum rules remain valid for any real numbers a, b from the domain of f, taking into account the following properties:

Integration Boundaries

(1) *Integration boundaries coincide:* $\displaystyle \int_a^a f(x)\, dx = 0.$

(2) *Swapping the integration limits:*

$$\int_b^a f(x)\, dx = - \int_a^b f(x)\, dx.$$

An important application of Note (2) is related to Example 8.9, moving the piston of a cylinder filled with gas. If we move the piston from a to b, we expand the gas and extract energy from the system. Otherwise, if we move the piston from b to a, we compress the gas, so we have to spend the same amount of energy, but the sign is negative.

Examples 8.12:

① $\displaystyle \int_2^2 \frac{1}{x}\, dx = \ln x \Big|_2^2 = \ln(2) - \ln(2) = 0.$

② $\displaystyle \int_{\pi/2}^0 \sin x\, dx = -\int_0^{\pi/2} \sin x\, dx = -\Big[-\cos x\Big]_0^{\pi/2} = -1.$ $\square$

Additivity of the Integral

For any point c in the integration domain $a \le c \le b$ of f, we have

$$\int_a^b f(x)\, dx = \int_a^c f(x)\, dx + \int_c^b f(x)\, dx.$$

This additivity of the integral is exploited when a function has different functional assignments on sub-intervals.

Example 8.13. Given is the function $f(x)$ defined by

$$f(x) = \begin{cases} x^2 & \text{for } 0 \le x \le 1 \\[2mm] -x + 2 & \text{for } 1 \le x \le 2 \end{cases}$$

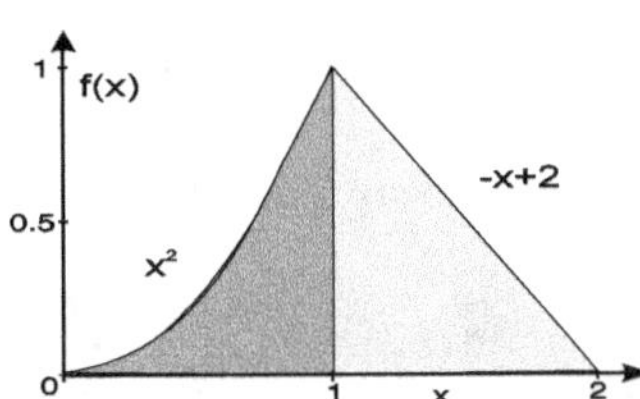

To calculate the area under the curve f in the range $[0, 2]$, the integral must be split into two parts: an integral over $[0, 1]$ and a second integral over $[1, 2]$, because the function has different assignments in the different sub-intervals. Therefore,

$$\begin{aligned}
A &= \int_0^2 f(x)\, dx = \int_0^1 x^2\, dx + \int_1^2 (-x+2)\, dx \\[2mm]
&= \left[\frac{x^3}{3}\right]_0^1 + \left[\frac{-x^2}{2} + 2x\right]_1^2 \\[2mm]
&= \frac{1}{3} + \frac{1}{2} = \frac{5}{6}.
\end{aligned}$$ $\square$

8.4 Integration Methods

In practice, the integration of functions is often more difficult than differentiation. While differentiation can be achieved by applying simple rules (product rule, quotient rule, chain rule), integration is more difficult. Nevertheless, in many cases an antiderivative function can be found using one of the following integration methods.

8.4.1 Integration by Parts

Integration by parts is the counterpart to the product rule of differentiation, which states that

$$\left(u\left(x\right)\cdot v\left(x\right)\right)' = u'\left(x\right)v\left(x\right) + u\left(x\right)v'\left(x\right).$$

We solve this equation in terms of $u\left(x\right)v'\left(x\right)$ and integrate

$$u\left(x\right)v'\left(x\right) = \left(u\left(x\right)v\left(x\right)\right)' - u'\left(x\right)v\left(x\right)$$

$$\int_a^b u\left(x\right)v'\left(x\right)\,dx = \int_a^b \left(u\left(x\right)v\left(x\right)\right)'\,dx - \int_a^b u'\left(x\right)v\left(x\right)\,dx.$$

According to the Fundamental Theorem of Calculus

$$\int_a^b \left(u\left(x\right)v\left(x\right)\right)'\,dx = \Big[u\left(x\right)\cdot v\left(x\right)\Big]_a^b,$$

so the following applies

Integration by Parts

$$\int_a^b u\left(x\right)v'\left(x\right)\,dx = \Big[u\left(x\right)v\left(x\right)\Big]_a^b - \int_a^b u'\left(x\right)v\left(x\right)\,dx.$$

Remarks:

(1) Whether integration by parts succeeds depends on the proper (appropriate) choice of $u\left(x\right)$ and $v'\left(x\right)$.

(2) Especially when integrating functions that contain a trigonometric function as a factor, it may happen that the integral to be calculated reappears on the right side after one or more partial integrations. In this case, the equation is solved with respect to the searched integral.

(3) In some cases, the integration procedure must be repeated several times before a basic integral is found.

(4) The formula for integration by parts also applies to indefinite integrals

$$\int u\left(x\right) v'\left(x\right) dx = u\left(x\right) v\left(x\right) - \int u'\left(x\right) v\left(x\right) dx.$$

Examples 8.14 (Integration by Parts):

① To find the integral $\int_1^2 x\,e^x\,dx$, we set

$$\begin{array}{lll} u\left(x\right) = x & \Rightarrow & u'\left(x\right) = 1 \\ v'\left(x\right) = e^x & \Rightarrow & v\left(x\right) = e^x \end{array}$$

and obtain

$$\int_1^2 x\,e^x\,dx = [x\,e^x]_1^2 - \int_1^2 1 \cdot e^x\,dx = \left[x\,e^x\right]_1^2 - \left[e^x\right]_1^2$$
$$= 2\,e^2 - e^1 - e^2 + e^1 = e^2.$$

So $\int x\,e^x\,dx = e^x\left(x - 1\right) + C.$

② To find the integral $\int x^2 \cos x\,dx$, we set

$$\begin{array}{lll} u\left(x\right) = x^2 & \Rightarrow & u'\left(x\right) = 2x \\ v'\left(x\right) = \cos x & \Rightarrow & v\left(x\right) = \sin x \end{array}$$

and obtain

$$\int x^2 \cos x\,dx = x^2 \sin x - \int 2x \sin x\,dx.$$

Integration by parts of $\int 2x \sin x\,dx$ gives

$$\begin{array}{lll} u\left(x\right) = 2x & \Rightarrow & u'\left(x\right) = 2 \\ v'\left(x\right) = \sin x & \Rightarrow & v\left(x\right) = -\cos x \end{array}$$

$$\int x^2 \cos x\,dx = x^2 \sin x - \left[2x\left(-\cos x\right) - \int 2\left(-\cos x\right) dx\right]$$

$$= x^2 \sin x + 2x \cos x - 2 \sin x + C. \qquad \square$$

Hint: Usually $u(x)$ is set to the power term in order to reduce this term by multiple integration by parts. But in some cases $v'(x) = 1 \Rightarrow v(x) = x$ will give the desired result, as the next example shows:

③ To calculate $\int \ln x \, dx$, we set

$$\boxed{\begin{aligned} u(x) &= \ln x & \Rightarrow & \quad u'(x) = \frac{1}{x} \\ v'(x) &= 1 & \Rightarrow & \quad v(x) = x \end{aligned}}$$

which gives

$$\int \ln x \, dx = x \ln x - \int \frac{1}{x} \cdot x \, dx = x \ln x - x + C$$
$$= x \left(\ln x - 1 \right) + C.$$

Tip: In some cases, the integral to be calculated appears on the right-hand side of the equation. Then solve the equation in terms of the integral:

④ To calculate $\int \cos^2 x \, dx = \int \cos x \cdot \cos x \, dx$ we set

$$\boxed{\begin{aligned} u(x) &= \cos x & \Rightarrow & \quad u'(x) = -\sin x \\ v'(x) &= \cos x & \Rightarrow & \quad v(x) = \sin x \end{aligned}}$$

which results in

$$\int \cos^2 x \, dx = \cos x \sin x - \int -\sin x \sin x \, dx = \cos x \sin x + \int \sin^2 x \, dx.$$

We replace $\sin^2 x = 1 - \cos^2 x$.

$$\Rightarrow \int \cos^2 x \, dx = \cos x \sin x + x - \int \cos^2 x \, dx.$$

We observe that the integral to be calculated appears also on the right-hand side of the equation. So we add $\int \cos^2 x \, dx$ on both sides and divide by the factor 2. Finally, we get

$$\int \cos^2 x \, dx = \tfrac{1}{2} \left(\sin x \cos x + x \right) + C. \qquad \square$$

8.4.2 Integration by Substitution

Besides integration by parts, there is another very important integration technique: *Integration by Substitution*. This integration method can be derived from the chain rule. For the derivation of the function $G\left(f\left(x\right)\right)$ with respect to x, we use the chain rule:

$$\frac{d}{dx} G\left(f\left(x\right)\right) = G'\left(f\left(x\right)\right) \cdot f'\left(x\right).$$

With $g(x) = G'(x)$ this result can be rewritten:

Substitution Rules for Indefinite Integrals

$$\int g\left(f\left(x\right)\right) f'\left(x\right)\, dx = G\left(f\left(x\right)\right) + C,$$

if G is an antiderivative of g.

This formula means that we can more or less only integrate the outer function g, if the inner function f also occurs with f' as a factor in the integrand. So f can be substituted. Table 8.3 shows special cases of this substitution rule. Note that the cases (A), (B) and (C) are included as special cases of the general form (D).

Table 8.3: Direct Integral Substitutions

	Integral type	Substitution	Solution
(A)	$\int g\left(a\,x + b\right)\, dx$	$y = a\,x + b$	$\frac{1}{a} G\left(a\,x + b\right) + C$
(B)	$\int f\left(x\right) f'\left(x\right)\, dx$	$y = f\left(x\right)$	$\frac{1}{2} f^2\left(x\right) + C$
(C)	$\int \dfrac{f'\left(x\right)}{f\left(x\right)}\, dx$	$y = f\left(x\right)$	$\ln\left|f\left(x\right)\right| + C$
(D)	$\int g\left(f\left(x\right)\right) f'\left(x\right)\, dx$	$y = f\left(x\right)$	$G\left(f\left(x\right)\right) + C$

In Table 8.3 we introduce direct substitutions, since we define the new integration variable $y = f(x)$ directly using the inner function $f(x)$. After discussing examples for these direct substitutions we will introduce indirect or implicit substitutions, see Table 8.4.

Examples 8.15 (Integral Substitutions according to Table 8.3):

(A1) $\displaystyle\int_2^3 (2x-3)^4\,dx =?$

We first determine an antiderivative and then use the upper and lower integration limits to calculate the definite integral. To do this, we substitute $\boxed{y = 2x-3}$ and replace each term of the integral containing the integration variable x by a corresponding term with y. In particular, the differential dx must also be replaced by a corresponding term dy.

From $y = 2x-3 \quad \hookrightarrow y' = \frac{dy}{dx} = 2 \quad \hookrightarrow dx = \frac{1}{2}\,dy$. So

$$\int (2x-3)^4\,dx = \int y^4\,\frac{1}{2}\,dy = \frac{1}{2}\int y^4\,dy = \frac{1}{2}\,\frac{1}{5}\,y^5 + C.$$

After the integral has been calculated, $y = 2x-3$ is replaced:

$$\int (2x-3)^4\,dx = \frac{1}{10}\,(2x-3)^5 + C.$$

The particular integral is therefore

$$\int_2^3 (2x-3)^4\,dx = \left[\frac{1}{10}\,(2x-3)^5\right]_2^3 = \frac{1}{10}\,[243-1] = 24.2\,.$$

(A1) Alternatively, we apply the substitution direct on the definite integral $\int_2^3 (2x-3)^4\,dx$. Then, the integration limits must also be substituted! After calculating the substituted integral, there is no more a back-substitution. Starting from $y = 2x-3$, follows for the lower boundary $x_u = 2 \hookrightarrow y_u = 1$ and for the upper boundary $x_o = 3 \hookrightarrow y_o = 3$:

$$\int_2^3 (2x-3)^4\,dx = \int_1^3 y^4\,\frac{1}{2}\,dy = \left[\frac{1}{10}\,y^5\right]_1^3 = \frac{1}{10}\,[243-1] = 24.2\,.$$

(A2) $\displaystyle\int_0^1 \frac{1}{1+4x}\,dx =?$

Substitute $\boxed{y = 1+4x} \quad \hookrightarrow y' = \frac{dy}{dx} = 4 \quad \hookrightarrow dx = \frac{1}{4}\,dy$.

$$\hookrightarrow \int \frac{1}{1+4x}\,dx = \int \frac{1}{y}\cdot\frac{1}{4}\,dy = \frac{1}{4}\int \frac{1}{y}\,dy = \frac{1}{4}\,\ln|y| + C.$$

By back-substituting $y = 1 + 4x$

$$\int \frac{1}{1+4x}\, dx = \frac{1}{4}\ln|1+4x| + C.$$

Finally, the result is

$$\int_0^1 \frac{1}{1+4x}\, dx = \left[\frac{1}{4}\ln|1+4x|\right]_0^1 = \frac{1}{4}\ln 5 - \frac{1}{4}\ln 1 = \frac{1}{4}\ln 5\,.$$

(B1) $\displaystyle\int \sin x \cos x\, dx = ?$

Substitute $\boxed{y = \sin x} \hookrightarrow y' = \dfrac{dy}{dx} = \cos x \hookrightarrow dx = \dfrac{1}{\cos x}\, dy.$

$$\hookrightarrow \int \sin x \cos x\, dx = \int y \cos x \frac{1}{\cos x}\, dy = \int y\, dy = \frac{1}{2}y^2 + C.$$

By back-substituting $y = \sin x$

$$\int \sin x \cos x\, dx = \frac{1}{2}\sin^2 x + C.$$

(B2) $\displaystyle\int_1^2 \frac{\ln x}{x}\, dx = ?$

Substitution: $\boxed{y = \ln x} \hookrightarrow y' = \dfrac{dy}{dx} = \dfrac{1}{x} \hookrightarrow dx = x\, dy.$

$$\hookrightarrow \int \frac{\ln x}{x}\, dx = \int \frac{y}{x}\cdot x\, dy = \int y\, dy = \frac{1}{2}y^2 + C.$$

Back-substitution: $\int \dfrac{\ln x}{x}\, dx = \frac{1}{2}\ln^2 x + C.$

Calculating the specific integral: $\displaystyle\int_1^2 \frac{\ln x}{x}\, dx = \frac{1}{2}\left[\ln^2 x\right]_1^2 = \frac{1}{2}\ln^2 2\,.$

(C1) $\displaystyle\int \frac{2x-3}{x^2-3x+1}\, dx = \int \frac{2x-3}{y}\cdot\frac{dy}{2x-3}$

$$= \int \frac{1}{y}\, dy = \ln|y| + C = \ln|x^2 - 3x + 1| + C$$

with the substitution $\boxed{y = x^2 - 3x + 1}$ and $dx = \dfrac{1}{2x-3}\, dy.$

(C2) $\displaystyle\int_0^1 \frac{e^x}{2\,e^x + 5}\,dx = ?$

Substitution: $\boxed{y = 2\,e^x + 5}$ $\hookrightarrow y' = \dfrac{dy}{dx} = 2\,e^x$ $\hookrightarrow dx = \dfrac{1}{2\,e^x}\,dy.$

Upper limit $x_o = 1 \hookrightarrow y_o = 2\,e + 5$

Lower limit $x_u = 0 \hookrightarrow y_u = 7.$

$$\int_0^1 \frac{e^x}{2\,e^x + 5}\,dx = \int_7^{2e+5} \frac{e^x}{y}\cdot\frac{1}{2\,e^x}\,dy = \frac{1}{2}\int_7^{2e+5} \frac{1}{y}\,dy$$

$$= \tfrac{1}{2}\,\ln y\,\Big|_7^{2e+5} = \tfrac{1}{2}\left[\ln\left(2\,e + 5\right) - \ln 7\right] = 0.1997\,.$$

(D1) $\displaystyle\int \left(x^3 + 2\right)^{\frac{1}{2}} x^2\,dx = ?$

Substitution: $\boxed{y = x^3 + 2}$ $\hookrightarrow \dfrac{dy}{dx} = 3x^2$ $\hookrightarrow dx = \dfrac{1}{3x^2}\,dy.$

$$\int \left(x^3 + 2\right)^{\frac{1}{2}} x^2\,dx = \int y^{\frac{1}{2}}\cdot x^2\cdot\frac{1}{3x^2}\,dy = \frac{1}{3}\int y^{\frac{1}{2}}\,dy = \frac{1}{3}\,\frac{2}{3}\,y^{\frac{3}{2}} + C.$$

Back-substitution:

$$\int \left(x^3 + 2\right)^{\frac{1}{2}} x^2\,dx = \frac{2}{9}\left(x^3 + 2\right)^{\frac{3}{2}} + C.$$

(D2) $\displaystyle\int \frac{e^x + x\,e^x}{\left(x\,e^x\right)^3}\,dx = ?$

Substitution: $\boxed{y = x\,e^x}$ $\hookrightarrow \dfrac{dy}{dx} = e^x + x\,e^x$ $\hookrightarrow dx = \dfrac{1}{e^x + x\,e^x}\,dy.$

$$\int \frac{e^x + x\,e^x}{\left(x\,e^x\right)^3}\,dx = \int \frac{e^x + x\,e^x}{y^3}\cdot\frac{1}{e^x + x\,e^x}\,dy = \int y^{-3}\,dy = -\frac{1}{2}\,y^{-2} + C.$$

Back-substitution:

$$\int \frac{e^x + x\,e^x}{\left(x\,e^x\right)^3}\,dx = -\frac{1}{2}\left(x\,e^x\right)^{-2} + C. \qquad \square$$

Note: If a substitution is made for a *definite* integral, the integration boundaries must also be replaced.

Besides the substitution rules given in Table 8.3, there are many others. Some of these are described in Table 8.4:

Table 8.4: Implicit Integral Substitutions

Integral type		Substitution
(E)	$\int g\left(x, \sqrt{a^2 - x^2}\right)\, dx$	$x = a \cdot \sin(y)$
(F)	$\int g\left(x, \sqrt{x^2 + a^2}\right)\, dx$	$x = a \cdot \sinh(y)$
(G)	$\int g\left(x, \sqrt{x^2 - a^2}\right)\, dx$	$x = a \cdot \cosh(y)$

Examples 8.16 (Integral Substitutions according to Table 8.4):

(E1) $\displaystyle \int \frac{dx}{\sqrt{4 - x^2}} = \int \frac{2\cos(y)}{2\cos(y)}\, dy = \int dy = y + C = \arcsin\left(\frac{1}{2}x\right) + C$

with the substitution $\boxed{x = 2\sin(y)}$ $\hookrightarrow \dfrac{dx}{dy} = 2\cos(y)$

$$\hookrightarrow dx = 2\cos(y)\, dy$$

and $\sqrt{4 - x^2} = \sqrt{4 - 4\sin^2(y)} = 2\cos(y)$, as $\cos^2(y) + \sin^2(y) = 1$.

(E2) $\displaystyle \int \frac{x}{\sqrt{4 - x^2}}\, dx = \int \frac{2\sin(y)\, 2\cos(y)}{2\cos(y)}\, dy = 2\int \sin(y)\, dy$

$$= -2\cos(y) + C = -2\sqrt{1 - \sin^2(y)} + C$$

$$= -2\sqrt{1 - \frac{x^2}{4}} + C = -\sqrt{4 - x^2} + C$$

with the substitution: $\boxed{x = 2\sin(y)}$
and $y = \arcsin\left(\frac{x}{2}\right)$.

(F) $\displaystyle \int \frac{dx}{\sqrt{1 + x^2}} = \int \frac{\cosh(y)}{\cosh(y)}\, dy = \int dy$

$$= y + C = ar\sinh(x) + C = \ln\left(x + \sqrt{1 + x^2}\right) + C$$

with the substitution $\boxed{x = \sinh(y)}$

$$\hookrightarrow \frac{dx}{dy} = \cosh(y) \hookrightarrow dx = \cosh(y)\, dy$$

and $\sqrt{1 + x^2} = \sqrt{1 + \sinh^2(y)} = \cosh(y)$,
since $\cosh^2(y) - \sinh^2(y) = 1$.

$$(G) \quad \int \frac{dx}{\sqrt{x^2 - 25}} = \int \frac{5\sinh(y)}{5\sinh(y)}\, dy = \int dy$$

$$= y + C = ar\cosh\left(\frac{x}{5}\right) + C = \ln\left(\frac{x}{5} + \sqrt{\left(\frac{x}{5}\right)^2 - 1}\right) + C$$

with the substitution $\boxed{x = 5\cosh(y)}$

$$\hookrightarrow \frac{dx}{dy} = 5\sinh(y) \hookrightarrow dx = 5\sinh(y)\, dy$$

and $\sqrt{x^2 - 25} = \sqrt{25\cosh^2(y) - 25} = 5\sqrt{\cosh^2(y) - 1} = 5\sinh(y)$,

as $\cosh^2(y) - \sinh^2(y) = 1$. The problem and the solution can be found
under the assumption that $|x| \geq 5$. $\qquad\square$

Despite the variety of substitutions, there are no general formulas for calculating integrals that will always lead to a result!

8.4.3 Decomposition into Partial Fraction

For rational functions $f(x) = \frac{Z(x)}{N(x)}$ ($Z(x)$, $N(x)$ Polynomials) there is a
special integration technique called *partial fractional decomposition*. This
method allows rational functions to be integrated in closed form. They must
be available in a proper rational representation to be integrated. A rational
function is called *properly rational* if the degree of the upper polynomial
is less than the degree of the lower polynomial, otherwise the function is
called *improperly rational*. In the case of an improper rational function,
polynomial division is used to ensure that the degree of the numerator is
then less than the degree of the denominator:

Example 8.17. $\dfrac{2x^3 - 2x^2 - 5x + 7}{x^2 - 3x + 2}$:

$$
\begin{array}{llll}
(2x^3 & -2x^2 & -5x & +7) : (x^2 - 3x + 2) = 2x + 4 + \dfrac{3x - 1}{x^2 - 3x + 2}\\[4pt]
-(2x^3 & -6x^2 & +4x)\\[2pt]
\hline
 & 4x^2 & -9x & +7\\[2pt]
 & -(4x^2 & -12x & +8)\\[2pt]
\hline
 & & 3x & -1 \qquad\qquad\qquad \square
\end{array}
$$

A properly rational function can be broken down into partial fractions. In the following we will always assume that

$$
\boxed{\,f(x) = \dfrac{p(x)}{q(x)}\,}
$$

is a proper rational function with degree $(p) <$ degree $(q) = n$.

Partial Fraction Decomposition (Single Zeros)

If $q(x) = a_n (x - x_1)(x - x_2) \cdot \ldots \cdot (x - x_n)$ has n **single real zeros**, then the rational function $f(x)$ can be decomposed in the form of

$$
f(x) = \frac{A_1}{x - x_1} + \frac{A_2}{x - x_2} + \cdots + \frac{A_n}{x - x_n}
$$

with **partial fractions** $\quad \dfrac{A_i}{x - x_i} \quad (i = 1, \ldots, n)$.

Example 8.18. $\displaystyle \int \frac{3x - 1}{x^2 - 3x + 2}\, dx = ?$

To perform a partial fractional decomposition, the zeros of the denominator must first be determined. From

$$
x^2 - 3x + 2 = 0
$$

follows with the p/q formula $x_{1/2} = \frac{3}{2} \pm \sqrt{\frac{9}{4} - 2}$. So $x_1 = 1$ and $x_2 = 2$ are the zeros of the denominator polynomial and $f(x) = \dfrac{3x - 1}{x^2 - 3x + 2} = \dfrac{3x - 1}{(x - 1)(x - 2)}$. By the decomposition

$$
f(x) = \frac{3x - 1}{(x - 1)(x - 2)} = \frac{A_1}{x - 1} + \frac{A_2}{x - 2}
$$

we get the partial decomposition. To calculate A_1 and A_2, the common denominator is formed as follows

$$\frac{A_1}{x-1} + \frac{A_2}{x-2} = \frac{A_1\,(x-2) + A_2\,(x-1)}{(x-1)\,(x-2)} = \frac{3x-1}{x^2 - 3x + 2}.$$

We compare the top with $3x - 1$:

$$A_1\,(x-2) + A_2\,(x-1) = 3x - 1 \qquad \text{for all} \quad x.$$

To determine the constants A_1 and A_2, either a coefficient comparison is performed or special values for x are used:

$$x = 1: \quad A_1\,(1-2) \;= 3-1 \quad \boxed{\Rightarrow A_1 = -2}$$
$$x = 2: \quad A_2\,(2-1) \;= 5 \quad \boxed{\Rightarrow A_2 = 5.}$$

Hence,

$$f(x) = \frac{-2}{x-1} + \frac{5}{x-2}$$

and we finally integrate the partial decomposition instead of the original representation of the function

$$\int f(x)\,dx = -2 \int \frac{1}{x-1}\,dx + 5 \int \frac{1}{x-2}\,dx$$
$$= -2 \ln|x-1| + 5 \ln|x-2| + C$$
$$= \ln|x-1|^{-2} + \ln|x-2|^{5} + C = \ln\left|\frac{(x-2)^5}{(x-1)^2}\right| + C. \quad \square$$

If the denominator polynomial has double or multiple zeros, then the partial fraction decomposition must be modified for these zeros:

Partial Fraction Decomposition (Multiple Zeros)

$q(x)$ has a **multiple real zero** x_l of order k, i.e. the term $(x - x_l)$ appears to the power k in the product representation of $q(x)$ alongside other zeros. Then, this zero of order k is to be considered alongside to the other zeros as follows:

$$f(x) = \cdots + \frac{B_1}{x - x_l} + \frac{B_2}{(x - x_l)^2} + \cdots + \frac{B_k}{(x - x_l)^k}.$$

Example 8.19. $\displaystyle\int \frac{2x^2 + 3x + 1}{x^3 - 5x^2 + 8x - 4}\, dx = ?$

$x = 1$ is a first-order zero and $x = 2$ is a second-order zero of the denominator polynomial, since $x^3 - 5x^2 + 8x - 4 = (x - 1)(x - 2)^2$. For the partial fractional decomposition of the integrand f we therefore choose the approach

$$
\begin{aligned}
f(x) &= \frac{A}{x - 1} + \frac{B_1}{x - 2} + \frac{B_2}{(x - 2)^2} \\
&= \frac{A(x - 2)^2 + B_1(x - 2)(x - 1) + B_2(x - 1)}{(x - 1)(x - 2)^2}.
\end{aligned}
$$

After multiplying by the common denominator, we get

$$
2x^2 + 3x + 1 \overset{!}{=} A(x - 2)^2 + B_1(x - 2)(x - 1) + B_2(x - 1).
$$

We use special x values to determine A, B_1 and B_2:

$$
x = 1 : \boxed{6 = A}
$$
$$
x = 2 : \boxed{15 = B_2}
$$
$$
x = 0 : 1 = 4A + 2B_1 - B_2 \Rightarrow 1 = 9 + 2B_1 \quad \Rightarrow \boxed{B_1 = -4}
$$

Therefore

$$
\begin{aligned}
\int \frac{2x^2 + 3x + 1}{x^3 - 5x^2 + 8x - 4}\, dx \\
= 6 \int \frac{1}{x - 1}\, dx - 4 \int \frac{1}{x - 2}\, dx + 15 \int \frac{1}{(x - 2)^2}\, dx \\
= 6 \ln|x - 1| - 4 \ln|x - 2| - 15 \frac{1}{x - 2} + C.
\end{aligned}
$$

$\square$

Example 8.20. $\displaystyle\int \frac{x^6 - 2x^5 + x^4 + 4x + 1}{x^4 - 2x^3 + 2x - 1}\, dx = ?$

(i) Using polynomial division, we decompose the integrand into a polynomial and a proper rational function:

$$
\begin{aligned}
\frac{x^6 - 2x^5 + x^4 + 4x + 1}{x^4 - 2x^3 + 2x - 1} &= (x^6 - 2x^5 + x^4 + 4x + 1) : (x^4 - 2x^3 + 2x - 1) \\
&= x^2 + 1 + \frac{x^2 + 2x + 2}{x^4 - 2x^3 + 2x - 1}.
\end{aligned}
$$

(ii) Since the denominator polynomial is of degree 4, the zeros of the denominator polynomial cannot be calculated by a formula. So it is necessary to guess a zero x_0 and then reduce the degree either by dividing the denominator polynomial by $(x - x_0)$ or by using the Horner scheme. For the denominator polynomial $x^4 - 2x^3 + 2x - 1$ we get:

$$x_1 = \quad 1 \quad \text{is a third-order zero}$$
$$x_2 = \quad -1 \quad \text{is a first-order zero}$$

This gives the linear factorization

$$x^4 - 2x^3 + 2x - 1 = (x-1)^3\,(x+1)$$

(iii) Partial fractions are assigned to the zeros according to their order:

$$x_1 = 1: \quad \frac{A_1}{x-1} + \frac{A_2}{(x-1)^2} + \frac{A_3}{(x-1)^3}\,;$$
$$x_2 = -1: \quad \frac{B}{x+1}\,.$$

Representation of $\frac{p(x)}{q(x)}$ by partial fractions:

$$\frac{x^2 + 2x + 2}{x^4 - 2x^3 + 2x - 1} = \frac{A_1}{(x-1)} + \frac{A_2}{(x-1)^2} + \frac{A_3}{(x-1)^3} + \frac{B}{x+1}$$

$$= \frac{A_1\,(x-1)^2\,(x+1) + A_2\,(x-1)\,(x+1) + A_3\,(x+1) + B\,(x-1)^3}{CD}$$

(iv) The coefficients are determined by multiplying by the common denominator (CD) and evaluating with special x-values:

$$
\begin{array}{llll}
x = 1: & 5 = A_3 \cdot 2 & \Rightarrow & A_3 = \frac{5}{2} \\
x = -1: & 1 = B \cdot (-2)^3 & \Rightarrow & B = -\frac{1}{8} \\
x = 0: & 2 = A_1 - A_2 + \frac{5}{2} + \left(-\frac{1}{8}\right)(-1) & \Rightarrow & A_1 - A_2 = -\frac{5}{8} \\
x = 2: & 10 = 3\,A_1 + 3\,A_2 + \frac{5}{2}\cdot 3 + \left(-\frac{1}{8}\right) & \Rightarrow & A_1 + A_2 = \frac{7}{8}
\end{array}
$$

Adding and subtracting the last two equations we get $A_1 = \frac{1}{8}$, $A_2 = \frac{3}{4}$.

(v) Finally, we integrate the polynomial and the partial fractions:

$$\int \frac{x^6 - 2x^5 + x^4 + 4x + 1}{x^4 - 2x^3 + 2x - 1}\,dx$$

$$= \int (x^2 + 1)\,dx + \frac{1}{8}\int \frac{1}{x-1}\,dx + \frac{3}{4}\int \frac{1}{(x-1)^2}\,dx$$

$$+ \frac{5}{2}\int \frac{1}{(x-1)^3}\,dx - \frac{1}{8}\int \frac{1}{x+1}\,dx$$

$$= \frac{1}{3}x^3 + x + \frac{1}{8}\ln|x-1| - \frac{3}{4}\frac{1}{(x-1)} - \frac{5}{4}\frac{1}{(x-1)^2} - \frac{1}{8}\ln|x+1| + C.\ \square$$

Integration by Decomposition into Partial Fractions

Any rational function $f(x) = \dfrac{Z(x)}{N(x)}$ ($Z(x)$, $N(x)$ polynomials) can be integrated by algebraic methods if $N(x)$ decomposes into linear factors:

(1) Decomposition of the function $f(x) = r(x) + \dfrac{p(x)}{q(x)}$ into a polynomial $r(x)$ and a proper fractional function $\frac{p(x)}{q(x)}$.
(Note: $N(x) = q(x)$).

(2) Determination of the real zeros of $q(x)$ and their multiplicity.

(3) Each zero x_0 of $q(x)$ is assigned partial fractions according to its multiplicity.

$$x_0 \text{ 1st order zero} \quad \rightarrow \quad \frac{A}{x - x_0}$$

$$x_0 \text{ 2nd order zero} \quad \rightarrow \quad \frac{A_1}{x - x_0} + \frac{A_2}{(x - x_0)^2}$$

$$\vdots$$

$$x_0 \text{ kth order zero} \quad \rightarrow \quad \frac{A_1}{x - x_0} + \frac{A_2}{(x - x_0)^2} + \cdots + \frac{A_k}{(x - x_0)^k} \, .$$

The proper rational function $\frac{p(x)}{q(x)}$ can then be represented as the sum of all partial fractions.

(4) Determination of the constants occurring in the partial fractions (either by comparing coefficients and solving the corresponding system of linear equations, or by using special x-values).

(5) Integration of $r(x)$ and all partial fractions with

$$\int \frac{1}{x - x_0}\, dx \quad = \ln |x - x_0| + C \, ;$$

$$\int \frac{1}{(x - x_0)^k}\, dx = -\frac{1}{k - 1}\, \frac{1}{(x - x_0)^{k-1}} + C \, .$$

Note: According to the Appendix to the Fundamental Theorem of Algebra (see Volume 1, Section 5.2.7), a real polynomial has n zeros, either real or occurring in pairs as complex and complex conjugated numbers. Complex zeros are therefore subject to a special decomposition into partial fractions:

(1) If the polynomial $q(x)$ has a complex zero at $x_0 = a + ib$, then $\bar{x}_0 = a - ib$ is also a zero and the product is

$$(x - x_0)(x - \bar{x}_0) = (x - a)^2 + b^2$$

cannot be decomposed within the real numbers. Such complex zeros have to be replaced with the other zeros by

$$f(x) = \cdots + \frac{Cx + D}{(x - a)^2 + b^2} + \cdots$$

(2) If the complex zeros are of order k, the approach changes to

$$f(x) = \cdots + \frac{C_1 x + D_1}{(x - a)^2 + b^2} + \frac{C_2 x + D_2}{\left[(x - a)^2 + b^2\right]^2} + \cdots + \frac{C_k x + D_k}{\left[(x - a)^2 + b^2\right]^k} + \cdots$$

Examples 8.21 (With MAPLE-Worksheet):

① $\displaystyle\int \frac{2x^3 + x^2 + 2x + 2}{x^4 + 2x^2 + 1}\, dx = ?$: $-i, -i, i, i$ are the zeros of the denominator $q(x)$. That is, $x = i$ and $x = -i$ are double zeros. The decomposition of the integrand into partial fractions is as follows

$$\frac{2x}{x^2 + 1} + \frac{1}{x^2 + 1} + \frac{1}{(x^2 + 1)^2}$$

and the subsequent integration gives

$$\ln(x^2 + 1) + \frac{3}{2}\arctan(x) + \frac{x}{2(x^2 + 1)} + C$$

② $\displaystyle\int \frac{x^3 + 2x^2 - 1}{x^4 - 2x^3 + 2x^2 - 2x + 1}\, dx = ?$:   1 (double) and $\pm i$ are the zeros of the denominator. A partial fraction decomposition results in

$$\frac{1 - 3x}{2(x^2 + 1)} + \frac{5}{2(x - 1)} + \frac{1}{(x - 1)^2}$$

with the following integration

$$-\frac{3}{4}\ln(x^2 + 1) + \frac{1}{2}\arctan(x) + \frac{5}{2}\ln(x - 1) - \frac{1}{x - 1} + C. \qquad \square$$

8.5 Improper Integrals

So far we have assumed that the integration limits are finite. However, there are applications where they are not finite and the integrals still exist. These integrals are considered to be a special case of the definite integral already introduced.

Definition: *Integrals, where the integration limits are $\pm\infty$, are called* **Improper Integrals:**

$$\int_a^\infty f(x)\,dx\,, \qquad \int_{-\infty}^b f(x)\,dx\,, \qquad \int_{-\infty}^\infty f(x)\,dx.$$

Application Example 8.22 **(Work Integral).**

In the gravitational field of the Earth a mass m is to be brought from a distance r_0 to infinity $(r = \infty)$. What work W_∞ has to be done and what velocity ($=$ *escape velocity*) does the mass need? The work W_R required to move the mass from $r = r_0$ to $r = R$ is based on the law of gravity

$$F(r) = G\,\frac{mM}{r^2}$$

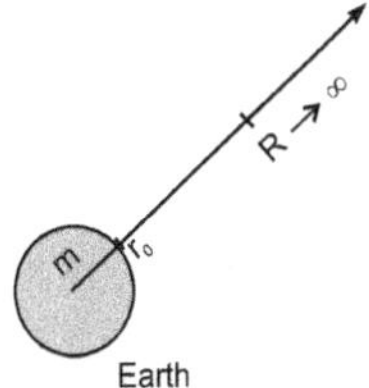

Figure 8.2. Work

given by

$$W_R = \int_{r_0}^R F(r)\,dr \;=\; \int_{r_0}^R G\,\frac{m\,M}{r^2}\,dr = G\,m\,M \int_{r_0}^R \frac{1}{r^2}\,dr$$

$$= G\,m\,M\,\left[-\frac{1}{r}\right]_{r_0}^R = G\,m\,M\,\left(\frac{1}{r_0} - \frac{1}{R}\right).$$

G is the gravitational constant and M is the mass of the Earth. For $R \to \infty$ we get

$$W_\infty = \lim_{R\to\infty} W_R = \lim_{R\to\infty} G\,m\,M\,\left[\frac{1}{r_0} - \frac{1}{R}\right] = \frac{G\,m\,M}{r_0}.$$

This work is equal to the kinetic energy $\frac{1}{2}\,m\,v^2$ that the mass must have at the beginning. So the escape velocity is

$$v_\infty = \sqrt{\frac{2\,G\,M}{r_0}}\,.$$

$\square$

The procedure chosen in the above example, i.e. first integrating r_0 to R and then letting R go to ∞, is the calculation method for improper integrals:

Calculation of Improper Integrals $\displaystyle\int_a^\infty f(x)\,dx$:

(1) Calculate the integral function $I(\lambda)$ as a function of the upper boundary $I(\lambda) = \displaystyle\int_a^\lambda f(x)\,dx$.

(2) Evaluate the limit value of the integral function as $\lambda \to \infty$:

$$\int_a^\infty f(x)\,dx = \lim_{\lambda\to\infty} I(\lambda) = \lim_{\lambda\to\infty} \int_a^\lambda f(x)\,dx.$$

Examples 8.23:

① $\displaystyle\int_1^\infty \frac{1}{x^3}\,dx = \lim_{\lambda\to\infty} \int_1^\lambda \frac{1}{x^3}\,dx = \lim_{\lambda\to\infty} \left[-\frac{1}{2}\frac{1}{x^2}\right]_1^\lambda$

$$= \lim_{\lambda\to\infty} \frac{1}{2}\left(-\frac{1}{\lambda^2}+1\right) = \frac{1}{2}.$$

② $\displaystyle\int_1^\infty \frac{1}{r}\,dr =?$ This improper integral does **not** exist:

$$\int_1^\infty \frac{1}{r}\,dr = \lim_{\lambda\to\infty} \int_1^\lambda \frac{1}{r}\,dr = \lim_{\lambda\to\infty} \ln r\Big|_1^\lambda = \lim_{\lambda\to\infty} \ln(\lambda) = \infty. \qquad \square$$

Application Example 8.24 (RL Circuit).

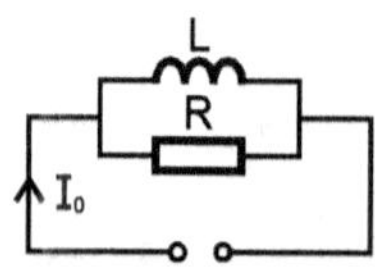

Figure 8.3. RL circuit

A coil (inductance L) and an ohmic resistor R are connected in parallel. A constant current I_0 flows. At the time $t_0 = 0$ the source is switched off and the current decreases according to $I(t) = I_0\, e^{-\frac{R}{L}t}$. The energy is given by

$$E = \int_0^\infty R\,I^2(t)\,dt = \lim_{T\to\infty} \int_0^T R\,I_0^2\, e^{-2\frac{R}{L}t}\,dt$$

$$= R\,I_0^2 \lim_{T\to\infty} \left[-\frac{L}{2R}\, e^{-2\frac{R}{L}t}\right]_0^T = \frac{1}{2}\,L\,I_0^2. \qquad \square$$

Remarks:

(1) Important *transformations* such as the *Laplace transform* ($\to$ Chapter 14) and *Fourier transform* (see Volume 3) are defined by improper integrals. The Laplace transform is defined by

$$\mathcal{L}\left(f\left(t\right)\right) = F\left(s\right) = \int_0^\infty f\left(t\right)\, e^{-st}\, dt.$$

and the Fourier transform is defined by

$$\mathcal{F}\left(f\left(t\right)\right) = F\left(\omega\right) = \int_{-\infty}^\infty f\left(t\right)\, e^{-i\,\omega\,t}\, dt.$$

(2) Integrals with *unrestricted* integrands are also called **improper integrals**:

$$\int_0^1 \frac{1}{\sqrt{1-t}}\, dt$$

is such an improper integral, since the integrand $\frac{1}{\sqrt{1-t}}$ is defined only for $0 \le t < 1$. Nevertheless, the integral has a finite value, since

$$\int_0^T \frac{1}{\sqrt{1-t}}\, dt = -2\sqrt{1-t}\,\Big|_0^T = 2 - 2\sqrt{1-T} \overset{T\to 1}{\longrightarrow} 2\,.$$

(3) There are three kinds of improper integrals:
  1. The integration interval is infinite.
  2. The integrand is infinite.
  3. Both the integration interval and the integrand are infinite.

8.6 Applications of Integral Calculus

This section gives some important applications of integral calculus. Other applications of integral calculus such as the averaging property, the arc length and the curvature behavior, and the calculation of rotational bodies together with the visualization can be found on the website.

8.6.1 Area calculations

By definition, the integral is first used to calculate areas where the area is bounded by $x = a$ and $x = b$, the x-axis and the function $f\left(x\right)$. The value of the area is then given by

$$\int_a^b f\left(x\right)\, dx.$$

Application Example 8.25 :

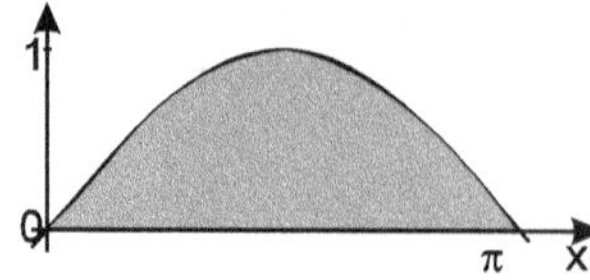

Figure 8.4. Area

We are looking for the area under a sine half-wave (see Fig. 8.4):

$$A = \int_0^\pi \sin x \, dx = -\cos x \Big|_0^\pi$$
$$= -(-1 - 1) = 2 \, .$$

The area under the curve is 2.   □

The **area between two curves** $y = f(x)$ and $y = g(x)$ is calculated from the difference of the individual integrals:

$$A = \int_a^b (f(x) - g(x)) \, dx = \int_a^b f(x) \, dx - \int_a^b g(x) \, dx.$$

Application Example 8.26 (Area between two curves).

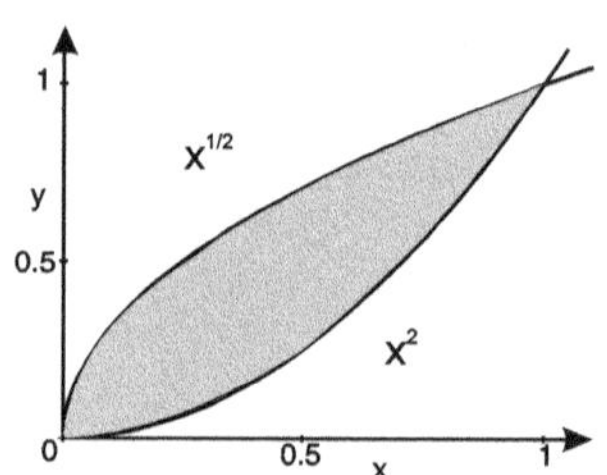

Figure 8.5. Area between curves

Find the grey area between the functions $y = \sqrt{x}$ and $y = x^2$ (see Fig. 8.5).

In order to calculate the grey hatched area, the intersection points of the curves must first be determined, as these are the integration limits:

$$\sqrt{x} = x^2 \quad \hookrightarrow x = 0 \text{ and } x = 1$$

The area enclosed by the curves is now calculated with the given integral over the difference of the functions:

$$A = \int_0^1 \left(\sqrt{x} - x^2 \right) dx$$
$$= \int_0^1 x^{\frac{1}{2}} \, dx - \int_0^1 x^2 \, dx = \frac{2}{3} x^{\frac{3}{2}} \Big|_0^1 - \frac{1}{3} x^3 \Big|_0^1 = \frac{1}{3} \, . \qquad □$$

8.6.2 Kinematics

For the motion of a mass point it holds

$$v(t) = \dot{s}(t) = \tfrac{d}{dt} s(t) \qquad\qquad \text{(velocity)},$$

$$a(t) = \tfrac{d}{dt} v(t) = \dot{v}(t) = \ddot{s}(t) \qquad \text{(acceleration)}.$$

If the acceleration is known as a function of time (e.g. by a law of force), the velocity $v(t)$ follows by integration and the space-time law $s(t)$ follows by a repeated integration:

$$v(t) = \int a(t)\, dt,$$

$$s(t) = \int v(t)\, dt.$$

Application Example 8.27 (Free Fall without Air Resistance).

Assuming a free fall without air resistance, the acceleration force is

$$m \cdot a = F_G = m g \quad \Rightarrow \quad a(t) = g = const.$$

So the velocity is

$$v(t) = \int a(t)\, dt = g\, t + C_1.$$

The integration constant is determined by the initial velocity

$$v(0) = v_0 \quad \hookrightarrow \quad C_1 = v_0 \quad \Rightarrow \quad v(t) = g\, t + v_0.$$

The space-time law follows by repeated integration

$$s(t) = \int v(t)\, dt = \int (g\, t + v_0)\, dt = \frac{g}{2}\, t^2 + v_0\, t + C_2.$$

The integration constant is determined from the initial position

$$s(0) = s_0 \quad \hookrightarrow \quad C_2 = s_0 \quad \Rightarrow \quad s(t) = \frac{1}{2}\, g\, t^2 + v_0\, t + s_0. \qquad \square$$

Application Example 8.28 (Equation of Motion of a Rocket).

A rocket rises vertically into the air with a constant thrust force F_0. The mass of the rocket decreases linearly due to the combustion of the fuel, i.e.

$$m(t) = m_0 - q\, t = m_0\,(1 - \alpha\, t) \qquad \text{with } \alpha = \frac{q}{m_0},$$

where m_0 the initial mass and q is the fuel consumption.

Assuming a constant gravitational acceleration of g and no air resistance, the accelerating force or acceleration is

$$m\,a = F_0 - m\,g \quad \hookrightarrow \quad a = \frac{F_0}{m_0\,(1 - \alpha\,t)} - g\,.$$

The velocity is

$$v\,(t) = \int a\,(t)\;dt = \frac{F_0}{m_0}\int \frac{dt}{1 - \alpha\,t} - g\int dt$$

$$= -\frac{F_0}{m_0}\,\frac{1}{\alpha}\,\ln\,(1 - \alpha\,t) - g\,t + C\,.$$

Note that the first integral is calculated with the substitution $y = 1 - \alpha\,t$. With the initial velocity $v\,(0) = 0$, we get $C = 0$.

The space-time behavior is obtained by repeated integration

$$s\,(t) = \int v\,(t)\;dt = -\frac{F_0}{m_0\,\alpha}\int \ln\,(1 - \alpha\,t)\;dt - g\int t\,dt\,.$$

With the substitution $y = 1 - \alpha\,t$ and the result from the Example 8.14 ③ $\int \ln x\,dx = x\cdot(\ln x - 1) + C$ we get

$$s\,(t) = \frac{F_0}{m_0\,\alpha}\,\frac{1}{\alpha}\,[(1 - \alpha\,t)\,\ln\,(1 - \alpha\,t) - (1 - \alpha\,t)] - \tfrac{1}{2}\,g\,t^2 + C\,.$$

With the initial condition $s\,(0) = 0$ follows $C = \frac{F_0}{m_0\,\alpha^2}$ and finally

$$s\,(t) = \frac{F_0}{m_0\,\alpha^2}\,[(1 - \alpha\,t)\,\ln\,(1 - \alpha\,t) + \alpha\,t] - \frac{1}{2}\,g\,t^2\,. \qquad \square$$

8.6.3 Electrodynamics

A point charge Q generates a radial **electric field** according to the formula

$$E\,(r) = \frac{1}{4\pi\,\varepsilon_0}\,\frac{Q}{r^2}$$

with the dielectric constant $\varepsilon_0 = 8.8542 \cdot 10^{-12}\,\frac{F}{m}$.

The *voltage* U_{12} between two points P_1 and P_2 at distances r_1 and r_2 from Q is given by

$$U_{12} = \int_{r_1}^{r_2} E\left(r\right) dr = \frac{Q}{4\pi\,\varepsilon_0} \int_{r_1}^{r_2} \frac{1}{r^2}\, dr = -\frac{Q}{4\pi\,\varepsilon_0}\,\frac{1}{r}\bigg|_{r_1}^{r_2}$$

$$= \frac{Q}{4\pi\,\varepsilon_0}\left(\frac{1}{r_1} - \frac{1}{r_2}\right).$$

8.6.4 Work Integrals

If a location-**independent** force F acts on a mass point m in the direction of the path, then the *work* done is

$$\Delta W := F \cdot \Delta s,$$

where $\Delta s = s_E - s_A$ is the distance by which the mass point is moved.

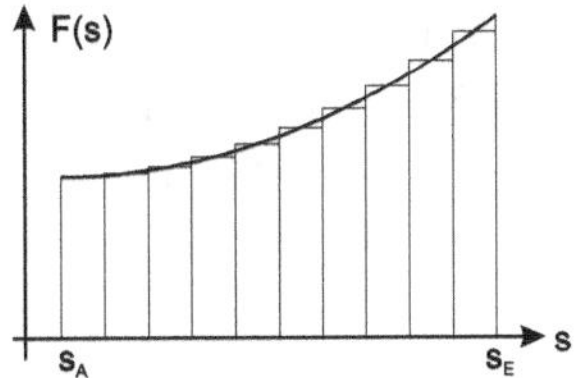

Figure 8.6. Non-constant force

However, if the force $F = F\left(s\right)$ is location-**dependent** (see Fig. 8.6), then the path must be divided into small intervals Δs. We assume a constant force in each sub-interval. In the interval Δs_i the work is then approximated by

$$\Delta W_i = F\left(s_i\right) \cdot \Delta s_i.$$

The total work W is the sum of all the individual contributions

$$W \approx \sum_{i=1}^{n} \Delta W_i = \sum_{i=1}^{n} F\left(s_i\right) \cdot \Delta s_i\,.$$

The exact value of the work is obtained by going to any fine sub-division ($\Delta s_i \to 0$ or $n \to \infty$):

$$W = \lim_{n\to\infty} \sum_{i=1}^{n} \Delta W_i = \lim_{n\to\infty} \sum_{i=1}^{n} F\left(s_i\right) \Delta s_i = \int_{s_A}^{s_E} F\left(s\right) ds.$$

Work Integral

The **work** of a location-dependent force is given by

$$W = \int_{s_A}^{s_E} F(s)\, ds,$$

if $F(s)$ is the force component in the direction of the path. Otherwise, $F(s)\, ds$ must be replaced by the scalar product

$$\vec{F}\vec{ds} = \left|\vec{F}\right| \cos\varphi\, ds$$

where φ is the angle between $\vec{F}$ and the direction $\vec{s}$.

Application Example 8.29 (**Tension of an Elastic Spring**).

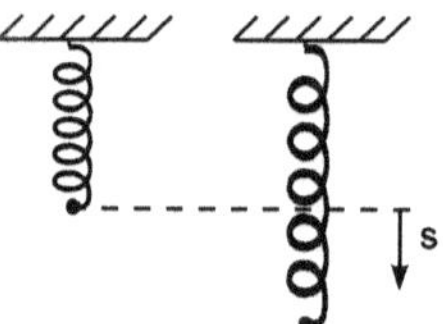

When an elastic spring is deflected by s from its rest position, a restoring force acts proportional to the deflection:

$$F(s) = -D \cdot s \qquad \text{(Hook's Law)}.$$

Figure 8.7. Elastic spring   To deflect a spring from its rest position by the distance s_0, the following work is done:

$$W = \int_0^{s_0} F(s)\, ds = D \int_0^{s_0} s\, ds = \tfrac{1}{2} D s_0^2. \qquad \square$$

Application Example 8.30 (**Work in an Electrostatic Field**).

Two point charges q_1 and q_2 exert a force on each other which is inversely proportional to the square of their distance

$$F = \frac{1}{4\pi\,\varepsilon_0} \frac{q_1 \cdot q_2}{r^2}.$$

If q_1 is in the origin and q_2 is moved from r_1 to r_2, we need to do the following work:

$$W = \int_{r_1}^{r_2} F(r)\, dr = \frac{q_1\, q_2}{4\pi\,\varepsilon_0} \int_{r_1}^{r_2} \frac{1}{r^2}\, dr = \frac{q_1\, q_2}{4\pi\,\varepsilon_0} \left(\frac{1}{r_1} - \frac{1}{r_2} \right). \qquad \square$$

8.6.5 Linear and Quadratic Mean Values

Problem: A two-way rectifier generates $i(t) = |i_0 \sin(\omega t)|$ from a sinusoidal alternating current of $\omega = \frac{2\pi}{T}$, shown in Fig. 8.8. We search for the *linear mean value*.

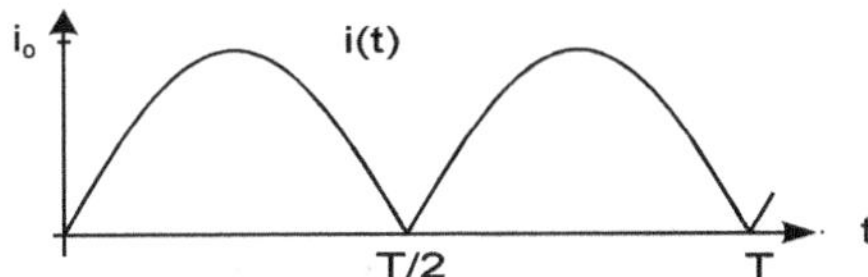

Figure 8.8. Two-way rectifier

Definition: *We define the **Linear Mean Value** of a function* $y = f(x)$ *in the interval* $[a, b]$ *as the value of*

$$\bar{y} = \frac{1}{b-a} \int_a^b f(x)\, dx.$$

Remarks:

According to the mean value theorem of the integral calculus there is always an intermediate value $\xi \in (a, b)$, such that $\bar{y} = f(\xi)$. Geometrically, the linear mean value means that the area under the curve is replaced by a rectangle of the same area.

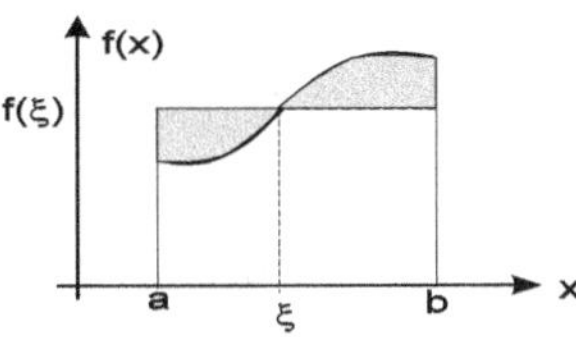

Figure 8.9. Mean value $f(\xi)$

Application Example 8.31 (Mean of a Two-way Rectifier).

The *linear* mean value of the two-way rectifier (see Fig. 8.8) is calculated using the integral

$$\bar{i} = \frac{1}{T/2} \int_0^{T/2} i_0 \sin(\omega t)\, dt = -\frac{2}{T} i_0 \frac{1}{\omega} \cos(\omega t)\Big|_0^{T/2}$$

$$= -\frac{2}{T} i_0 \frac{T}{2\pi} \left[\cos(\frac{2\pi}{T} \cdot \frac{T}{2}) - 1\right] = \frac{2}{\pi} i_0 .$$

The linear mean value of a sinusoidal alternating current, however, is 0. □

In electrical applications, therefore, the *root mean square value (RMS)* is *introduced.*

Definition: *We define the* **Root Mean Square (RMS) Value** *of a function* $y = f(x)$ *in the interval* $[a, b]$ *as the value of*

$$\bar{y}_q = \left(\frac{1}{b-a} \int_a^b f^2(x) \, dx \right)^{\frac{1}{2}} .$$

Application Example 8.32   (Effective Value of a AC).

The **Effective Value** I_{eff} of an alternating current is the *root mean square value* over a period T:

$$
\begin{aligned}
I_{\text{eff}} &= \left(\frac{1}{T} \int_0^T i^2(t) \, dt \right)^{\frac{1}{2}} = \left(\frac{1}{T} \int_0^T i_0^2 \sin^2(\omega t) \, dt \right)^{\frac{1}{2}} \\
&= \left(\frac{i_0^2}{T} \int_0^T \sin^2(\omega t) \, dt \right)^{\frac{1}{2}} = \left(\frac{i_0^2}{T} \left[\tfrac{1}{2} t - \frac{1}{4\omega} \sin(2\omega t) \right]_0^T \right)^{\frac{1}{2}} \\
&= \left(\frac{i_0^2}{T} \left[\tfrac{1}{2} T - \frac{1}{4\omega} \sin(2\omega T) \right] \right)^{\frac{1}{2}} = \frac{i_0}{\sqrt{2}} .
\end{aligned}
$$

Similarly, the effective value of an alternating voltage is $U_{\text{eff}} = \frac{U_0}{\sqrt{2}}$. $\qquad \square$

8.6.6 Center of Gravity of a Plane Surface

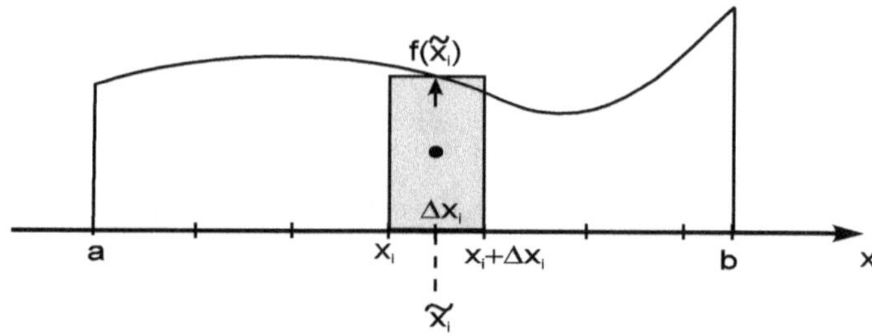

Figure 8.10. Center of gravity of a plane surface

If $(x_1, y_1), (x_2, y_2), \ldots, (x_n, y_n)$ are the coordinates of n mass points with masses $m_1, m_2, \ldots, m_n$, then the coordinates of the *center of gravity* are obtained by

$$x_s = \frac{\sum\limits_{i=1}^{n} m_i \, x_i}{\sum\limits_{i=1}^{n} m_i} \quad \text{and} \quad y_s = \frac{\sum\limits_{i=1}^{n} m_i \, y_i}{\sum\limits_{i=1}^{n} m_i}.$$

To calculate the center of gravity of a plane surface, we assume that the surface is bounded by a curve $y = f(x)$ and the x-axis between $x = a$ and $x = b$ (Fig. 8.10). We occupy the surface homogeneously with a mass density of 1. To calculate the center of gravity we divide the interval $[a, b]$ into n sub-intervals, $a = x_0 < x_1 < \cdots < x_n = b$, select for each sub-interval Δx_i an intermediate point $\tilde{x}_i \in [x_i, x_i + \Delta x_i]$ and determine $f(\tilde{x}_i)$. The center of gravity of each rectangle $A_i = \Delta x_i \, f(\tilde{x}_i)$ is

$$x_{s,i} = x_i + \frac{1}{2}\Delta x_i \quad , \quad y_{s,i} = \frac{1}{2} f(\tilde{x}_i)$$

with the masses $m_i = \frac{A_i}{A} = A_i = \Delta x_i \, f(\tilde{x}_i)$ $(i = 1, \ldots, n)$. The coordinates of the center of gravity of the n masses are given by $(x_{s,1}, y_{s,1}), \ldots, (x_{s,n}, y_{s,n})$. According to the formulas above we get

$$x_s = \frac{\sum\limits_{i=1}^{n} m_i \, x_{s,i}}{\sum\limits_{i=1}^{n} m_i} = \frac{\sum\limits_{i=1}^{n} \Delta x_i \, f(\tilde{x}_i) \cdot \left(x_i + \frac{1}{2}\Delta x_i\right)}{\sum\limits_{i=1}^{n} \Delta x_i \, f(\tilde{x}_i)},$$

$$y_s = \frac{\sum\limits_{i=1}^{n} m_i \, y_{s,i}}{\sum\limits_{i=1}^{n} m_i} = \frac{\sum\limits_{i=1}^{n} \Delta x_i \, f(\tilde{x}_i) \cdot \frac{1}{2} f(\tilde{x}_i)}{\sum\limits_{i=1}^{n} \Delta x_i \, f(\tilde{x}_i)}.$$

As $n \to \infty$, the subtotals merge into the corresponding integrals.

Center of Gravity

The coordinates of the **Center of Gravity** $S = (x_s, y_s)$ for the area under a graph $y = f(x)$ between $x = a$ and $x = b$ are

$$x_s = \frac{\int_a^b x \, f(x) \, dx}{\int_a^b f(x) \, dx} \quad \text{and} \quad y_s = \frac{\frac{1}{2}\int_a^b f^2(x) \, dx}{\int_a^b f(x) \, dx}.$$

Examples 8.33:

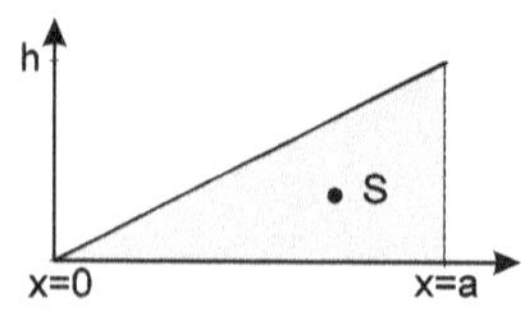

① The coordinates of the center of gravity for the adjacent triangle below the straight line $y = f(x) = \frac{h}{a} x$ are given by

$$x_s = \frac{1}{A} \int_0^a x \, \frac{h}{a} x \, dx = \frac{1}{A} \frac{h}{a} \left[\frac{x^3}{3} \right]_0^a = \frac{1}{A} \frac{h}{3} a^2,$$

$$y_s = \frac{1}{2A} \int_0^a \left(\frac{h}{a} x \right)^2 dx = \frac{1}{2A} \frac{h^2}{a^2} \left[\frac{x^3}{3} \right]_0^a = \frac{1}{2A} \frac{h^2}{3} a.$$

With

$$A = \int_0^a f(x) \, dx = \int_0^a \frac{h}{a} x \, dx = \frac{h}{a} \frac{x^2}{2} \Big|_0^a = \frac{1}{2} h a$$

we get the result

$$\boxed{x_s = \tfrac{2}{3} a} \quad \text{and} \quad \boxed{y_s = \tfrac{1}{3} h.}$$

② We calculate the center of gravity of the adjacent semicircle with radius R: For reasons of symmetry, the center of gravity is on the y-axis, so that $x_s = 0$. For the y-coordinate of the center of gravity we calculate

$$y_s = \tfrac{1}{2A} \int_{-R}^{R} y^2 \, dx$$

$$= \tfrac{1}{2A} \int_{-R}^{R} (R^2 - x^2) \, dx$$

$$= \tfrac{1}{2A} \left[R^2 x - \tfrac{1}{3} x^3 \right]_{-R}^{R} = \tfrac{1}{2A} \tfrac{4}{3} R^3.$$

With $A = \frac{\pi}{2} R^2$ we get $\quad y_s = \dfrac{4}{3\pi} R.$

③ The coordinates of the center of gravity of the area A, which is bounded by the two functions $y_2 = f(x)$ and $y_1 = g(x)$ with $f \geq g$ and the lines $x = a$ and $x = b$, are given by the difference of the individual centers of gravity:

$$x_s = \frac{1}{A} \int_a^b x \left[f(x) - g(x) \right] dx$$

$$y_s = \frac{1}{2A} \int_a^b \left[f^2(x) - g^2(x) \right] dx$$

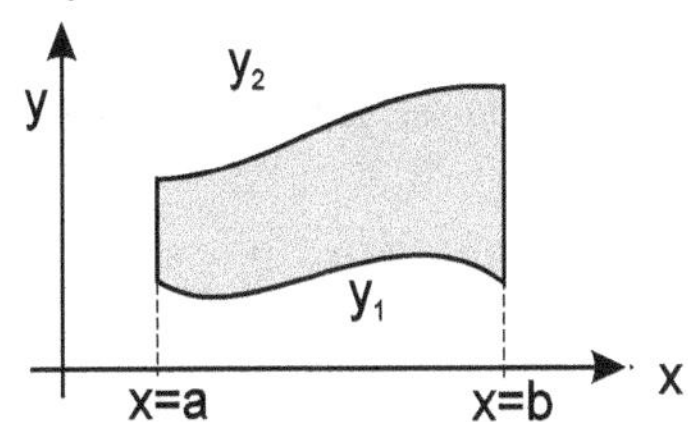

and

$$A = \int_a^b \left[f(x) - g(x) \right] dx.$$
□

8.6.7 Averaging property

An important property of the integral is its smoothing behavior, because

$$\frac{1}{(b-a)} \int_a^b f(x)\, dx$$

is the mean value of a function in the interval $[a, b]$. This property is used when interpreting measurement results, as these are usually distorted by noise.

Application Example 8.34 . We look at the function

$$f(x) = x^2 \left(1 + \frac{1}{20} \sin(200\,x) \right) + \frac{1}{20} \cos(50\,x)$$

which corresponds on average to an x^2-function, but with high frequency noise superimposed.

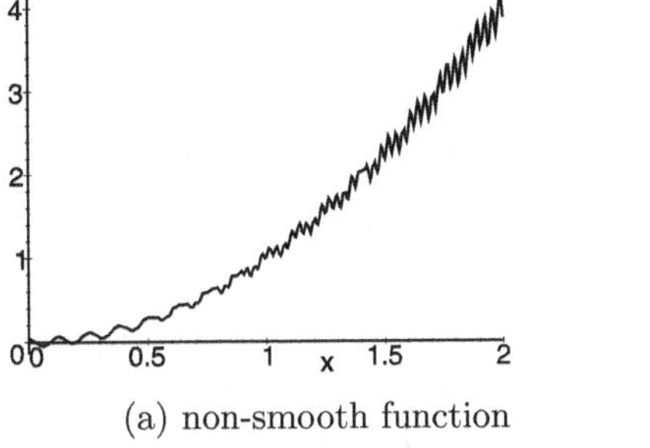

(a) non-smooth function   (b) after averaging

Figure 8.11. Non-smooth function

Linear averaging with an appropriate interval length $h = 0.1$ produces a smooth curve: But if h is too small, oscillations are still observed (e.g. for $h = 0.05$); if h is too large, the resulting function becomes angular (e.g. for $h = 0.5$). The appropriate h is determined by the occurring frequencies. Here, the lowest frequency is $\omega = 50 = 2\pi/T$, corresponding to a period of $T = 2\pi/50 = 0.125$. The appropriate h is therefore about 0.1. In practice, however, the interfering frequencies are not known. To reconstruct them from the signal, methods of the Fourier analysis (see Volume 3) must be applied. $\qquad\square$

8.6.8 Arc length

The *arc length* of a curve $\overset{\frown}{AB}$ is calculated by first dividing the interval $[a, b]$ into n sub-intervals $a = x_0 < x_1 < x_2 < \cdots < x_n = b$.

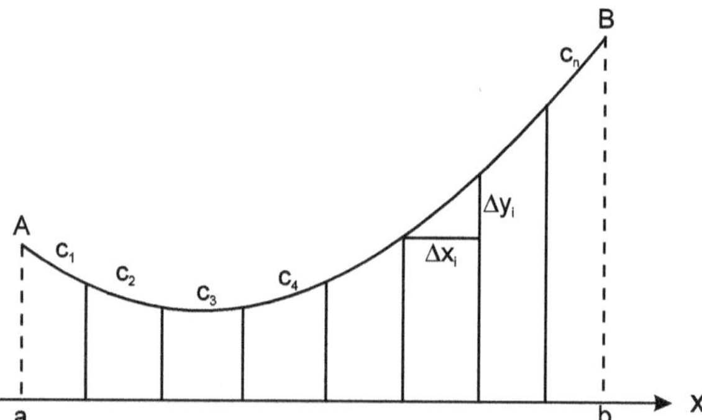

Figure 8.12. Arc length of a curve

For each sub-interval, the curve is replaced by a straight line $c_1, c_2, \ldots, c_n$ with the individual lengths

$$|c_i| = \sqrt{(\Delta x_i)^2 + (\Delta y_i)^2} = \sqrt{1 + \left(\frac{\Delta y_i}{\Delta x_i}\right)^2} \cdot \Delta x_i.$$

The total length of all the pieces is

$$S_n = \sum_{i=1}^{n} |c_i| = \sum_{i=1}^{n} \sqrt{1 + \left(\frac{\Delta y_i}{\Delta x_i}\right)^2} \cdot \Delta x_i.$$

By refining the decomposition of the interval $[a, b]$ as $n \to \infty$, the graph of f is exactly approximated by the stretch. If the limit value $\lim\limits_{n \to \infty} S_n$ exists, f is called *rectifiable* and the limit value is the *arc length* of the graph.

To calculate the limit value, the mean value theorem of differential calculus is used. According to this theorem, for each interval there is $\tilde{x}_i \in [x_i, x_i + \Delta x_i]$, so that $\frac{\Delta y_i}{\Delta x_i} = f'(\tilde{x}_i)$. For $n \to \infty$, then $\Delta x_i \to 0$ and $\frac{\Delta y_i}{\Delta x_i} \to f'(x_i)$. So the arc length is

$$S = \lim_{n \to \infty} S_n = \lim_{n \to \infty} \sum_{i=1}^{n} \sqrt{1 + (f'(\tilde{x}_i))^2} \cdot \Delta x_i = \int_a^b \sqrt{1 + (f'(x))^2} \, dx.$$

Arc Length

Let f be a function that is continuously differentiable on the interval $[a, b]$. Then the **arc length** S of $y = f(x)$ between $x = a$ and $x = b$ is

$$S = \int_a^b \sqrt{1 + (f'(x))^2} \, dx = \int_a^b \sqrt{1 + (y')^2} \, dx.$$

Example 8.35. Find the arc length of the function $y = \cosh(x)$ between $x = 0$ and $x = 1$:

The derivative of $y = \cosh(x)$ is $y' = \sinh(x)$. This gives the integrand to be

$$\hookrightarrow \sqrt{1 + (y')^2} = \sqrt{1 + \sinh^2(x)} = \cosh(x),$$

because $\cosh^2(x) - \sinh^2(x) = 1$. So the arc length is

$$S = \int_0^1 \cosh(x) \, dx = \sinh(x) \Big|_0^1 = \sinh(1) = 1.175.$$ $\square$

Example 8.36. Calculate the arc length of a quarter circle (see Figure 8.13).

The circle equation is $x^2 + y^2 = 1$. We solve this equation for

$$y = f(x) = \sqrt{1 - x^2}$$

and obtain the derivative

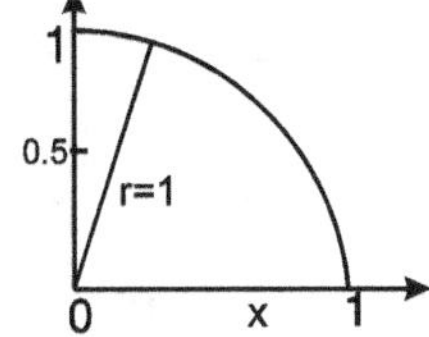

Figure 8.13. Circle

$$y' = \frac{1}{2} \frac{1}{\sqrt{1 - x^2}} \cdot (-2x) = -\frac{x}{\sqrt{1 - x^2}}.$$

The integrand for calculating the arc length is therefore

$$\sqrt{1+y'^2} = \sqrt{1+\frac{x^2}{1-x^2}} = \sqrt{\frac{1-x^2+x^2}{1-x^2}} = \frac{1}{\sqrt{1-x^2}} = (1-x^2)^{-\frac{1}{2}}.$$

To calculate the integral

$$S = \int_0^1 \sqrt{1+y'^2}\,dx = \int_0^1 (1-x^2)^{-\frac{1}{2}}\,dx$$

we choose the substitution

$$x = \sin(t).$$

Because then

$$(1-x^2)^{-\frac{1}{2}} = (1-\sin^2(t))^{-\frac{1}{2}} = (\cos^2(t))^{-\frac{1}{2}} = (\cos(t))^{-1}$$

and

$$\frac{dx}{dt} = \cos(t) \Rightarrow dx = \cos(t)dt.$$

Inserted into the integral, we get the results taking the limits into account $t_u = \arcsin(x_u) = \arcsin(0) = 0$ and $t_o = \arcsin(x_o) = \arcsin(1) = \pi/2$:

$$S = \int_0^1 (1-x^2)^{-1/2}\,dx = \int_0^{\pi/2} \frac{1}{\cos(t)}\cos(t)\,dt = \Big[t\Big]_0^{\pi/2} = \frac{\pi}{2}. \qquad \square$$

8.6.9 Curvature

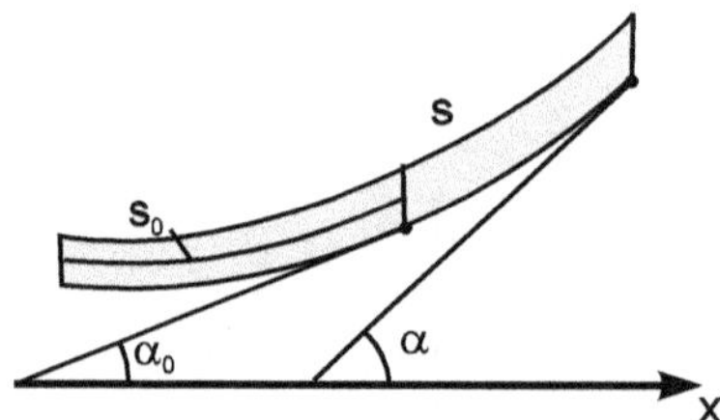

Figure 8.14. Curvature of a curve

The *Curvature* κ of a curve is a measure of how much the helix angle α changes as a function of the arc length s:

$$\kappa := \frac{d\alpha}{ds} \qquad \text{(Curvature)}$$

This initially very unwieldy quantity is qualitatively easy to understand: it says that if the angle α changes more for the same arc length, the curve has a greater curvature. The arc length s is given as a function of x by

$$S = \int_{x_0}^x \sqrt{1+(f'(\tilde{x}))^2}\,d\tilde{x}.$$

The tilt angle α is implicitly given as a function of x by the derivative of the function f:

$$\tan \alpha = f'(x) \quad \Rightarrow \quad \alpha(x) = \arctan(f'(x)) \ .$$

According to the chain rule

$$\kappa = \frac{d\alpha}{ds} = \frac{d\alpha}{dx} \cdot \frac{dx}{ds}$$

or according to the formula for the derivation of the inverse function

$$\kappa = \frac{d\alpha}{dx} \Big/ \frac{ds}{dx} \ .$$

Because of

$$\frac{d\alpha}{dx} = \frac{d}{dx} \arctan f'(x) = \frac{1}{(f'(x))^2 + 1} \cdot f''(x)$$

and

$$\frac{ds}{dx} = \frac{d}{dx} \int_{x_0}^{x} \sqrt{1 + (f'(\tilde{x}))^2} \ d\tilde{x} = \sqrt{1 + (f'(x))^2}$$

we finally get the expression

Curvature of a Function f

$$\kappa = \frac{f''(x)}{\left(1 + (f'(x))^2\right)^{\frac{3}{2}}} \ .$$

Examples 8.37:

① Curvature of a straight line: $y = a\,x + b \quad \Rightarrow \quad y'' = 0 \quad \Rightarrow \quad \kappa = 0.$

② Curvature of a parabola:
$$y = a\,x^2 \quad \Rightarrow \quad y'' = 2\,a \quad \Rightarrow \quad \kappa = \frac{2\,a}{(1 + 4\,a^2\,x^2)^{\frac{3}{2}}} \ .$$

Especially at the point $x = 0$ the curvature is $\kappa = 2\,a.$

③ Curvature of a circle with radius R:
$$y = \sqrt{R^2 - x^2} \quad \Rightarrow \quad \kappa = -\frac{1}{\sqrt{R^2 - x^2}} \ \frac{1}{\sqrt{\frac{R^2}{R^2 - x^2}}} = -\frac{1}{R} \ . \qquad \square$$

8.6.10 Volume and Surface Area of Bodies of Revolution

A body created by rotating a plane surface about an axis is called a *rotation body* or *solid of revolution*. In this section we only consider solids of revolution created by rotating the surface between a function $y = f(x)$ and the x-axis.

Bodies of revolution created by rotation about the y-axis are traced back to the case discussed by considering the inverse function.

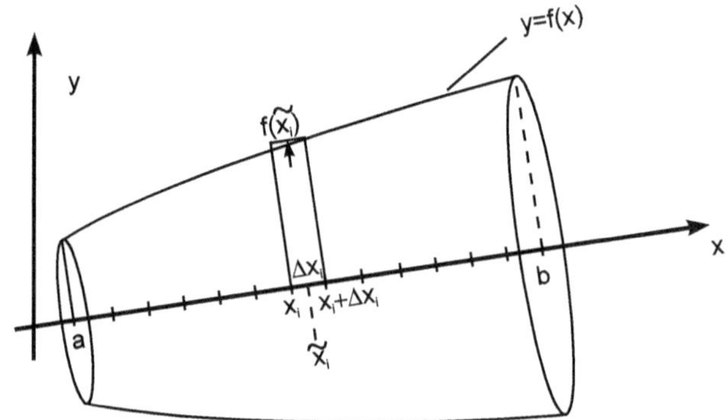

Figure 8.15. Volume of a body of revolution

(1) Volume of a body of revolution. To calculate the volume, we divide the interval $[a,\ b]$ into n sub-intervals $a = x_0 < x_1 < \cdots < x_n = b$ with interval lengths Δx_i. For each sub-interval Δx_i we choose an intermediate value $\tilde{x}_i$ and calculate the functional value $f(\tilde{x}_i)$.

The volume of the corresponding cylinder with height Δx_i and radius $f(\tilde{x}_i)$ is

$$V_i = \pi\, f(\tilde{x}_i)^2\, \Delta x_i .$$

We add all the partial cylinders and obtain the subtotal S_n

$$S_n = \sum_{i=1}^{n} V_i = \pi \sum_{i=1}^{n} f(\tilde{x}_i)^2\, \Delta x_i .$$

By refining the subdivision ($n \to \infty$ or $\Delta x_i \to 0$) the subtotal S_n becomes the integral.

Volumes of Solids of Revolution

The **volume of a body (solids of revolution)**, which is obtained by rotating the surface under the graph $y = f(x)$ about the x-axis with limits $x = a$ and $x = b$, is calculated by the expression

$$V = \pi \int_a^b \left(f(x) \right)^2 \, dx = \pi \int_a^b y^2 \, dx.$$

Examples 8.38:

① **Volume of a cone:** By rotating the straight line $y = \frac{r}{h} \cdot x$ around the x-axis, we obtain a cone with a volume of

$$V_{cone} = \pi \int_0^h y^2 \, dx = \pi \int_0^h \frac{r^2}{h^2} x^2 \, dx = \pi \frac{r^2}{h^2} \left[\frac{x^3}{3} \right]_0^h = \frac{\pi}{3} r^2 h.$$

② **Volume of a sphere:** By rotating the function $y = \sqrt{R^2 - x^2}$ around the x-axis, we obtain a sphere with the volume

$$V_{sphere} = \pi \int_{-R}^R y^2 \, dx = \pi \int_{-R}^R \left(R^2 - x^2 \right) \, dx$$

$$= \pi \left[R^2 x - \frac{x^3}{3} \right]_{-R}^R = \frac{4}{3} \pi R^3. \qquad \square$$

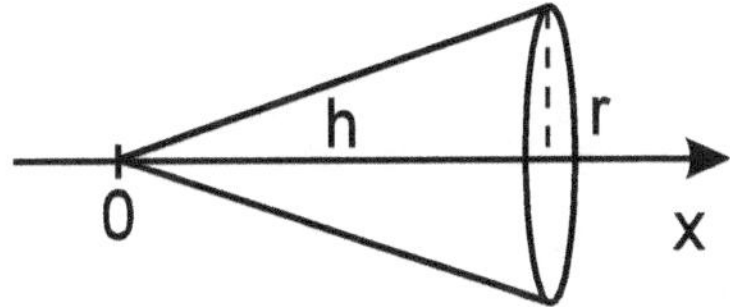

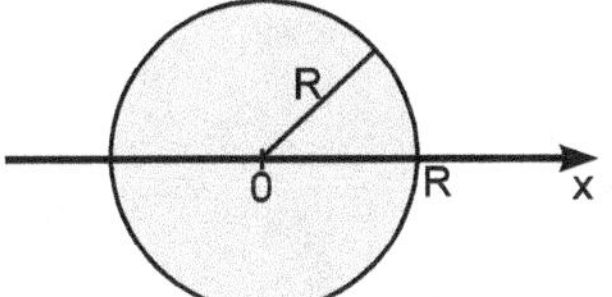

(2) Lateral surfaces of rotating bodies. According to the procedure for calculating the volume of a body of revolution from Fig. 8.15, its lateral surface is obtained by calculating the arc length of the curve segments c_i according to Section 8.6.8

$$|c_i| = \sqrt{1 + \left(\frac{\Delta y_i}{\Delta x_i} \right)^2} \, \Delta x_i.$$

The surface of the truncated cone has the mean radius $f(\tilde{x}_i)$, where $\tilde{x}_i = \frac{1}{2}(x_i + x_{i+1})$, with a side length $|c_i|$ of

$$M_i = 2\pi f(\tilde{x}_i) \sqrt{1 + \left(\frac{\Delta y_i}{\Delta x_i}\right)^2} \, \Delta x_i .$$

Summing over all surfaces gives

$$S_n = \sum_{i=1}^{n} M_i = 2\pi \sum_{i=1}^{n} f(\tilde{x}_i) \sqrt{1 + \left(\frac{\Delta y_i}{\Delta x_i}\right)^2} \, \Delta x_i .$$

By refining the subdivision ($n \to \infty$ or $\Delta x_i \to 0$), we get

Lateral Surfaces

The **surface of a body in revolution**, obtained by rotating the graph $y = f(x)$ around the x-axis with the limits $x = a$ and $x = b$, is given by

$$M = 2\pi \int_a^b f(x) \sqrt{1 + (f'(x))^2} \, dx .$$

The total surface area is obtained by adding the base area $\pi f^2(a)$ and the top area $\pi f^2(b)$ to M. In the next table we summarize the rotation expressions:

Rotation about x-axis	Rotation about y-axis
$V_x = \int_a^b \pi f^2(x) \, dx$	$V_y = \int_a^b 2\pi x \, f(x) \, dx$
$M_x = \int_a^b 2\pi f(x) \sqrt{1 + f'^2(x)} \, dx$	$M_y = \int_a^b 2\pi x \sqrt{1 + f'^2(x)} \, dx$

Note that the formulas are only valid if the graph of the function $y = f(x)$ does not intersect the axis of rotation. Otherwise, y must be replaced by the absolute value of y.

8.7 Problems on Integral Calculus

8.1   Take the function $f(x) = \sqrt{x}$ in the $x \in [0, 2]$ range and calculate for $n = 10$ the right sum.

8.2   The following integrals are to be found:

a) $\displaystyle\int x^5 \, dx$ b) $\displaystyle\int \frac{dx}{x^2}$ c) $\displaystyle\int \sqrt[3]{z} \, dz$

d) $\displaystyle\int \frac{dx}{\sqrt[3]{x^2}}$ e) $\displaystyle\int \left(2x^2 - 5x + 3\right) dx$ f) $\displaystyle\int (1-x)\sqrt{x} \, dx$

8.3   Calculate the following definite integrals:

a) $\displaystyle\int_0^{\frac{\pi}{2}} \sin x \, dx$ b) $\displaystyle\int_{\frac{1}{2}\pi}^{\frac{3}{2}\pi} \left(2x + \sin x - \cos x\right) dx$ c) $\displaystyle\int_1^a \frac{1}{x} \, dx$

8.4   Use integration by parts to determine the following integrals:

a) $\displaystyle\int x \cos x \, dx$ b) $\displaystyle\int \sin x \cos x \, dx$ c) $\displaystyle\int x^2 \sin x \, dx$

d) $\displaystyle\int x^2 \ln x \, dx$ e) $\displaystyle\int x \, e^x \, dx$ f) $\displaystyle\int x^2 \, e^x \, dx$

8.5   Determine the following integrals by substitution:

a) $\displaystyle\int \frac{1}{x+2} \, dx$ b) $\displaystyle\int \frac{x}{x^2 - 1} \, dx$ c) $\displaystyle\int \frac{x^2}{1 - 2x^3} \, dx$

d) $\displaystyle\int (3s + 4)^8 \, ds$ e) $\displaystyle\int \sin(\omega t + \varphi) \, dt$ f) $\displaystyle\int \cos(3t) \, dt$

g) $\displaystyle\int e^{-x} \, dx$ h) $\displaystyle\int \frac{\sin t}{\cos t} \, dt$ i) $\displaystyle\int \frac{e^x + x \, e^x}{x \, e^x} \, dx$

j) $\displaystyle\int \sin x \cos x \, dx$ k) $\displaystyle\int \sqrt{4 + 3x} \, dx$

8.6   Show the validity of the following expressions:

a) $\displaystyle\int e^{-x}(1-x) \, dx = x \, e^{-x} + C$

b) $\displaystyle\int \frac{\sqrt{x^2 - 4}}{x} \, dx = \sqrt{x^2 - 4} - 2 \arccos\left(\frac{2}{x}\right) + C$

c) $\displaystyle\int \cos(x) \, e^{\sin(x)} \, dx = e^{\sin(x)} + C$

d) $\displaystyle\int \cos(3x) \cdot \sin(3x) \, dx = \frac{1}{6} \sin^2(3x) + C$

8.7   a) Solve the integral $\displaystyle\int \frac{2-x}{1+\sqrt{x}} \, dx$ with the substitution $u = 1 + \sqrt{x}$.

     b) Solve the integral $\displaystyle\int x\sqrt{1 - x^2} \, dx$ with the substitution $x = \sin u$.

8.8   Calculate the following integrals using a suitable substitution:

a) $\displaystyle\int \frac{x^2}{\sqrt{1 + x^3}} \, dx$ b) $\displaystyle\int (5x + 12)^{\frac{1}{2}} \, dx$ c) $\displaystyle\int \sqrt[3]{1 - t} \, dt$

d) $\displaystyle\int_0^\pi \cos^3 x \cdot \sin x \, dx$ e) $\displaystyle\int \frac{\arctan z}{1 + z^2} \, dz$ f) $\displaystyle\int \frac{2x + 6}{x^2 + 6x - 12} \, dx$

g) $\displaystyle\int \frac{dx}{x \ln x}$
h) $\displaystyle\int x \cdot \sin\left(x^2\right) dx$
i) $\displaystyle\int \frac{3\,x^2 - 2}{2\,x^3 - 4\,x + 2}\,dx$

j) $\displaystyle\int_{-1}^{1} \frac{t}{\sqrt{1 + t^2}}\, dt$
k) $\displaystyle\int_{0}^{\frac{\pi}{2}} \sin\left(3\,t - \frac{\pi}{4}\right) dt$
l) $\displaystyle\int_{-1}^{1} \frac{5 + x}{5 - x}\, dx$

m) $\displaystyle\int x^2\, e^{x^3 - 2}\, dx$
n) $\displaystyle\int \frac{\tan\left(z + 5\right)}{\cos^2\left(z + 5\right)}\, dz$
o) $\displaystyle\int \frac{\sqrt{4 - x^2}}{x^2}\, dx$

8.9 Solve the following integrals by integration by parts:

a) $\displaystyle\int x\, \ln x\, dx$
b) $\displaystyle\int x\, \cos x\, dx$
c) $\displaystyle\int \ln t\, dt$

d) $\displaystyle\int x\, \sin\left(3\,x\right) dx$
e) $\displaystyle\int \arctan x\, dx$
f) $\displaystyle\int \sin^2\left(\omega t\right) dt$

g) $\displaystyle\int e^x\, \cos x\, dx$
h) $\displaystyle\int x^2\, e^{-x}\, dx$

8.10 Solve the following integrals by partial fractional decomposition:

a) $\displaystyle\int \frac{1}{x^2 - a^2}\, dx$
b) $\displaystyle\int \frac{4\,x^3}{x^3 + 2\,x^2 - x - 2}\, dx$

c) $\displaystyle\int \frac{3\,z}{z^3 + 3\,z^2 - 4}\, dz$
d) $\displaystyle\int \frac{4\,x - 2}{x^2 - 2\,x - 63}\, dx$

e) $\displaystyle\int \frac{2\,x + 1}{x^3 - 6\,x^2 + 9\,x}\, dx$

8.11 Calculate the following integrals:

a) $\displaystyle\int \frac{\sqrt{\ln x}}{x}\, dx$
b) $\displaystyle\int \cot x\, dx$
c) $\displaystyle\int x\, \cosh x\, dx$

d) $\displaystyle\int \sin x\, e^{\cos x}\, dx$
e) $\displaystyle\int \frac{x^3}{\left(x^2 - 1\right)\left(x + 1\right)}\, dx$
f) $\displaystyle\int \frac{x - 4}{x + 1}\, dx$

g) $\displaystyle\int \frac{\left(\ln x\right)^3}{x}\, dx$
h) $\displaystyle\int \frac{12\,x^2}{2\,x^3 - 1}\, dx$
i) $\displaystyle\int x \cdot \arctan x\, dx$

8.12 Determine $\displaystyle\int \ln(x + \sqrt{1 + x^2})\, dx$ by first performing the substitution $u^2 = 1 + x^2$ and then integrating it by parts.

8.13 Calculate the indefinite integral

$$\int \frac{1}{\sqrt{1 + \sqrt{x}}}\, dx$$

by the substitution $u^2 = 1 + \sqrt{x}$.

8.14 Perform a partial fractional decomposition and then integrate

a) $\displaystyle\int \frac{x^4 - x^3 - x - 1}{x^3 - x^2}\, dx$

b) $\displaystyle\int \frac{x^4 - x^3 + 3\,x}{x^2\left(x + 2\right)\left(x - 3\right)}\, dx$

c) $\displaystyle\int \frac{x^4 - x^3 + 3\,x^2 - 2\,x + 1}{x^3 - x^2 - x + 1}$

8.15 Calculate the following definite integrals (n, $m \in \mathbb{N}$) by using the addition theorems for sine and cosine.

a) $\displaystyle\int_0^{2\pi} \sin(n\,x)\,dx$

b) $\displaystyle\int_0^{2\pi} \cos(n\,x)\,dx$

c) $\displaystyle\int_0^{2\pi} \cos(n\,x)\cos(m\,x)\,dx$ for $m = n$ and for $m \neq n$

d) $\displaystyle\int_0^{2\pi} \sin(n\,x)\sin(m\,x)\,dx$ for $m = n$ and for $m \neq n$

e) $\displaystyle\int_0^{2\pi} \sin(n\,x)\cos(m\,x)\,dx$

8.16 Determine the value of the improper integrals:

a) $\displaystyle\int_2^{\infty} x^{-2}\,dx$ b) $\displaystyle\int_0^1 \frac{1}{\sqrt{1-x^2}}\,dx$ c) $\displaystyle\int_0^{\infty} \left(3\,e^{-2\,x} - e^{-x}\right)\,dx$

d) $\displaystyle\int_0^{\infty} e^{at}e^{-st}\,dt$ e) $\displaystyle\int_0^{\infty} \cos(at)\,e^{-st}\,dt$ f) $\displaystyle\int_0^{\infty} t^n\,e^{-st}\,dt$

8.17 Calculate the area between the parabola
$f(x) = x^2 - 2\,x - 1$ and the straight line $g(x) = 3\,x - 1$.

⊚ Additional Problems on Integral Calculus

8.a1 Calculate the average power of a sinusoidal alternating current

$$\bar{P} = \frac{1}{T} \int_0^T P(t)\, dt,$$

if $P(t) = U(t) \cdot I(t) = u_0 \sin(\omega t) \cdot i_0 \sin(\omega t + \varphi)$.

8.a2 Three mean values are defined for an alternating current $I(t)$ with period T:

$$I_{eff} = \sqrt{\frac{1}{T} \int_0^T I^2(t)\, dt} \qquad \text{(Effective value)}$$

$$\hat{I} = \frac{1}{T} \int_0^T I(t)\, dt \qquad \text{(Linear average value)}$$

$$|\hat{I}| = \frac{1}{T} \int_0^T |I(t)|\, dt \qquad \text{(Rectified value)}$$

Calculate these quantities for
a) for $I(t) = I_0 \sin(\frac{2\pi}{T} t)$
b) for a sawtooth current these three average values.

8.a3 Determine the arc length and the curvature of the curve $y = x^3$ between $x = 0$ and $x = 5$.

8.a4 Calculate the arc length of the chain line $y = a \cosh\left(\frac{x}{a}\right)$ between $x = 0$ and $x = b$.

8.a5 Calculate the lengths and curvatures of the curves
a) $y = \frac{1}{2} x^2$, $-1 \le x \le 1$ b) $y = \cosh(x)$, $-a \le x \le a$.

8.a6 Show that the volume of a truncated cone is

$$V = \frac{1}{3} \pi h \left(R^2 + R r + r^2\right)$$

by dividing the linear equation by the points $(0, r)$ and (h, R) and rotating the graph around the x axis.

8.a7 Show that the arc length of a semicircle with radius $r = 1$ is π.

8.a8 Determine the arc length of the graph of the function $f(x) = \frac{2}{3} x^{\frac{3}{2}}$ between the points $(0, 0)$ and $(3, f(3))$. What is the curvature of the curve? What is its volume of rotation?

8.a9 Determine the volume of the body, which is generated by rotating the curve $y = a x^2$ $(x \in [0, r])$ around the y axis.

8.a10 Calculate the center of gravity coordinates of the area between the graph $f(x) = h$ and $g(x) = \frac{h}{a^2} x^2$ for $x \in [0, a]$.

Chapter 9
Taylor Series

Important functions, especially those occurring in applications, can be described as *power series* in the form of $\sum_{n=0}^{\infty} a_n \, (x - x_0)^n$, the so-called *Taylor series*. This chapter provides a way to explicitly calculate the Taylor series of e^x, $\sin(x)$, $\tan(x)$, $\sqrt{x}$, $\ln(x)$ or $\arctan(x)$. With this kind of representation of the functions, they can be evaluated at any point x_0 only using the basic arithmetic operations $+ - * /$. In addition, it is important for many applications that approximation formulas are available for complicated expressions.

9

9 Taylor Series

Important functions, especially those occurring in applications, can be described as *power series* in the form of $\sum_{n=0}^{\infty} a_n \, (x - x_0)^n$, the so-called *Taylor series*. This chapter provides a way to explicitly calculate the Taylor series of e^x, $\sin(x)$, $\tan(x)$, $\sqrt{x}$, $\ln(x)$ or $\arctan(x)$. With this kind of representation of the functions, they can be evaluated at any point x_0 only using the basic arithmetic operations $+ \; - \; * \; /$. In addition, it is important for many applications that approximation formulas are available for complicated expressions.

Application Example 9.1 (Headlight Control).

When adjusting headlights, the law requires the height of the dipped beam to be reduced by a specified amount $H_{opt} = 0.1\,m$ over a distance of $10\,m$. This specification gives a target inclination of the headlamps for the cut-off line $\beta_{ab} = \arctan \frac{H_{opt}}{10} \approx 0.009999 \mathrel{\hat{=}} 0.5729°$.

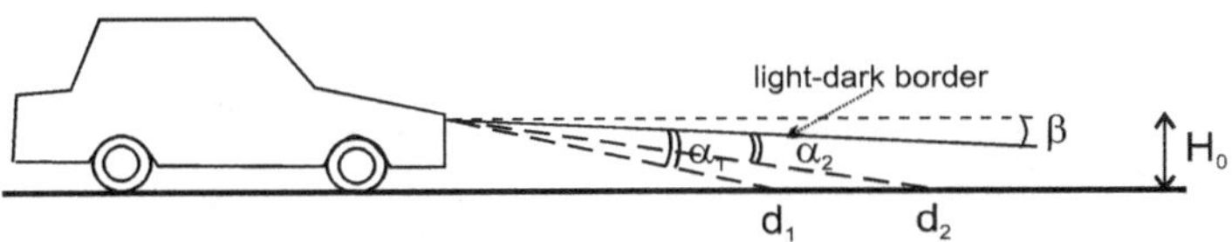

Figure 9.1. Geometric arrangement of the two measuring beams

Since the actual inclination angle β of the headlight cannot be determined directly, we use an optical measurement of two distances d_1 and d_2, which are related to the actual headlight angle β by means of

$$d_1 = \frac{H_0}{\sin(\alpha_1 + \beta)} \quad \text{and} \quad d_2 = \frac{H_0}{\sin(\alpha_2 + \beta)} \, ,$$

where α_1 and α_2 are the fixed angles. When we take the quotient, the equations reduce to

$$\frac{d_1}{d_2} = \frac{\sin(\alpha_2 + \beta)}{\sin(\alpha_1 + \beta)} = q(\beta).$$

To determine the actual inclination angle β of the headlight, we need to solve the equation with respect to β. This task cannot be solved analytically. Therefore, we need an approximate solution. The idea is to apply an approximation formula for the quotient to simplify the equation such that it can then be solved with respect to β. $\hfill \square$

9.1 Series of Numbers

Before we introduce power series and Taylor series in general, we will first discuss number series and their convergence criteria. This will be necessary when we examine the convergence properties of the Taylor series.

According to Volume 1, Chapter 6.1, an ordered set of real numbers

$$(a_n)_{n \in \mathbb{N}} = a_1, \, a_2, \, a_3, \ldots, \, a_n, \ldots$$

is called *sequence of numbers*. This sequence is *convergent* if there exists a real number $a \in \mathbb{R}$ such that for every $\varepsilon > 0$ there is a $n_0 \in \mathbb{N}$ with

$$|a_n - a| < \varepsilon \qquad \text{for} \quad n \geq n_0.$$

Examples 9.2: In the first column of the table sequences are given by the first few elements. The second column represents the general term of each sequence. All sequences can be explicitly described by a construction rule. The last column discusses the question whether the sequences converge or diverge.

Sequence	General term	Convergence?
$(a_n)_n = 1, \, 2, \, 3, \, 4, \ldots$	$a_n = n$	No
$(a_n)_n = 1, \, \frac{1}{2}, \, \frac{1}{3}, \, \frac{1}{4}, \ldots$	$a_n = \dfrac{1}{n}$	Yes: $\quad a_n \to 0$
$(a_n)_n = q^0, \, q^1, \, q^2, \, q^3, \ldots$	$a_n = q^{n-1}$	$\begin{cases} \text{for} \ \ |q| < 1: a_n \to 0 \\ \text{for} \ \ |q| > 1: \text{divergent} \\ \text{for} \ \ q = 1: \quad a_n \to 1 \\ \text{for} \ \ q = -1: \text{divergent} \end{cases}$
$(a_n)_n = 1, \, \frac{1}{2!}, \, \frac{1}{3!}, \, \frac{1}{4!}, \ldots$	$a_n = \dfrac{1}{n!}$	Yes: $\quad a_n \to 0$
$(a_n)_n = -1, \, \frac{1}{2}, \, -\frac{1}{3}, \, \frac{1}{4}, \ldots$	$a_n = (-1)^n \dfrac{1}{n}$	Yes: $\quad a_n \to 0 \qquad \square$

Transition to Series. We consider the sequence of numbers

$$(a_n)_n = \frac{1}{0!}, \, \frac{1}{1!}, \, \frac{1}{2!}, \, \frac{1}{3!}, \, \frac{1}{4!}, \ldots, \, \frac{1}{(n-1)!}, \ldots$$

with the general expression $a_n = \frac{1}{(n-1)!}$. (Note that $0! = 1$.) From the elements of this sequence we form *sums* (= *partial sums*), by adding the first elements:

$$
\begin{aligned}
S_1 &= 1 & &= 1 \\
S_2 &= 1 + 1 & &= 2 \\
S_3 &= 1 + 1 + \tfrac{1}{2!} & &= 2.5 \\
S_4 &= 1 + 1 + \tfrac{1}{2!} + \tfrac{1}{3!} & &= 2.66666 \\
S_5 &= 1 + 1 + \tfrac{1}{2!} + \tfrac{1}{3!} + \tfrac{1}{4!} & &= 2.70833 \\
S_6 &= 1 + 1 + \tfrac{1}{2!} + \tfrac{1}{3!} + \tfrac{1}{4!} + \tfrac{1}{5!} & &= 2.71666 \\
S_7 &= 1 + 1 + \tfrac{1}{2!} + \tfrac{1}{3!} + \tfrac{1}{4!} + \tfrac{1}{5!} + \tfrac{1}{6!} & &= 2.71804 \\
S_8 &= 1 + 1 + \tfrac{1}{2!} + \tfrac{1}{3!} + \tfrac{1}{4!} + \tfrac{1}{5!} + \tfrac{1}{6!} + \tfrac{1}{7!} & &= 2.71823
\end{aligned}
$$

We summarize the partial sums in a sequence $(S_n)_{n\in\mathbb{N}}$.

$$
S_n = 1 + 1 + \frac{1}{2!} + \frac{1}{3!} + \cdots + \frac{1}{(n-1)!} = \sum_{k=1}^{n} \frac{1}{(k-1)!} \, .
$$

$(S_n)_n$ is called a *series*.

Definition: (Series). *Let* $(a_k)_{k\in\mathbb{N}}$ *be a sequence of numbers. Then*

$$
S_n = a_1 + a_2 + a_3 + \cdots + a_n = \sum_{k=1}^{n} a_k
$$

is a **Partial Sum** *and the sequence of partial sums* $(S_n)_{n\in\mathbb{N}}$ *is an* **infinite sum,** *(in short:* **series***):*

$$
(S_n)_{n\in\mathbb{N}} = \left(\sum_{k=1}^{n} a_k \right)_{n\in\mathbb{N}} = (a_1 + a_2 + \cdots + a_n)_{n\in\mathbb{N}} \, .
$$

Remarks:

(1) Often the sum of a series starts at $k = 0$.

(2) Depending on the terms, the construction of the sum can vary:

$$
a_2 + a_3 + \cdots = \sum_{k=2}^{\infty} a_k = \sum_{k=1}^{\infty} a_{k+1} \, .
$$

(3) The name of the summation index is arbitrary:

$$
\sum_{k=0}^{\infty} a_k = \sum_{i=0}^{\infty} a_i \, .
$$

Examples 9.3:

General term	Partial sum
$a_n = n$	$\sum_{k=1}^{n} k = 1 + 2 + 3 + \cdots + n$
$a_n = \dfrac{1}{n}$	$\sum_{k=1}^{n} \frac{1}{k} = 1 + \frac{1}{2} + \frac{1}{3} + \cdots + \frac{1}{n}$
$a_n = q^n$	$\sum_{k=0}^{n} q^k = 1 + q + q^2 + \cdots + q^n$
$a_n = \dfrac{1}{n!}$	$\sum_{k=0}^{n} \frac{1}{k!} = 1 + 1 + \frac{1}{2!} + \cdots + \frac{1}{n!}$
$a_n = (-1)^n \dfrac{1}{n}$	$\sum_{k=1}^{n} (-1)^k \frac{1}{k} = -1 + \frac{1}{2} - \frac{1}{3} \pm \cdots (-1)^n \frac{1}{n}$ $\square$

So a series is the sequence of the partial sums $\left(\sum_{k=1}^{n} a_k\right)_{n\in\mathbb{N}}$. The question arises whether these series converge, i.e. whether $\lim_{n\to\infty} S_n = \sum_{k=1}^{\infty} a_k$ has a finite value.

Definition:

(1) *A series $\left(\sum_{k=1}^{n} a_k\right)_{n\in\mathbb{N}}$* **converges**, *if the sequence of the partial sums $S_n := \sum_{k=1}^{n} a_k$ is a convergent sequence. In the case of convergence, the limit value is*

$$\lim_{n\to\infty} S_n = \lim_{n\to\infty} \sum_{k=1}^{n} a_k = \sum_{k=1}^{\infty} a_k.$$

This is called the value of the series.

(2) *A series $\left(\sum_{k=1}^{n} a_k\right)_{n\in\mathbb{N}}$ is* **divergent**, *if it has no limit.*

(3) *A series $\left(\sum_{k=1}^{n} a_k\right)_{n\in\mathbb{N}}$ is called* **absolutely convergent**, *if the partial sum of the absolute values $S_n := \sum_{k=1}^{n} |a_k|$ converges.*

Remarks:

(1) A convergent series always has a finite, uniquely determined sum value.

(2) An absolutely convergent series is always convergent. The inversion does not apply $\to$ see Example 9.14!

(3) A series is called *definitely divergent*, if $\sum_{k=1}^{\infty} a_k$ is either $+\infty$ or $-\infty$.

(4) In some rare cases it is possible to evaluate the partial sum as a closed expression. Then the sum value can be calculated. Otherwise, the sum value is unknown and we have to use convergence criteria to show the convergence of the series.

We will first discuss series whose partial sums can be evaluated. Then we will introduce important convergence criteria.

Example 9.4. The **geometric series**

$$\sum_{k=0}^{\infty} q^k = 1 + q + q^2 + \cdots + q^k + \cdots$$

converges for $|q| < 1$ and diverges for $|q| \geq 1$.

We apply the geometric sum

$$S_n = \sum_{k=0}^{n} q^k = \frac{1 - q^{n+1}}{1 - q} \qquad \text{for} \quad q \neq 1.$$

For $|q| < 1$ it is $\lim_{n \to \infty} q^{n+1} = 0$ and the sequence of partial sums has the value

$$S = \lim_{n \to \infty} S_n = \lim_{n \to \infty} \frac{1 - q^{n+1}}{1 - q} = \frac{1}{1 - q}.$$

Consequently,

$$\sum_{k=0}^{\infty} q^k = \frac{1}{1 - q} \qquad \text{for } |q| < 1.$$

For $|q| > 1$ the power q^{n+1} diverges and so does S_n. For $q = 1$ we have $S_n = \sum_{k=0}^{n} 1 = n + 1$ which diverges. For $q = -1$ the series is also divergent, as the following example shows. $\qquad \square$

Example 9.5. The series $\sum_{n=0}^{\infty} (-1)^n$ is **divergent**.

The sequence of the partial sums is

$$S_0 = 1, \quad S_1 = 1 - 1 = 0, \quad S_2 = 1 - 1 + 1 = 1, \quad S_3 = S_2 - 1 = 0,$$
$$S_4 = 1, \quad S_5 = 0, \qquad\qquad S_6 = 1, \qquad\qquad\qquad S_7 = 0, \qquad \ldots$$

So the sequence (S_n) has no limit and the series $\sum_{n=0}^{\infty} (-1)^n$ is divergent. This example also shows that a divergent series does not necessarily end in $+\infty$ or $-\infty$. $\qquad \square$

Example 9.6. The **arithmetic series**

$$\sum_{k=1}^{\infty} k = 1 + 2 + 3 + \cdots + n + \cdots$$

is **divergent**.

By mathematical induction we have proved (Volume 1, Example 1.3) that

$$S_n = \sum_{k=1}^{n} k = 1 + 2 + 3 + \cdots + n = \frac{n\,(n+1)}{2}.$$

So the sum value is

$$S = \lim_{n\to\infty} S_n = \lim_{n\to\infty} \frac{1}{2}\,n\,(n+1) = \infty.$$

The arithmetic series is therefore a **definite divergent** series.   □

Example 9.7. The series

$$\sum_{k=1}^{\infty} \frac{1}{k\,(k+1)} = \frac{1}{1\cdot 2} + \frac{1}{2\cdot 3} + \frac{1}{3\cdot 4} + \cdots + \frac{1}{k\,(k+1)} + \cdots$$

is convergent.

By mathematical induction we check that

$$S_n = \sum_{k=1}^{n} \frac{1}{k\,(k+1)} = \frac{1}{1\cdot 2} + \frac{1}{2\cdot 3} + \cdots + \frac{1}{n\,(n+1)} = \frac{n}{n+1}.$$

Consequently,

$$\lim_{n\to\infty} S_n = \lim_{n\to\infty} \frac{n}{n+1} = 1 \quad\Rightarrow\quad \sum_{k=1}^{\infty} \frac{1}{k\,(k+1)} = 1.$$ □

⚠ **Theorem/ Example 9.8.**

The harmonic series

$$\sum_{k=1}^{\infty} \frac{1}{k} = 1 + \frac{1}{2} + \frac{1}{3} + \frac{1}{4} + \cdots + \frac{1}{n} + \cdots$$

diverges.

Note: We compare the harmonic series with a series whose sequence elements are smaller than those of the harmonic series; but the comparison series already diverges.

Harmonic series:

$$1 + \frac{1}{2} + \left(\frac{1}{3} + \frac{1}{4}\right) + \left(\frac{1}{5} + \frac{1}{6} + \frac{1}{7} + \frac{1}{8}\right) + \cdots + \left(\frac{1}{2^n + 1} + \cdots + \frac{1}{2^{n+1}}\right) + \cdots$$

The brackets are arranged in such a way that the terms

$$\frac{1}{2^n + 1} + \cdots + \frac{1}{2^{n+1}}$$

are combined. We replace all the terms in a bracket by the value $\frac{1}{2^{n+1}}$ indicated by the arrow. In this way, we reduce the value of the sum and obtain the comparison series

$$1 + \frac{1}{2} + \left(\frac{1}{4} + \frac{1}{4}\right) + \left(\frac{1}{8} + \frac{1}{8} + \frac{1}{8} + \frac{1}{8}\right) + \left(\frac{1}{16} + \cdots + \frac{1}{16}\right) + \cdots$$

For this series

$$\sum_{i=1}^{n} \frac{1}{2} = n\frac{1}{2} \to \infty \quad \text{for} \quad n \to \infty.$$

Since the comparison series diverges to ∞, the harmonic series whose members are greater than those of the comparison series must also diverge.   □

With these considerations we have implicitly introduced the *minor criterion (comparison test)*. It means that a series $\sum_{i=1}^{\infty} b_i$ diverges if there exists a divergent comparison series $\sum_{i=1}^{\infty} a_i$ (minor) whose elements a_i are smaller than those b_i of the series under investigation.

Comparison Test

Let $0 < a_i \le b_i$ for all $i \ge m$ (stating at a fixed $m \in \mathbb{N}$). Then the following applies:

$$\sum_{i=1}^{\infty} a_i \text{ divergent} \quad \Rightarrow \quad \sum_{i=1}^{\infty} b_i \text{ divergent.}$$

Conclusions from the Divergence of the Harmonic Series

(1) ⚠ $\lim\limits_{n\to\infty} a_n = 0$ does not ensure the convergence of $\sum\limits_{k=1}^{\infty} a_k$.

(2) If a_n is convergent with $\lim\limits_{n\to\infty} a_n \neq 0$, then the series $\sum\limits_{k=1}^{\infty} a_k$ is divergent.

(3) If $\sum\limits_{k=1}^{\infty} a_k$ is convergent, then $\lim\limits_{n\to\infty} a_n = 0$.

⚠ **Caution:** A numerical calculation of a series is not sufficient to check for convergence or to determine the sum value in case of convergence! The harmonic series $\sum_{n=1}^{\infty} \frac{1}{n}$ is numerically **always** convergent, which contradicts the Example 9.8. The reason is that the numerical accuracy is finite. Therefore, from a certain N numerically it is

$$\sum_{i=1}^{N} \frac{1}{i} + \frac{1}{N+1} = \sum_{i=1}^{N} \frac{1}{i} \qquad \text{(numerically!)},$$

because $\frac{1}{N+1}$ no longer contributes to the sum. After this N, the series no longer changes its numerical value.

Example 9.9 (With MAPLE-Worksheet). To illustrate this effect, we calculate the harmonic series with an accuracy of 5 digits. Depending on N, we get the following results

N	10	100	1000	10000	15000	20000	30000
sum	2.9290	5.1873	7.4847	9.7509	10.000	10.000	10.000

At about $N = 15000$ the sum value does not change anymore, although the series diverges! If we change the order of the summation, almost any value greater than 10 can be obtained as the sum. □

Only in very few cases can the partial sum be simplified into a closed expression. Criteria are therefore needed to decide whether a series converges or not. This leads to the concept of *convergence criteria*. Only the three most important ones are introduced.

9.1.1 Majorant Criterion

A very descriptive criterion is the *majorant criterion*, which states that a series $\sum_{i=1}^{\infty} a_i$ converges, if we know that a larger series $\sum_{i=1}^{\infty} A_i$ already converges (comparison test). In short

---

Majorant Criterion (Comparison Test)

$$|a_i| \leq A_i \ (i \geq m) \ \text{ and } \ \sum_{i=1}^{\infty} A_i \text{ convergent} \ \Rightarrow \ \sum_{i=1}^{\infty} a_i \text{ convergent.}$$

The expression $\sum_{i=1}^{\infty} A_i$ is then called the *majorant*.

Example 9.10. For $p \geq 2$, the series $\sum_{n=1}^{\infty} \dfrac{1}{n^p}$ converges.

The convergent majorant is the series discussed in Example 9.7:

$$\sum_{k=1}^{\infty} \frac{1}{k \, (k+1)} = 1.$$

For $p \geq 2$ we have

$$\frac{1}{k^p} \leq \frac{1}{k^2} \leq \frac{2}{k \, (k+1)} \, .$$

Therefore, it holds

$$\sum_{k=1}^{N} \frac{1}{k^p} \leq \sum_{k=1}^{N} \frac{2}{k \, (k+1)} \leq 2 \sum_{k=1}^{\infty} \frac{1}{k \, (k+1)} = 2$$

and $\sum_{k=1}^{\infty} \dfrac{1}{k^p}$ converges to a sum value of ≤ 2. $\qquad\qquad\square$

More generally, the following statement applies:

Power Series:

The series $\sum_{n=1}^{\infty} \dfrac{1}{n^p}$ converges for $p > 1$ and diverges for $p \leq 1$.

Examples 9.11:

① The series $\displaystyle\sum_{n=1}^{\infty} \frac{1}{\sqrt{n}} = \sum_{n=1}^{\infty} \frac{1}{n^{\frac{1}{2}}}$ diverges because $p = \frac{1}{2} < 1$.

② The series $\displaystyle\sum_{n=1}^{\infty} \frac{n}{\sqrt{n^5}} = \sum_{n=1}^{\infty} \frac{1}{n^{\frac{3}{2}}}$ converges because $p = \frac{3}{2} > 1$.

③ The series $\displaystyle\sum_{n=1}^{\infty} \frac{\sin(n)}{\sqrt{n^3}}$ converges: Because of $\left|\frac{\sin(n)}{\sqrt{n^3}}\right| \leq \frac{1}{n^{\frac{3}{2}}}$ and the series $\sum_{n=1}^{\infty} \frac{1}{n^{\frac{3}{2}}}$ is a convergent major. $\qquad\square$

9.1.2 Quotient criterion

The quotient criterion proves to be extraordinary successful for power series.

Quotient Criterion

The series $\displaystyle\sum_{n=1}^{\infty} a_n$ converges if there exist an integer $N \in \mathbb{N}$ and a non-negative real number $q < 1$ such that

$$\left|\frac{a_{n+1}}{a_n}\right| \leq q < 1 \qquad \text{for all } n \geq N.$$

Proof: For the convergence discussion, all elements a_n $(n \leq N)$ are not relevant. So we assume that $|a_{n+1}| \leq q\,|a_n|$ for all n. Then inductively, we get

$$|a_n| \leq q\,|a_{n-1}| \leq q^2\,|a_{n-2}| \leq q^3\,|a_{n-3}| \leq \cdots \leq q^n\,|a_0|.$$

We use the geometric series from the Example 9.4. Since $q < 1$

$$\left|\sum_{n=0}^{\infty} a_n\right| \leq \sum_{n=0}^{\infty} |a_n| \leq |a_0| \sum_{n=0}^{\infty} q^n = |a_0|\,\frac{1}{1-q}. \qquad\square$$

Remarks:

(1) Since the geometric series diverges for $|q| > 1$, the following statement is analogous to the argument of the previous proof:

If there is a $N \in \mathbb{N}$ such that $\left|\frac{a_{n+1}}{a_n}\right| > 1$ for all $n \geq N$, then the series $\sum_{n=1}^{\infty} a_n$ diverges.

(2) For applications with the power and Taylor series, it is often easier to use the limit form of the quotient criterion. Here we calculate $\lim_{n \to \infty} \left|\frac{a_{n+1}}{a_n}\right|$. It is then equivalent to the quotient criterion:

Limit Form of the Quotient Criterion

$$\text{For} \quad \lim_{n \to \infty} \left|\frac{a_{n+1}}{a_n}\right| < 1 \quad \Rightarrow \quad \sum_{n=1}^{\infty} a_n \quad \text{converges.}$$

$$\text{For} \quad \lim_{n \to \infty} \left|\frac{a_{n+1}}{a_n}\right| > 1 \quad \Rightarrow \quad \sum_{n=1}^{\infty} a_n \quad \text{diverges.}$$

For $\lim_{n \to \infty} \left|\frac{a_{n+1}}{a_n}\right| = 1$, **no** statement about convergence is possible. In this case, other convergence criteria have to be applied.

Examples 9.12:

① The series

$$\sum_{n=1}^{\infty} \frac{n^2}{2^n}$$

converges. This follows directly from the limit form of the quotient criterion, since

$$\left|\frac{a_{n+1}}{a_n}\right| = \left|\frac{\frac{(n+1)^2}{2^{n+1}}}{\frac{n^2}{2^n}}\right| = \frac{2^n (n+1)^2}{2^{n+1} n^2}$$

$$= \frac{1}{2}\left(1 + \frac{1}{n}\right)^2 \xrightarrow{n \to \infty} \frac{1}{2} < 1.$$

② The series

$$\sum_{n=0}^{\infty} \frac{1}{n!} x^n$$

converges for every $x \in \mathbb{R}$. This follows from the limit form of the quotient criterion, since

$$\left|\frac{a_{n+1}}{a_n}\right| = \left|\frac{x^{n+1}}{(n+1)!} \cdot \frac{n!}{x^n}\right| = \frac{|x|}{n+1} \xrightarrow{n \to \infty} 0 < 1. \qquad \square$$

Comments on the Quotient Criterion

(1) $\triangle$ **Caution: If** $\lim\limits_{n\to\infty}\left|\dfrac{a_{n+1}}{a_n}\right| = 1$ **no statement is possible:**

For $a_n = \frac{1}{n}$ the harmonic series is $\sum_{n=1}^{\infty} \frac{1}{n}$. This is

$$\left|\frac{a_{n+1}}{a_n}\right| = \frac{n}{n+1} \overset{n\to\infty}{\longrightarrow} 1.$$

For $a_n = \frac{1}{n^2}$ we get the series $\sum_{n=1}^{\infty} \frac{1}{n^2}$. Here is also

$$\left|\frac{a_{n+1}}{a_n}\right| = \frac{n^2}{(n+1)^2} \overset{n\to\infty}{\longrightarrow} 1.$$

In both cases, the limit form of the quotient criterion returns the value 1. According to Example 9.8 the first series *diverges* and according to Example 9.10 the second series *converges*. So the quotient criterion cannot be used in these cases!

(2) The quotient criterion is therefore only a sufficient but not a necessary condition for the convergence of a series.

(3) Note that the convergence criterion only indicates whether a series converges or not; it does not give any indication of the sum. In particular, $\lim\limits_{n\to\infty}\left|\dfrac{a_{n+1}}{a_n}\right|$ does not match the value of the series!

9.1.3 Leibniz Criterion

For *alternating* series, there is a simple criterion by Leibniz $(1646 - 1716)$. Alternating series have the form $\sum_{n=1}^{\infty} (-1)^{n+1} a_n = a_1 - a_2 + a_3 - a_4 \pm \cdots$ with $a_n > 0$. The sign $(-1)^{n+1}$ changes throughout.

Leibniz Criterion

An **alternating series**

$$\sum_{n=1}^{\infty} (-1)^{n+1} a_n = a_1 - a_2 + a_3 - a_4 \pm \cdots \qquad \text{is convergent,}$$

if $a_1 > a_2 > a_3 > a_4 > \cdots > 0$ and $\lim\limits_{n\to\infty} a_n = 0$.

An alternating series is always convergent if the absolute values of the elements are a strictly decreasing zero sequence.

Example 9.13. $\displaystyle\sum_{n=1}^{\infty} (-1)^{n+1}\frac{1}{n!}$ is convergent.

The elements of the series are alternating and the absolute value of the elements

$$\frac{1}{1!} > \frac{1}{2!} > \frac{1}{3!} > \cdots > \frac{1}{n!} > \frac{1}{(n+1)!} > \cdots > 0$$

form a strictly decreasing zero sequence. The series converges according to the Leibniz criterion. $\square$

Example 9.14. The *alternating* harmonic series

$$\sum_{n=1}^{\infty} (-1)^{n+1}\frac{1}{n} = 1 - \tfrac{1}{2} + \tfrac{1}{3} - \tfrac{1}{4} \pm \cdots$$

converges.

The elements of the series are alternating and their absolute value

$$1 > \frac{1}{2} > \frac{1}{3} > \cdots > \frac{1}{n} > \frac{1}{n+1} > \cdots > 0$$

form a strictly decreasing zero sequence. The series converges according to the Leibniz criterion. $\square$

Example 9.15. $\displaystyle\sum_{n=1}^{\infty} (-1)^{n+1}$ diverges according to Example 9.5. The Leibniz criterion can **not** be applied. $|a_n| = 1$ is not a zero sequence. $\square$

Remarks:

(1) Absolutely convergent series are also convergent in the usual sense. But the reverse is not true! The alternating harmonic series is convergent ($\rightarrow$ Example 9.14) but not absolutely convergent, since the harmonic series $\sum_{n=1}^{\infty} \left|(-1)^{n+1}\frac{1}{n}\right| = \sum_{n=1}^{\infty}\frac{1}{n}$ diverges according to Example 9.8.

(2) When using the Leibniz criterion, it is not sufficient to check only the alternating property! Even if the members of the series have an alternating sign and form a zero sequence, the convergence can **not** be concluded, as the next series shows $\displaystyle\sum_{k=1}^{\infty} (-1)^{k} \left\{ \frac{1}{\sqrt{k+1}} + \frac{(-1)^{k}}{k+1} \right\}$. The elements of the series alternate but do not form a strictly decreasing zero sequence.

9.2 Power Series

This section presents the transition from the number series to the Taylor series. The convergence criteria of the number series are transferred to the power series to determine the domain of the power series.

If the terms in a series are themselves functions of a variable x, the expression $\sum_{n=0}^{\infty} a_n (x)$ represents a function, a so-called **function series**. An important special case of such function series are the *power series*.

Definition: *A function*

$$\sum_{n=0}^{\infty} a_n x^n = a_0 + a_1 x + \cdots + a_n x^n + \cdots$$

is called a **Power Series.** *The domain of a power series consists of all real numbers x for which $\sum_{n=0}^{\infty} a_n x^n$ converges. Therefore, the set is called*

$$K := \left\{ x \in \mathbb{R} : \sum_{n=0}^{\infty} a_n x^n \quad \text{is convergent} \right\}$$

the **Convergence Range** *of the power series.*

Remarks:

(1) $a_0, a_1, a_2, \ldots, a_n, \ldots$ are called the *coefficients* of the power series.

(2) For any fixed x, a power series reduces to a series of numbers.

(3) A more general representation of power series is given by expressions of the form

$$\sum_{n=0}^{\infty} a_n (x - x_0)^n = a_0 + a_1 (x - x_0) + \cdots + a_n (x - x_0)^n + \cdots \ .$$

The point x_0 is then called the *expansion point* of the series.

Example 9.16. $\displaystyle\sum_{n=0}^{\infty} n x^n = 1 x + 2 x^2 + 3 x^3 + \cdots + n x^n + \cdots .$ □

Example 9.17. $\displaystyle\sum_{n=0}^{\infty} \frac{1}{n!} x^n = 1 + x + \frac{1}{2!} x^2 + \frac{1}{3!} x^3 + \cdots + \frac{1}{n!} x^n + \cdots .$ □

Example 9.18. Let f be an infinitely differentiable function at $x_0 \in \mathbb{D}$ as often as desired. Then

$$\sum_{n=0}^{\infty} \frac{1}{n!} f^{(n)}(x_0) (x - x_0)^n$$

is a power series at the expansion point x_0 with the coefficients

$$a_n = \frac{1}{n!} f^{(n)}(x_0).$$

Such a series is called a *Taylor series* of the function f at the expansion point x_0 (see Section 9.3). $\qquad\square$

Example 9.19 (Geometric Power Series): According to the Example 9.4, the power series

$$\sum_{n=0}^{\infty} x^n = 1 + x + x^2 + \cdots + x^n + \cdots = \frac{1}{1 - x}$$

converges for $|x| < 1$ and diverges for $|x| \geq 1$. The **convergence range** is therefore $K = (-1, 1)$.

Example 9.20. We calculate the convergence range of the power series

$$\sum_{n=1}^{\infty} \frac{1}{n} x^n = x + \frac{1}{2} x^2 + \frac{1}{3} x^3 + \cdots + \frac{1}{n} x^n + \cdots$$

for a fixed $x \in \mathbb{R}$. For this, we use the quotient criterion with $b_n = \frac{1}{n} x^n$:

$$\left| \frac{b_{n+1}}{b_n} \right| = \left| \frac{\frac{x^{n+1}}{n+1}}{\frac{x^n}{n}} \right| = \left| \frac{x^{n+1}}{n+1} \frac{n}{x^n} \right| = \frac{n}{n+1} |x| \overset{n \to \infty}{\longrightarrow} |x|.$$

The series converges for $|x| < 1$ and diverges for $|x| > 1$. For $|x| = 1$, separate investigations must be made by inserting both values into the series:

$x = 1$: Then $\displaystyle\sum_{n=1}^{\infty} \frac{1}{n} x^n = \sum_{n=1}^{\infty} \frac{1}{n} 1^n = \sum_{n=1}^{\infty} \frac{1}{n}$ is the harmonic series. According to Example 9.8 this series is divergent.

$x = -1$: Then $\displaystyle\sum_{n=1}^{\infty} \frac{1}{n} x^n = \sum_{n=1}^{\infty} \frac{1}{n} (-1)^n$. The alternating harmonic series is convergent according to the Example 9.14.

So the convergence range is $K = [-1, 1)$. $\qquad\square$

⊙ Convergence Behavior of Power Series

If we apply the quotient criterion according to the procedure in Example 9.20 to any power series $\sum_{n=0}^{\infty} a_n\, x^n$, the convergence behavior can be characterized as follows:

Convergence Behavior of Power Series

Every power series

$$\sum_{n=0}^{\infty} a_n\, x^n = a_0 + a_1\, x + a_2\, x^2 + \cdots + a_n\, x^n + \cdots$$

has a well-defined **convergence radius** ρ $(0 \le \rho \le \infty)$ with the properties:

(1) The series converges for all x with $|x| < \rho$.

(2) The series diverges for all x with $|x| > \rho$.

(3) No general statement is possible for $|x| = \rho$.

Remark: To find ρ, we apply the quotient criterion to the series

$$\sum_{n=0}^{\infty} b_n$$

with $b_n = a_n\, x^n$:

$$\left| \frac{b_{n+1}}{b_n} \right| = \left| \frac{a_{n+1}\, x^{n+1}}{a_n\, x^n} \right| = \left| \frac{a_{n+1}}{a_n} \right| |x| \overset{n \to \infty}{\longrightarrow} \lim_{n \to \infty} \left| \frac{a_{n+1}}{a_n} \right| \cdot |x| .$$

According to the limit form of the quotient criterion, the series converges for

$$\lim_{n \to \infty} \left| \frac{a_{n+1}}{a_n} \right| \cdot |x| < 1 \;\hookrightarrow\; |x| < \frac{1}{\lim\limits_{n \to \infty} \left| \frac{a_{n+1}}{a_n} \right|} = \lim_{n \to \infty} \left| \frac{a_n}{a_{n+1}} \right|$$

and it diverges for

$$|x| > \frac{1}{\lim\limits_{n \to \infty} \left| \frac{a_{n+1}}{a_n} \right|} = \lim_{n \to \infty} \left| \frac{a_n}{a_{n+1}} \right| .$$

We define

$$\rho := \lim_{n \to \infty} \left| \frac{a_n}{a_{n+1}} \right| .$$

So the statements are verified and we calculated the radius of convergence. $\square$

Convergence Radius

The convergence radius ρ of a power series $\sum_{n=0}^{\infty} a_n x^n$ is

$$\rho = \lim_{n \to \infty} \left| \frac{a_n}{a_{n+1}} \right|.$$

Remarks:

(1) The special case $\rho = \infty$ is allowed because then $|x| < \rho$ is always satisfied and the series converges for all $x \in \mathbb{R}$.

(2) There are power series with $\rho = 0$. These series converge at most for $x = 0$, otherwise for no other $x \in \mathbb{R}$.

(3) No statement is possible if $\lim\limits_{n \to \infty} \left| \frac{a_n}{a_{n+1}} \right|$ does not exist.

(4) ⚠ **Caution:** When using the quotient criterion, we have to calculate the quotient $\left| \frac{b_{n+1}}{b_n} \right|$ whereas for the calculation of the radius of convergence the ratio $\left| \frac{a_n}{a_{n+1}} \right|$ has to be considered!

Example 9.21. The series $\quad \sum_{n=0}^{\infty} \frac{1}{n!} x^n = 1 + x + \frac{1}{2!} x^2 + \cdots + \frac{1}{n!} x^n + \cdots$

converges for all $x \in \mathbb{R}$ because the convergence radius is

$$\rho = \lim_{n \to \infty} \left| \frac{a_n}{a_{n+1}} \right| = \lim_{n \to \infty} \left| \frac{\frac{1}{n!}}{\frac{1}{(n+1)!}} \right| = \lim_{n \to \infty} \frac{(n+1)!}{n!} = \lim_{n \to \infty} (n+1) = \infty. \quad \square$$

Example 9.22. The series

$$\sum_{n=0}^{\infty} \frac{(-1)^n}{(2n+1)!} x^{2n+1} = x - \frac{1}{3!} x^3 + \frac{1}{5!} x^5 - \frac{1}{7!} x^7 \pm \cdots$$

converges for all $x \in \mathbb{R}$.

With $a_n = \frac{(-1)^n}{(2n+1)!}$, we get

$$\left| \frac{a_n}{a_{n+1}} \right| = \left| \frac{(-1)^n}{(2n+1)!} \frac{(2n+3)!}{(-1)^{n+1}} \right| = \frac{(2n+3)!}{(2n+1)!} = (2n+2)\,(2n+3)$$

With this series we have to take into account that two subsequent terms $b_n = a_n x^{2n+1}$ and $b_{n+1} = a_{n+1} x^{2n+3}$ will give us a ratio of x^2 so that the computation of $\lim\limits_{n \to \infty} \left| \frac{a_n}{a_{n+1}} \right|$ gives ρ^2:

$$\Rightarrow \; \rho^2 = \lim_{n \to \infty} \left| \frac{a_n}{a_{n+1}} \right| = \infty \; \Rightarrow \rho = \infty \; \Rightarrow K = \mathbb{R}. \qquad \square$$

⊘ Notes on the Convergence Range

A power series $\sum_{n=0}^{\infty} a_n x^n$ always converges symmetrically to zero within the interval $|x| < \rho$ and diverges outside. Figure 9.2 shows the graphical representation of the convergence range.

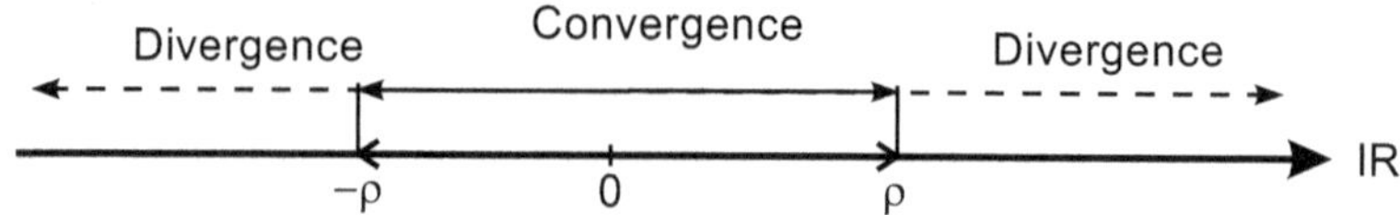

Figure 9.2. Convergence range of a power series

⚠ **Caution:** For $|x| = \rho$, i.e. $x = \pm \rho$, no general statement can be made. Separate investigations must be carried out at these limits by inserting $x = \rho$ and $x = -\rho$ into the power series. The convergence / divergence of the number series must then be discussed using the criteria from Section 9.1:

Example 9.23. We examine the power series

$$\sum_{n=1}^{\infty} \frac{(-1)^{n+1}}{n} (x-1)^n$$

in terms of its convergence properties.

To do this, we apply the quotient criterion to the series with the terms $b_n = \frac{(-1)^{n+1}}{n} (x-1)^n$:

$$\frac{b_{n+1}}{b_n} = \frac{(-1)^{n+2}}{n+1} (x-1)^{n+1} \cdot \frac{n}{(-1)^{n+1} (x-1)^n} = \frac{n}{n+1} (-1) (x-1)$$

$$\hookrightarrow \; \lim_{n \to \infty} \left| \frac{b_{n+1}}{b_n} \right| = \lim_{n \to \infty} \frac{n}{n+1} |x-1| = |x-1|.$$

So the series converges for $|x - 1| < 1$ and diverges for $|x - 1| > 1$. For the case $|x - 1| = 1$ separate investigations are carried out. From $|x - 1| = 1$ follows either $x = 2$ or $x = 0$:

For $x = 2$

$$\sum_{n=1}^{\infty} \frac{(-1)^{n+1}}{n} (2-1)^n = \sum_{n=1}^{\infty} \frac{(-1)^{n+1}}{n}$$

is the alternating harmonic series and therefore convergent.

For $x = 0$

$$\sum_{n=1}^{\infty} \frac{(-1)^{n+1}}{n} (0-1)^n = \sum_{n=1}^{\infty} \frac{(-1)^{n+1} (-1)^n}{n} = -\sum_{n=1}^{\infty} \frac{1}{n}$$

is the harmonic series and thus divergent.

In summary we obtain the range of convergence $K = (0, 2]$. $\square$

Remark: For power series in the form of $\sum_{n=0}^{\infty} a_n (x - x_0)^n$ the radius of convergence is also calculated by the formula

$$\rho = \lim_{n \to \infty} \left| \frac{a_n}{a_{n+1}} \right|.$$

(1) The series converges for $|x - x_0| < \rho$.

(2) The series diverges for $|x - x_0| > \rho$.

(3) For $|x - x_0| = \rho$ $(\hookrightarrow x = \pm \rho + x_0)$ no general statement can be made.

Animation: The **convergence range** of a power series is visualized graphically by displaying the power series with increasing order in the form of an animation.

Remark: For power series in the form of $\sum_{n=0}^{\infty} a_n (x - x_0)^{kn}$ with $k \in \mathbb{N}$, the radius of convergence is calculated by the formula

$$\rho = \sqrt[k]{\lim_{n \to \infty} \left| \frac{a_n}{a_{n+1}} \right|}.$$

Examples 9.24 (With Maple-Worksheet):

① The convergence radius of the power series $\sum_{i=1}^{\infty} \frac{1}{4^i} x^i$ is $\rho = 4$. The convergence range is the open interval $(-4, 4)$.

② For the power series $\sum_{i=1}^{\infty} (-1)^i \frac{1}{i} (x-1)^i$ we compute the convergence radius $\rho = 1$. The convergence range is the semi-open interval $(0, 2]$.

The animations show the partial sum up to $n = 25$ and $n = 26$, respectively.

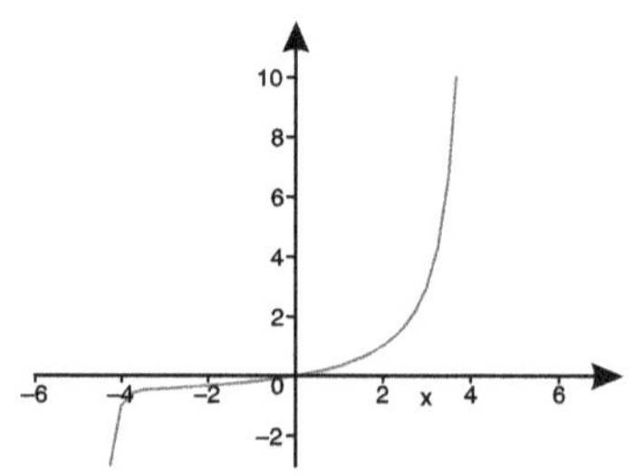
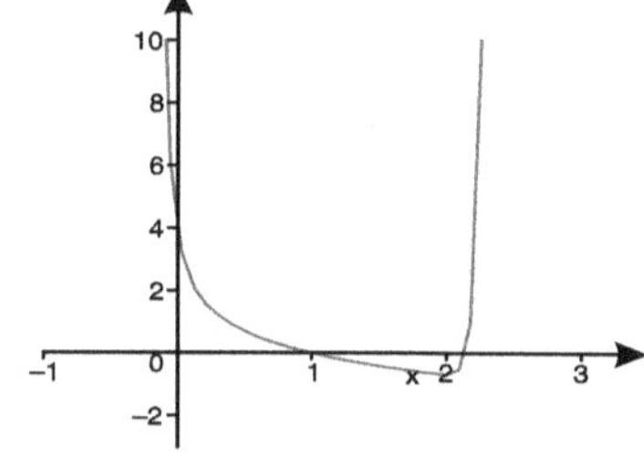

Partial sum $\sum_{i=1}^{25} \frac{1}{4^i} x^i$ Partial sum $\sum_{i=1}^{26} (-1)^i \frac{1}{i} (x-1)^i$

Animation: The animation clearly shows that the series stabilize inside the convergence range, while outside they go to infinity. In the first series, the convergence range is between -4 and 4, while in the second series it varies from 0 to 2. □

Finally, three important properties of power series are summarized.

Important Properties of Power Series

(1) A power series is absolutely convergent within its convergence range.

(2) A power series can be differentiated and integrated within its convergence range. The resulting power series have the same radius of convergence as the original series.

(3) Two power series can be added and multiplied in the common convergence range of the series.

9.3 Taylor Series

Now we come to the main topic of this chapter: Taylor series. Taylor's theorem states that almost any elementary function can be approximated by polynomials at any point x_0 of its convergence domain. It even shows that these functions can be represented by a power series of the form

$$\sum_{n=0}^{\infty} a_n \left(x - x_0\right)^n.$$

In addition to the determination of the coefficients a_n, there is information about the maximum error that occurs when this series is truncated after a finite number of summation elements. Thus, on the one hand, we obtain a method for calculating the elementary functions

$$e^x, \ \sin x, \ \sqrt{x}, \ \ln x \quad \text{etc.}$$

with arbitrary precision and, on the other hand, approximate formulas for these functions.

Example 9.25 (Introduction): Applying Example 9.19 to the geometric power series gives

$$1 + x + x^2 + \cdots + x^n + \cdots = \sum_{n=0}^{\infty} x^n = \frac{1}{1 - x}$$

for $|x| < 1$. So the power series $\sum_{n=0}^{\infty} x^n$ corresponds to the function $\frac{1}{1-x}$ for all $x \in (-1, 1)$. Outside this open interval, $\frac{1}{1-x}$ is still defined $(x \neq 1)$, but no longer the power series. A heuristic formula is now derived which allows a transfer to the power series for elementary functions. $\qquad \square$

Derivation of Taylor Polynomials. Given is a function $f(x)$, see Fig. 9.3. We are looking for an approximation of the function in an environment of the point $x_0 \in \mathbb{D}$. We assume that the function f can be differentiated several times in this environment.

(0.) The zero order approximation p_0 to the function is given by the constant function

$$\boxed{p_0\left(x\right) = f\left(x_0\right).}$$

The function p_0 has only the functional value at the position x_0 in common with f.

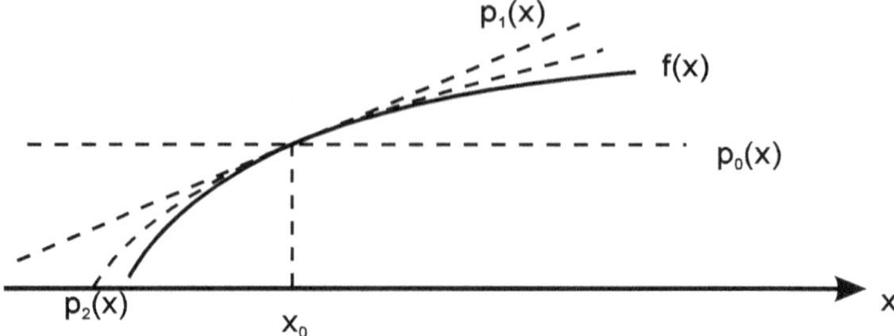

Figure 9.3. Function f and approximations in the environment of x_0

(1.) The linear approximation p_1 to the function is obtained by selecting the tangent line at x_0:

$$p_1(x) = f(x_0) + f'(x_0)(x - x_0).$$

The tangent has both the function value and the derivative at x_0 in common with the function.

(2.) The quadratic function p_2 is to be found, with *additionally* the same curvature as f at the point x_0:

Approach: $p_2(x) = f(x_0) + f'(x_0)(x - x_0) + c(x - x_0)^2.$

Condition: $p_2''(x_0) = f''(x_0).$

Because $p_2''(x) = 1 \cdot 2 \cdot c$ follows $p_2''(x_0) = 1 \cdot 2 \cdot c = f''(x_0)$

$$\Rightarrow \quad c = \frac{1}{2!} f''(x_0)$$

$$\Rightarrow \quad p_2(x) = f(x_0) + f'(x_0)(x - x_0) + \frac{f''(x_0)}{2!}(x - x_0)^2.$$

(3.) The cubic function p_3 is to be found, which *additionally* has the 3rd derivative in common with f in x_0:

Approach:

$$p_3(x) = f(x_0) + f'(x_0)(x - x_0) + \frac{1}{2!} f''(x_0)(x - x_0)^2 + d(x - x_0)^3.$$

Condition: $p_3'''(x_0) = f'''(x_0).$

Because $\qquad p_3'''(x_0) = 1 \cdot 2 \cdot 3 \cdot d = f'''(x_0)$

$$\Rightarrow d = \frac{1}{3!}\, f'''(x_0)$$

$$\Rightarrow \boxed{\begin{aligned} p_3(x) &= f(x_0) + f'(x_0)(x-x_0) + \tfrac{1}{2!} f''(x_0)(x-x_0)^2 \\ &\quad + \tfrac{1}{3!} f'''(x_0)(x-x_0)^3. \end{aligned}}$$

$$\vdots$$

(n.) A better approximation of the function f in an environment of point x_0 can be obtained by adding terms of the form

$$\frac{1}{n!}\, f^{(n)}(x_0)(x-x_0)^n\,,$$

so that the n-th approximation polynomial (**Taylor's polynomial of degree** n) is given by

$$p_n(x) = f(x_0) + f'(x_0)(x-x_0) + \cdots + \frac{1}{n!}\, f^{(n)}(x_0)(x-x_0)^n$$

$$= \sum_{i=0}^{n} \frac{1}{i!}\, f^{(i)}(x_0)(x-x_0)^i\,.$$

Visualization: To illustrate the convergence of the Taylor polynomials p_n to the function f we choose an animation for $f(x) = \sqrt{6-(x-2.5)^2}$ at point $x_0 = 1$. For this purpose, we determine the first 10 Taylor polynomials.

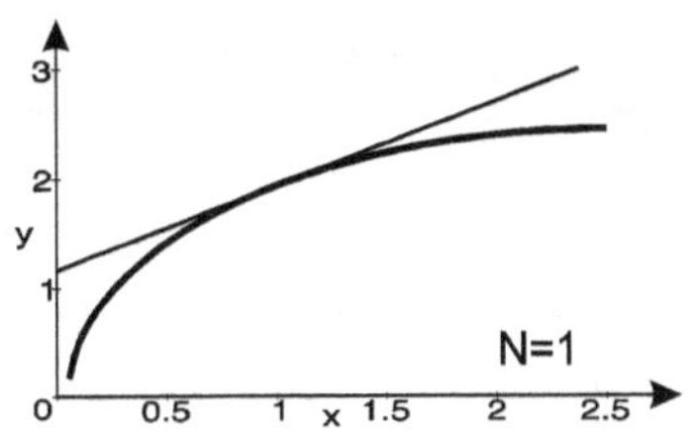

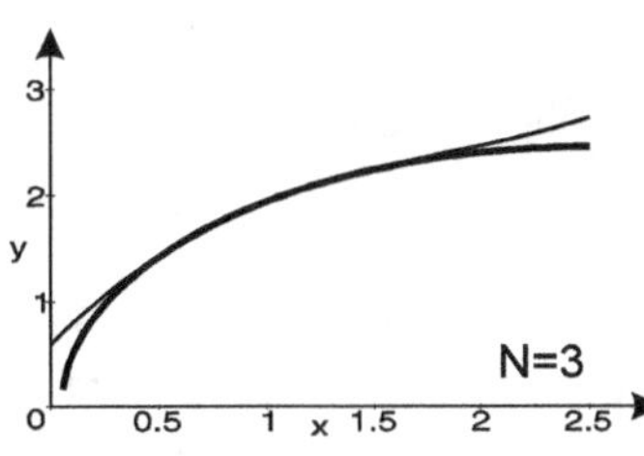

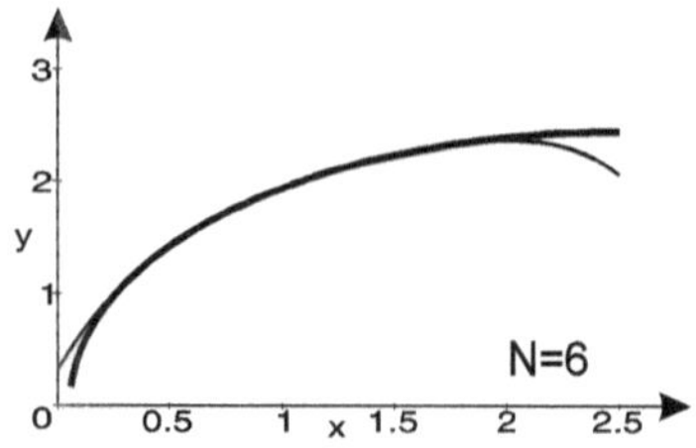

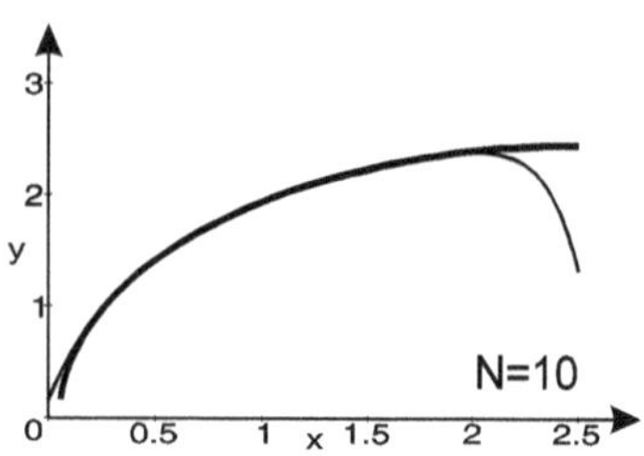

The animation clearly shows that as the degree of the Taylor polynomial increases, the range in which the function and the Taylor polynomial graphically coincide increases. For $N = 10$ in the range $0.5 \leq x \leq 1.7$ there is no graphical difference between the function f and the approximation polynomial p_{10}. So the question arises, how large is the deviation of the approximation function $p_n(x)$ from the function f in a certain environment of x_0. This is explained in the following theorem.

Theorem on Taylor Polynomials

Given is a function f that can be continuously differentiated in $x_0 \in \mathbb{D}$ $(m + 1)$ times. Then we get the **Taylor polynomial**

$$f(x) = f(x_0) + f'(x_0)(x - x_0) + \cdots + \frac{1}{m!} f^{(m)}(x_0)(x - x_0)^m$$
$$+ R_m(x)$$

$$= \sum_{n=0}^{m} \frac{1}{n!} f^{(n)}(x_0)(x - x_0)^n + R_m(x)$$

with the **Residual**

$$R_m(x) = \frac{1}{(m+1)!} f^{(m+1)}(\xi)(x - x_0)^{m+1} \qquad (x \in \mathbb{D})$$

where ξ is an unknown value between x and x_0.

The theorem of Taylor $(1685 - 1731)$ does not specify the intermediate point ξ between x and x_0. Therefore, it is not possible to specify the exact deviation of the approximation function $p_n(x)$ from the function f. For concrete applications, however, this will not play an important role, since an upper bound for the residual $R_m(x)$ can be specified. If the residual $R_m(x) \xrightarrow{m \to \infty} 0$ is fulfilled, the result is

Theorem on Taylor Series

If f is an infinitely differentiable function at $x_0 \in \mathbb{D}$ and if the residual satisfies $R_m(x) \to 0$ for $m \to \infty$, then the following holds

$$
\begin{aligned}
f(x) \;=\;& f(x_0) + f'(x_0)(x - x_0) + \frac{1}{2!} f''(x_0)(x - x_0)^2 + \cdots \\
& \cdots + \frac{1}{n!} f^{(n)}(x_0)(x - x_0)^n + \cdots \\[2mm]
\;=\;& \sum_{n=0}^{\infty} \frac{1}{n!} f^{(n)}(x_0)(x - x_0)^n .
\end{aligned}
$$

This power series is called the **Taylor Series for the function** f **at the expansion point** x_0.

Remarks:

(1) The convergence radius of the Taylor series is not necessarily > 0.

(2) When the Taylor series converges, it does not necessarily converge to $f(x)$.

(3) The Taylor series converges to $f(x)$ exactly when the residual of $R_m(x)$ goes to zero as $m \to \infty$. In this case, the function and the Taylor series coincide for all x from the convergence range of the power series.

(4) If the expansion point is $x_0 = 0$, the Taylor series is often also called *MacLaurin's series*.

(5) If f is an even function, then only terms with even powers appear in the Taylor series. If f is an odd function, then only terms with odd powers appear.

We will now determine the Taylor series of important functions, including the exponential, logarithmic and trigonometric functions. However, finding the Taylor series of functions is not enough. We also need to know the range of convergence. Only within this range can the series be evaluated and only in this range does the function match its Taylor series.

Example 9.26 (Exponential Function): The **Taylor series of** e^x at the expansion point $x_0 = 0$ is $e^x = \sum_{n=0}^{\infty} \dfrac{1}{n!} x^n$ for all $x \in \mathbb{R}$:

Because of

$$
\begin{aligned}
f(x) &= e^x \\
f'(x) &= e^x \\
f''(x) &= e^x \\
f'''(x) &= e^x \\
&\vdots \\
f^{(n)}(x) &= e^x
\end{aligned}
$$

follows

$$
\begin{aligned}
f(0) &= 1 \\
f'(0) &= 1 \\
f''(0) &= 1 \\
f'''(0) &= 1 \\
&\vdots \\
f^{(n)}(0) &= 1
\end{aligned}
$$

This makes the Taylor series for e^x:

$$
1 + x + \frac{1}{2!} x^2 + \frac{1}{3!} x^3 + \cdots + \frac{1}{n!} x^n + \cdots = \sum_{k=0}^{\infty} \frac{1}{k!} x^k.
$$

The convergence radius of this power series is $\rho = \infty$ ($\rightarrow$ Example 9.21), $K = \mathbb{R}$. The following applies to the residual with $\xi \in [-x, x]$

$$
R_m(x) = \frac{1}{(m+1)!} f^{(m+1)}(\xi) (x - x_0)^{m+1} = \frac{x^{m+1}}{(m+1)!} e^{\xi}
$$

$$
\Rightarrow \quad |R_m(x)| \leq \frac{|x|^{m+1}}{(m+1)!} e^{\xi} \to 0 \quad \text{as} \quad m \to \infty.
$$

So the Taylor series equals e^x for all $x \in \mathbb{R}$ and it holds

$$
e^x = \sum_{n=0}^{\infty} \frac{1}{n!} x^n. \qquad \Box
$$

 **Visualization:** The animation shows how the Taylor polynomials approach the exponential function e^x uniformly with increasing order.

Example 9.27 (Sinus Function): The **Taylor series of** $\sin(x)$ at the expansion point $x_0 = 0$:

$$
\begin{array}{lll}
\text{Because of} & f(x) = \sin x & \text{follows} \quad f(0) = 0 \\
& f'(x) = \cos x & f'(0) = 1 \\
& f''(x) = -\sin x & f''(0) = 0 \\
& f'''(x) = -\cos x & f'''(0) = -1 \\
& f^{(4)}(x) = \sin x & f^{(4)}(0) = 0 \\
& f^{(5)}(x) = \cos x & f^{(5)}(0) = 1 \\
& f^{(6)}(x) = -\sin x & f^{(6)}(0) = 0 \\
& \vdots & \vdots
\end{array}
$$

So it is $f^{(2n)}(0) = 0$ and $f^{(2n+1)}(0) = (-1)^n$. Only the odd exponents in the Taylor series occur with alternating signs:

$$
x - \frac{1}{3!}\, x^3 + \frac{1}{5!}\, x^5 - \frac{1}{7!}\, x^7 \pm \cdots = \sum_{n=0}^{\infty} \frac{(-1)^n}{(2n+1)!}\, x^{2n+1}.
$$

In the Example 9.22 the radius of convergence is $\rho = \infty$ and analogous to 9.26 $R_m(x) \to 0$ for $m \to \infty$. So the Taylor series matches for all $x \in \mathbb{R}$ with the final result:

$$
\sin(x) = \sum_{n=0}^{\infty} \frac{(-1)^n}{(2n+1)!}\, x^{2n+1}. \qquad \square
$$

Example 9.28 (Cosine Function): The **Taylor series of** $f(x) = \cos(x)$ at the expansion point $x_0 = 0$:

The Taylor series of $f(x) = \cos(x)$ results immediately from the example above. Since the power series can be differentiated term by term within the convergence range, the following applies for all $x \in \mathbb{R}$

$$
\cos(x) = \sin'(x) = \sum_{n=0}^{\infty} \frac{(-1)^n}{(2n+1)!}\, (2n+1)\, x^{2n}
$$

$$
\Rightarrow \quad \cos(x) = \sum_{n=0}^{\infty} \frac{(-1)^n}{(2n)!}\, x^{2n}. \qquad \square
$$

Example 9.29 (Logarithmic Function): The **Taylor series of** $\ln(x)$ at the expansion point $x_0 = 1$:

$$
\begin{aligned}
f(x) &= \ln x & f(1) &= 0 \\
f'(x) &= x^{-1} & f'(1) &= 1 \\
f''(x) &= (-1)\, x^{-2} & f''(1) &= (-1) \\
f'''(x) &= (-1)\,(-2)\, x^{-3} & f'''(1) &= (-1)\,(-2) \\
f^{(4)}(x) &= (-1)\,(-2)\,(-3)\, x^{-4} & f^{(4)}(1) &= (-1)\,(-2)\,(-3) \\
f^{(5)}(x) &= (-1)\,(-2)(-3)(-4)\, x^{-5} & f^{(5)}(1) &= (-1)\,(-2)(-3)(-4)
\end{aligned}
$$

$$
\vdots \qquad\qquad\qquad\qquad \vdots
$$

$$
f^{(n)}(x) = (-1)^{n+1}\,\frac{(n-1)!}{x^n} \qquad\qquad f^{(n)}(1) = (-1)^{n+1}\,(n-1)!
$$

This gives the Taylor coefficients for $n \geq 1$ as

$$
\frac{f^{(n)}(1)}{n!} = \frac{(-1)^{n+1}\,(n-1)!}{n!} = \frac{(-1)^{n+1}}{n}.
$$

Since the residual $R_m(x) \to 0$ as $m \to \infty$, the Taylor series at point $x_0 = 1$ is given by

$$
\ln x = (x-1) - \tfrac{1}{2}\,(x-1)^2 + \tfrac{1}{3}\,(x-1)^3 \pm \cdots \pm \frac{(-1)^{n+1}}{n}\,(x-1)^n \pm \cdots
$$

$$
\Rightarrow \quad \ln x = \sum_{n=1}^{\infty} \frac{(-1)^{n+1}}{n}\,(x-1)^n \qquad \text{for} \quad x \in (0,\,2].
$$

According to Example 9.23 the convergence range is $K = (0,\,2]$.

Also: If we evaluate the series at $x = 2$, then we get

$$
\boxed{\ln 2 = \sum_{n=1}^{\infty} \frac{(-1)^{n+1}}{n}.}
$$

The sum of the alternating harmonic series has the sum value $\ln 2$. $\quad\square$

Example 9.30 (Binomial Series): The **Taylor series of** $(1+x)^\alpha$ at the expansion point $x_0 = 0$ for any $\alpha \in \mathbb{R}$:

$$
(1+x)^\alpha = \sum_{k=0}^{\infty} \binom{\alpha}{k}\, x^k \qquad \text{for} \quad x \in (-1,\,1)\,,
$$

where the generalized binomial coefficients are defined as

$$\binom{\alpha}{0} := 1 \quad \text{and} \quad \binom{\alpha}{k} := \frac{\alpha\,(\alpha-1)\,(\alpha-2)\cdot\ldots\cdot(\alpha-k+1)}{k!}\,.$$

Because of

$$
\begin{aligned}
f(x) &= (1+x)^{\alpha} & f(0) &= 1 \\
f'(x) &= \alpha\,(1+x)^{\alpha-1} & f'(0) &= \alpha \\
f''(x) &= \alpha\,(\alpha-1)\,(1+x)^{\alpha-2} & f''(0) &= \alpha\,(\alpha-1) \\
f'''(x) &= \alpha\,(\alpha-1)\,(\alpha-2)\,(1+x)^{\alpha-3} & f'''(x) &= \alpha\,(\alpha-1)\,(\alpha-2) \\
&\;\;\vdots & &\;\;\vdots \\
f^{(n)}(x) &= \alpha\,(\alpha-1)\cdot\ldots & f^{(n)}(0) &= \alpha\,(\alpha-1)\cdot\ldots \\
&\quad \ldots\cdot(\alpha-n+1)\,(1+x)^{\alpha-n} & &\quad \ldots\cdot(\alpha-n+1)
\end{aligned}
$$

it follows for the Taylor coefficients

$$\frac{f^{(k)}(x_0)}{k!} = \frac{\alpha\,(\alpha-1)\,(\alpha-2)\cdot\ldots\cdot(\alpha-k+1)}{k!} = \binom{\alpha}{k}$$

and for the Taylor series

$$(1+x)^{\alpha} = \sum_{k=0}^{\infty} \binom{\alpha}{k}\, x^k\,.$$

The convergence range is calculated using the quotient criterion, resulting in: $K = (-1,\,1)$.

Special Cases:

① $\alpha = -1$ (Geometric Series):

$$\frac{1}{1+x} = 1 - x + x^2 - x^3 \pm \cdots = \sum_{k=0}^{\infty} (-1)^k\, x^k\,.$$

② $\alpha = -2$ (Derivation of the geometric series):

$$\frac{1}{(1+x)^2} = 1 - 2x + 3x^2 \mp \cdots = \sum_{k=0}^{\infty} (-1)^k\,(k+1)\, x^k\,.$$

③ $\alpha = \dfrac{1}{2}$:

$$\sqrt{1+x} = 1 + \frac{1}{2}\,x - \frac{1}{2\cdot4}\,x^2 + \frac{1\cdot3}{2\cdot4\cdot6}\,x^3 - \frac{1\cdot3\cdot5}{2\cdot4\cdot6\cdot8}\,x^4 \pm \cdots \quad .$$

④ $\alpha = -\dfrac{1}{2}$:

$$\frac{1}{\sqrt{1+x}} = 1 - \frac{1}{2}\,x + \frac{1 \cdot 3}{2 \cdot 4}\,x^2 - \frac{1 \cdot 3 \cdot 5}{2 \cdot 4 \cdot 6}\,x^3 + \frac{1 \cdot 3 \cdot 5 \cdot 7}{2 \cdot 4 \cdot 6 \cdot 8}\,x^4 \mp \cdots \qquad . \ \square$$

Tip: Sometimes the calculation of the Taylor series is based on known power series and then differentiated or integrated, as the following two examples show.

Example 9.31 (Arcus Tangent Function): The **Taylor series of** $f(x) = \arctan(x)$ at the expansion point $x_0 = 0$:

From $f(x) = \arctan(x)$

$$\Rightarrow \quad f'(x) = \frac{1}{1+x^2}\,.$$

According to Example 9.19 it is for $|x| < 1$

$$\frac{1}{1+x^2} = \frac{1}{1-(-x^2)} = \sum_{n=0}^{\infty} \left(-x^2\right)^n = \sum_{n=0}^{\infty} (-1)^n\, x^{2n}\,.$$

Since power series can be integrated term-wise, we get

$$\arctan(x) \;=\; f(0) + \int_0^x f'(\tilde{x})\, d\tilde{x} = 0 + \sum_{n=0}^{\infty} (-1)^n \int_0^x \tilde{x}^{2n}\, d\tilde{x}$$

$$=\; \sum_{n=0}^{\infty} \frac{(-1)^n}{2n+1}\, x^{2n+1}\,.$$

According to the Leibniz criterion, the power series also converges for $x = \pm 1$, so that in total:

$$\arctan(x) = \sum_{n=0}^{\infty} \frac{(-1)^n}{2n+1}\, x^{2n+1} \qquad \text{for } x \in [-1, 1]. \qquad \square$$

Example 9.32 (Area Functions): The **Taylor series of** $\operatorname{artanh}(x)$ at the expansion point $x_0 = 0$. We determine the Taylor series of the area functions by tracing them back to the binomial series:

① We differentiate $f(x) = \operatorname{artanh}(x)$

$$f'(x) = \operatorname{artanh}'(x) = \frac{1}{1 - x^2} = \sum_{n=0}^{\infty} x^{2n}$$

and integrate this result

$$f(x) = f(0) + \sum_{n=0}^{\infty} \frac{1}{2n + 1} x^{2n+1}.$$

Since $f(0) = \operatorname{artanh}(0) = 0$, we finally get

$$\operatorname{artanh}(x) = \sum_{n=0}^{\infty} \frac{1}{2n + 1} x^{2n+1} \qquad \text{for} \quad |x| < 1.$$

② The Taylor series of $\operatorname{arsinh}(x)$, $\operatorname{arcosh}(x)$ and $\operatorname{arcoth}(x)$ are calculated in the same way because they have the following derivatives

$$\operatorname{arsinh}'(x) = \frac{1}{\sqrt{1 + x^2}} \quad \text{for } x \in \mathbb{R},$$

$$\operatorname{arcosh}'(x) = \frac{1}{\sqrt{x^2 - 1}} \quad \text{for } |x| > 1 \text{ and}$$

$$\operatorname{arcoth}'(x) = \frac{1}{1 - x^2} \quad \text{for } |x| > 1. \qquad \qquad \square$$

 **Visualization of Convergence Range:** Using the procedure **taylor_poly** the animation shows the function $f(x)$ along with the Taylor polynomials as n increases. It can be seen that the function adapts smoothly as the order of the polynomials increases. It is clear that for a finite n the functions cannot be completely described because the Taylor polynomials p_n have the property that for any n it is $|p_n| \to \infty$ as $x \to -\infty$!

The Table 9.1 shows the Taylor series of important functions together with their convergence range.

Table 9.1: Taylor's Series:

Function	Power Series Expansion	Range
$(1+x)^{\alpha}$	$\displaystyle\sum_{k=0}^{\infty} \binom{\alpha}{k} x^k$	$|x| < 1$
$(1 \pm x)^{\frac{1}{2}}$	$1 \pm \frac{1}{2}\,x - \frac{1}{2\cdot4}\,x^2 \pm \frac{1\cdot3}{2\cdot4\cdot6}\,x^3 - \frac{1\cdot3\cdot5}{2\cdot4\cdot6\cdot8}\,x^4 \pm \cdots$	$|x| \leq 1$
$(1 \pm x)^{-\frac{1}{2}}$	$1 \mp \frac{1}{2}\,x + \frac{1\cdot3}{2\cdot4}\,x^2 \mp \frac{1\cdot3\cdot5}{2\cdot4\cdot6}\,x^3 + \frac{1\cdot3\cdot5\cdot7}{2\cdot4\cdot6\cdot8}\,x^4 \mp \cdots$	$|x| < 1$
$(1 \pm x)^{-1}$	$1 \mp x + x^2 \mp x^3 + x^4 \mp \cdots$	$|x| < 1$
$(1 \pm x)^{-2}$	$1 \mp 2\,x + 3\,x^2 \mp 4\,x^3 + 5\,x^4 \mp \cdots$	$|x| < 1$
$\sin x$	$x - \frac{x^3}{3!} + \frac{x^5}{5!} - \frac{x^7}{7!} + \frac{x^9}{9!} \mp \cdots$	$|x| < \infty$
$\cos x$	$1 - \frac{x^2}{2!} + \frac{x^4}{4!} - \frac{x^6}{6!} + \frac{x^8}{8!} \mp \cdots$	$|x| < \infty$
$\tan x$	$x + \frac{1}{3}\,x^3 + \frac{2}{15}\,x^5 + \frac{17}{315}\,x^7 + \frac{62}{2835}\,x^9 + \cdots$	$|x| < \frac{\pi}{2}$
e^x	$1 + \frac{x}{1!} + \frac{x^2}{2!} + \frac{x^3}{3!} + \frac{x^4}{4!} + \cdots$	$|x| < \infty$
$\ln x$	$(x-1) - \frac{1}{2}\,(x-1)^2 + \frac{1}{3}\,(x-1)^3 \mp \cdots$	$0 < x \leq 2$
$\arcsin x$	$x + \frac{1}{2\cdot3}\,x^3 + \frac{1\cdot3}{2\cdot4\cdot5}\,x^5 + \frac{1\cdot3\cdot5}{2\cdot4\cdot6\cdot7}\,x^7 + \cdots$	$|x| < 1$
$\arccos x$	$\frac{\pi}{2} - \left[x + \frac{1}{2\cdot3}\,x^3 + \frac{1\cdot3}{2\cdot4\cdot5}\,x^5 + \frac{1\cdot3\cdot5}{2\cdot4\cdot6\cdot7}\,x^7 + \cdots\right]$	$|x| < 1$
$\arctan x$	$x - \frac{1}{3}\,x^3 + \frac{1}{5}\,x^5 - \frac{1}{7}\,x^7 + \frac{1}{9}\,x^9 \mp \cdots$	$|x| \leq 1$
$\sinh x$	$x + \frac{1}{3!}\,x^3 + \frac{1}{5!}\,x^5 + \frac{1}{7!}\,x^7 + \cdots$	$|x| < \infty$
$\cosh x$	$1 + \frac{1}{2!}\,x^2 + \frac{1}{4!}\,x^4 + \frac{1}{6!}\,x^6 + \cdots$	$|x| < \infty$
$\tanh x$	$x - \frac{1}{3}\,x^3 + \frac{2}{15}\,x^5 - \frac{17}{315}\,x^7 + \frac{62}{2835}\,x^9 \mp \cdots$	$|x| < \frac{\pi}{2}$

9.4 Applications

⊙ Approximation Polynomials of a Function

In many applications complicated functions are approximated by Taylor polynomials $p_n(x)$. On the one hand to evaluate the functions in a simple way with a given accuracy, on the other hand to obtain a simpler physical relationship, e.g. with a linear approximation.

According to Taylor's theorem, the error between the function $f(x)$ and the Taylor polynomial $p_n(x)$ is given by the residual

$$R_n(x) = \frac{1}{(n+1)!} f^{(n+1)}(\xi)(x - x_0)^{n+1},$$

where x_0 is the expansion point and ξ is an unknown intermediate value between x and x_0. For most functions occurring in practice, the residual goes to zero as $n \to \infty$. Thus, for a sufficiently large n, an arbitrarily high accuracy is achieved. In practice, however, functions near their expansion point are often replaced only by the Taylor polynomial $p_1(x)$ or $p_2(x)$!

Example 9.33 (Calculating the Number e): We want to calculate the number e to 6 decimal places. We start with the Taylor expansion of the exponential function at $x_0 = 0$.

$$e^x = \sum_{n=0}^{\infty} \frac{1}{n!} x^n = 1 + x + \frac{1}{2!} x^2 + \cdots + \frac{1}{n!} x^n + \cdots$$

and calculate e^1 using the Taylor polynomial of order n

$$e^1 \approx p_n(1) = 1 + 1 + \frac{1}{2!} 1^2 + \cdots + \frac{1}{n!} 1^n.$$

The error according to the residual is

$$R_n(1) = \frac{1}{(n+1)!} e^\xi \leq \frac{1}{(n+1)!} e^1 < \frac{3}{(n+1)!}$$

(since $e^\xi \leq e^1 < 3$). To reduce the error to be less than 6 decimals, the following condition must be satisfied

$$R_n(1) < \frac{3}{(n+1)!} < 0.9 \cdot 10^{-6} \Rightarrow (n+1)! > \frac{3}{0.9 \cdot 10^{-6}} \approx 3\,333\,333.$$

So $n \geq 9$, because $(9+1)! = 3628800$. For $n = 9$ the number e^1 is calculated to 6 decimal places:

$$e^1 \approx \sum_{n=0}^{9} \frac{1}{n!} = 2.71828. \qquad \square$$

Remarks:

(1) The same error estimate applies to the evaluation of e^x in the range of $-1 \leq x \leq 1$. So e^x is approximated with the Taylor polynomial $p_9(x)$ upto 6 digits for all $|x| \leq 1$.

(2) If we compare this method of calculating the number e with the sequence $\left(1 + \frac{1}{n}\right)^n$ from the Example 6.3 in Volume 1, the series converges very quickly. For an accuracy of 6 decimal places, only 9 summation members are needed. Using the sequence representation we need at least $n > 10^5 = 100000$.

Application Example 9.34 (Linearization of Model Equations).

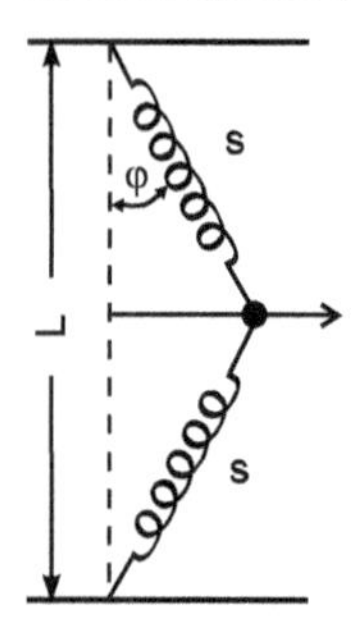

A mass is attached to the end of two opposing springs with spring constant D. The rest deflection of the springs is $l_0 < L/2$ (large spring preload). The mass m is deflected in the x-direction by the value x. We look for an effective D^* which represents the complete system.

The restoring force in the direction of the loaded springs is

$$F_D = -D\,(s - l_0);$$

so the restoring force F in x-direction is

$$F = 2\,F_D\,\sin\varphi = -2\,D\,(s - l_0)\,\sin\varphi = -2\,D\,(s - l_0)\,\frac{x}{s}.$$

Because of $s = \sqrt{x^2 + \left(\frac{L}{2}\right)^2}$ it is

$$F = -2\,D\,x\,\left(1 - \frac{l_0}{s}\right) = -2\,D\,x\,\left(1 - \frac{l_0}{\sqrt{x^2 + (L/2)^2}}\right)$$

$$= -2\,D\,x\,\left(1 - \frac{l_0}{L/2}\,\frac{1}{\sqrt{1 + (2x/L)^2}}\right).$$

For small deflections $\left(x << \frac{L}{2}\right)$ it is almost $\frac{2x}{L} \approx 0$. Near the expansion point $x_0 = 0$ we approximate the fraction according to Table 9.1

$$\frac{1}{\sqrt{1+x}} \approx 1 - \frac{1}{2}\,x \quad \text{or} \quad \frac{1}{\sqrt{1+(2x/L)^2}} \approx 1 - \frac{1}{2}\left(\frac{2x}{L}\right)^2 .$$

If the function is replaced by the **quadratic** Taylor polynomial, the restoring force is simplified to

$$F \approx -2\,D\,x\left(1 - \frac{2\,l_0}{L} + 4\,\frac{l_0\,x^2}{L^3}\right) .$$

In that case we choose the **constant** Taylor polynomial, $\frac{1}{\sqrt{1+(2x/L)^2}} \approx 1$, we get

$$F \approx -2\,D\,x\left(1 - \frac{2\,l_0}{L}\right) = -2\,D\left(1 - \frac{2\,l_0}{L}\right) x .$$

This is Hooke's law. The restoring force is proportional to the deflection with the effective spring constant $D^* = 2\,D\left(1 - \frac{2\,l_0}{L}\right)$.   □

Application Example 9.35 (Relativistic Particles).

According to Einstein, the total energy of a particle is

$$E = m\,c^2,$$

where c is the speed of light and m is the mass depending on the particle's velocity v:

$$m = \frac{m_0}{\sqrt{1 - (v/c)^2}},$$

m_0 is the rest mass of the particle. If $E_0 = m_0\,c^2$ denotes the *rest mass energy*, then the *kinetic energy* is

$$E_{kin} = E - E_0 = m\,c^2 - m_0\,c^2 = m_0\,c^2\left(\frac{1}{\sqrt{1-(v/c)^2}} - 1\right).$$

For a non-relativistic particle $v << c$, i.e. $0 \approx \frac{v}{c} << 1$. Then $\left(\frac{v}{c}\right)^2$ is close to the expansion point $x_0 = 0$ of the function $\frac{1}{\sqrt{1-x}}$. Therefore, if we substitute $x = \left(\frac{v}{c}\right)^2$ according to Table 9.1, we replace

$$\frac{1}{\sqrt{1-x}} \approx 1 + \frac{1}{2}\,x \quad \text{or} \quad \frac{1}{\sqrt{1-(v/c)^2}} \approx 1 + \frac{1}{2}\left(\frac{v}{c}\right)^2 .$$

The kinetic energy is then

$$E_{kin} = m_0\,c^2 \left(\frac{1}{\sqrt{1-(v/c)^2}} - 1\right) \approx m_0\,c^2 \left(1 + \frac{1}{2}\left(\frac{v}{c}\right)^2 - 1\right) = \frac{1}{2}\,m_0\,v^2.$$

The expression $\frac{1}{2}\,m_0\,v^2$ represents the kinetic energy of a particle in the limit case $v << c$ ($= $ *classical case*).   □

Application Example 9.36 (Headlight Adjustment).

Let us return to the introductory example of headlight adjustment. To deduce the actual inclination angle β from the quotient of the distance values d_1 and d_2, this quotient must be resolved for β. This is done by defining the function $q(\beta)$, which is expanded into a Taylor series.

$$q(\beta) := \frac{d_1}{d_2} = \frac{\sin(\alpha_2 + \beta)}{\sin(\alpha_1 + \beta)}. \tag{*}$$

Assuming the parameter values $\beta_{ab} = 0.0099996$, $\alpha_1 = 0.20337$ and $\alpha_2 = 0.097913$, the quotient q_0 of the angle β_{ab} between the horizontal and the cut-off line for the stationary vehicle is

$$q_0 := q(\beta_{ab}) = 0.5086238522.$$

To solve the quotient of $(*)$ for β, we now expand the right side into a Taylor series up to the order 2

$$q_2(\beta) := 0.4851497843 + 2.347500693\,\beta - 10.83456844\,(\beta - 0.0099996)^2$$

and solve equation $(*)$ for any left side $\frac{d_1}{d_2}$ with the approximation for the right side $\frac{\sin(\alpha_2+\beta)}{\sin(\alpha_1+\beta)} \sim q_2(\beta)$ to

$$\beta_{1/2} := \quad 0.11833343 \pm 0.36918867 \sqrt{0.43052561 - 0.67716052\,\frac{d_1}{d_2}}.$$

Of the two solutions found, only the one that gives the correct deflection angle β_{ab} for the size $q_0 = \frac{d_1}{d_2}$ is valid. This is true for β_2. The approximation function is drawn as the dashed line and the original, implicitly given function, is the solid line.

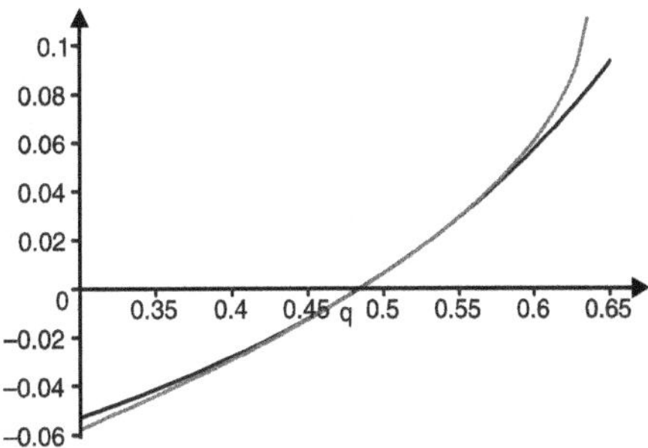

Figure 9.4. Function and approximation

The graph shows that the approximation formula for q between 0.4 and 0.58 corresponds quite well with the implicit function. This provides an angular range from -0.03 (-1.71°) to 0.05 (2.864°) in which the approximation can be used. However, to get an approximation formula without roots at all, β_2 is expanded once again into a Taylor series with expansion point $q = q_0$.

$$\beta_2 = -2.373259209 + 23.42846966\, q - 96.31243152\, q^2$$
$$+200.8851230\, q^3 - 210.4285911\, q^4 + 89.10928185\, q^5. \qquad \square$$

⊙ Integration using Power Series

Power series, and therefore Taylor series, can be easily differentiated and integrated in their range of convergence. For the power series

$$f(x) = \sum_{n=0}^{\infty} a_n \, (x - x_0)^n$$

we get its derivative

$$f'(x) = \frac{d}{dx} \sum_{n=0}^{\infty} a_n \, (x - x_0)^n = \sum_{n=0}^{\infty} a_n \, \frac{d}{dx} \, (x - x_0)^n = \sum_{n=1}^{\infty} n \, a_n \, (x - x_0)^{n-1} \, .$$

Note that differentiating the constant a_0 gives zero and the Taylor series starts at $n = 1$. To integrate the power series, we obtain

$$\int f(x) \, dx = \int \sum_{n=0}^{\infty} a_n \, x^n \, dx = \sum_{n=0}^{\infty} a_n \int x^n \, dx = \sum_{n=0}^{\infty} \frac{a_n}{n+1} \, x^{n+1} + C.$$

Note that for a given integral, the integration limits must be within the convergence range of the power series.

Example 9.37. Find the integral function $F(x) = \int_0^x e^{-t^2}\, dt$ that cannot be represented by an elementary function

With the power series approach

$$e^x = \sum_{n=0}^{\infty} \frac{1}{n!}\, x^n$$

follows

$$e^{-t^2} = \sum_{n=0}^{\infty} \frac{1}{n!}\, \left(-t^2\right)^n = \sum_{n=0}^{\infty} \frac{1}{n!}\, (-1)^n\, t^{2n}.$$

$$\Rightarrow \quad F(x) = \int_0^x e^{-t^2}\, dt = \sum_{n=0}^{\infty} \frac{1}{n!}\, (-1)^n \int_0^x t^{2n}\, dt$$

$$= \sum_{n=0}^{\infty} \frac{1}{n!}\, (-1)^n\, \frac{1}{2n+1}\, x^{2n+1} \quad (x \in \mathbb{R}). \qquad \square$$

⊘ **Solving Differential Equations with Power Series**

A method often used in physics to solve differential equations is to develop the desired function into a power series. This power series contains the coefficients a_n as unknown quantities. By inserting the power series into the differential equation, the a_n can be determined by comparing the coefficients (see Section 13.5.3).

9.5 Complex Functions

Describing complex numbers, the Euler formula is used

$$e^{i\varphi} = \cos(\varphi) + i\,\sin(\varphi), \qquad \varphi \in [0,\, 2\pi],$$

as an abbreviation. In this section we will show the equality of the function e^z and the function $\cos(z) + i\,\sin(z)$ for any complex number $z \in \mathbb{C}$. However, we first have to define e^z, $\cos(z)$ and $\sin(z)$ ($z \in \mathbb{C}$) as complex functions $f : \mathbb{C} \to \mathbb{C}$ with $z \longmapsto f(z)$ such that for $z \in \mathbb{R}$ the conventional real functions are included as a special case.

In the complex domain, the basic arithmetic operations $+, -, *, /$ are available. Therefore, complex functions should be defined only using these basic operations. As has been introduced in the last section, the exponential, the sine and the cosine functions are examples of the representation of a function by its Taylor series. Since only the basic operations mentioned above are needed to evaluate the Taylor series, the complex functions e^z, $\sin(z)$ and $\cos(z)$ are defined by their Taylor series.

9.5.1 Complex Power Series

The properties of the real power series apply analogously to the complex case. The convergence of a complex power series can be described as follows:

Complex Power Series

Let us consider a **complex power series**

$$\sum_{n=0}^{\infty} a_n \, (z - z_0)^n$$

with $a_n \in \mathbb{C}$ and the expansion point $z_0 \in \mathbb{C}$. The **real** power series $\sum_{n=0}^{\infty} |a_n| \, |z - z_0|^n$ is a majorant and the radius of convergence is

$$\rho = \lim_{n \to \infty} \frac{|a_n|}{|a_{n+1}|}.$$

Convergence exists for $z \in \mathbb{C}$ with $|z - z_0| < \rho$ and divergence for $|z - z_0| > \rho$. No general statements can be made for $|z - z_0| = \rho$.

Proof: In the complex domain the following calculation rules apply

$$|z_1 + z_2| \leq |z_1| + |z_2| \quad \text{and} \quad |a\,z| = |a| \, |z|.$$

Therefore, we get

$$\left| \sum_{n=0}^{N} a_n \, (z - z_0)^n \right| \leq \sum_{n=0}^{N} |a_n \, (z - z_0)^n| = \sum_{n=0}^{N} |a_n| \, |z - z_0|^n.$$

So $\sum_{n=0}^{\infty} |a_n| \, |z - z_0|^n$ is a majorant of $\sum_{n=0}^{\infty} a_n \, (z - z_0)^n$. So the complex power series has the same radius of convergence as the real majorant, namely $\rho = \lim_{n \to \infty} \frac{|a_n|}{|a_{n+1}|}$. $\qquad \square$

Interpretation:

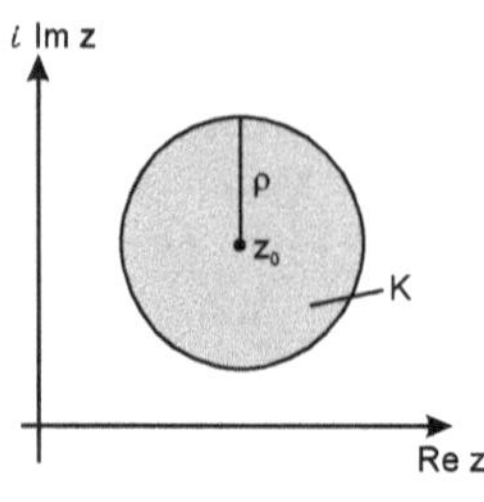

Figure 9.5.

Only in the complex domain the term *convergence radius* gets its full meaning, because the set $K = \{z \in \mathbb{C} : |z - z_0| < \rho\}$ corresponds to a circle around z_0 with radius ρ (see Fig. 9.5). Inside the circle the power series converges, outside it diverges. No general statements can be made on the border of the circle which is defined by $|z - z_0| = \rho$.

Examples 9.38. By representing the exponential and trigonometric functions by the Taylor series, the definition of the associated complex functions is obtained directly:

① **Complex Exponential Function**

$$e^z := \sum_{n=0}^{\infty} \frac{1}{n!} z^n = 1 + \frac{1}{1!} z^1 + \frac{1}{2!} z^2 + \frac{1}{3!} z^3 + \cdots \qquad \text{for} \ \ z \in \mathbb{C}.$$

It is

$$\left| \sum_{n=0}^{N} \frac{1}{n!} z^n \right| \leq \sum_{n=0}^{N} \frac{1}{n!} |z|^n \leq \sum_{n=0}^{\infty} \frac{1}{n!} |z|^n = e^{|z|}.$$

So $e^{|z|}$ is a convergent majorant and e^z converges for all $z \in \mathbb{C}$.

② **Complex Sinus Function**

$$\sin(z) := \sum_{n=0}^{\infty} \frac{(-1)^n}{(2n+1)!} z^{2n+1} = z - \frac{1}{3!} z^3 + \frac{1}{5!} z^5 - \frac{1}{7!} z^7 \pm \cdots \ \ \text{for} \ \ z \in \mathbb{C}.$$

③ **Complex Cosine Function**

$$\cos(z) := \sum_{n=0}^{\infty} \frac{(-1)^n}{(2n)!} z^{2n} = 1 - \frac{1}{2!} z^2 + \frac{1}{4!} z^4 - \frac{1}{6!} z^6 \pm \cdots \qquad \text{for} \ \ z \in \mathbb{C}.$$

The absolute convergence of the power series ② and ③ is guaranteed by the majorant criterion for all $z \in \mathbb{C}$, because the majorants are the real power series $\displaystyle\sum_{n=0}^{\infty} \frac{1}{(2n+1)!} |z|^{2n+1}$ and $\displaystyle\sum_{n=0}^{\infty} \frac{1}{(2n)!} |z|^{2n}$. $\qquad\square$

9.5.2 Euler's Formula

According to the preliminary remarks, e^z, $\cos(z)$, $\sin(z)$ are defined as independent functions for each $z \in \mathbb{C}$. In this context, we conclude the following:

> **Theorem: Euler's Formula**
>
> For each $z \in \mathbb{C}$ the following applies
>
> $$e^{iz} = \cos(z) + i\,\sin(z) \qquad \textit{(Euler's formula)}$$

Proof: We represent $\cos(z)$ and $i\,\sin(z)$ by their Taylor series. We then add the two series and identify the sum as e^{iz}. With $i^0 = 1$, $i^1 = i$, $i^2 = -1$, $i^3 = -i$, $i^4 = 1$ and $i^5 = i$, $i^6 = -1$, $i^7 = -i$, $i^8 = 1$, and so on, we obtain:

$$
\begin{aligned}
\cos(z) \quad &= 1 & &- \frac{z^2}{2!} & &+ \frac{z^4}{4!} & &\mp \cdots\\
&= 1 & &+ \frac{i^2 z^2}{2!} & &+ \frac{i^4 z^4}{4!} & &+ \cdots\\
&= 1 & &+ \frac{(i z)^2}{2!} & &+ \frac{(i z)^4}{4!} & &+ \cdots
\end{aligned}
$$

$$
\begin{aligned}
i\,\sin(z) \quad &= & i\frac{z}{1!} & & -i\frac{z^3}{3!} & & +i\frac{z^5}{5!}\ \mp \cdots\\
&= & \frac{i z}{1!} & & +\frac{i^3 z^3}{3!} & & +\frac{i^5 z^5}{5!}\ + \cdots\\
&= & \frac{(i z)}{1!} & & +\frac{(i z)^3}{3!} & & +\frac{(i z)^5}{5!}\ + \cdots
\end{aligned}
$$

$$
\cos(z) + i\,\sin(z) \;= 1\; + \frac{(i z)}{1!} + \frac{(i z)^2}{2!} + \frac{(i z)^3}{3!} + \frac{(i z)^4}{4!} + \frac{(i z)^5}{5!} + \cdots
$$

Therefore,

$$\cos(z) + i\,\sin(z) = \sum_{n=0}^{\infty} \frac{1}{n!}\,(i z)^n = e^{iz}. \qquad \square$$

This simple argument proves Euler's formula for all $z \in \mathbb{C}$. In particular, for $z = \varphi \in \mathbb{R}$ we obtain $e^{i\varphi} = \cos\varphi + i\,\sin\varphi$ for $\varphi \in \mathbb{R}$. If the angle φ is replaced by $\varphi = \omega t$, the identity of the functions is as follows

$$e^{i\omega t} = \cos(\omega t) + i\,\sin(\omega t)\,, \qquad t \in \mathbb{R}.$$

This equality of functions will be used in the following section to differentiate and integrate $e^{i\omega t}$.

9.5.3 Properties of the Complex Exponential Function

E1:
$$e^{z_1+z_2} = e^{z_1} \cdot e^{z_2}, \qquad z_1, z_2 \in \mathbb{C}.$$

As in the real domain, the theorem of addition applies to the exponential function in the complex domain. The proof of this central formula is the same as in the real domain. We compute the product of the series

$$e^{z_1} \cdot e^{z_2} = \sum_{k=0}^{\infty} \frac{1}{k!} z_1^k \cdot \sum_{k=0}^{\infty} \frac{1}{k!} z_2^k = \sum_{k=0}^{\infty} \sum_{j=0}^{k} \frac{1}{j!(k-j)!} z_1^j z_2^{k-j}$$

$$= \sum_{k=0}^{\infty} \frac{1}{k!} \sum_{j=0}^{k} \frac{k!}{j!(k-j)!} z_1^j z_2^{k-j} = \sum_{k=0}^{\infty} \frac{1}{k!} \sum_{j=0}^{k} \binom{k}{j} z_1^j z_2^{k-j}$$

$$= \sum_{k=0}^{\infty} \frac{1}{k!} (z_1 + z_2)^k = e^{z_1+z_2}$$

using the binomial theorem (see Volume 1, Section 1.2.5) which is also valid for complex numbers. It is worth noting the conclusion:

E2:
$$e^{z+2\pi i} = e^z, \qquad z \in \mathbb{C}.$$

We substitute $z_2 = 2\pi i$ into $(E1)$ and use the identity $e^{z_2} = e^{2\pi i} = 1$. The complex exponential function is therefore periodic with the complex period $2\pi i$. $\qquad\qquad\square$

E3:
$$\sin(-z) = -\sin(z), \qquad z \in \mathbb{C}.$$

$$\cos(-z) = \cos(z), \qquad z \in \mathbb{C}.$$

As in the real domain, the sine is an odd function, because for the sine only odd powers occur

$$\sin(-z) = \sum_{n=0}^{\infty} \frac{(-1)^n}{(2n+1)!} (-z)^{2n+1} = -\sum_{n=0}^{\infty} \frac{(-1)^n}{(2n+1)!} z^{2n+1} = -\sin(z).$$

Since by definition $\cos(z)$ has only even powers z^{2n}, $\cos(z)$ is an even function. $\qquad\qquad\square$

E4:
$$\cos(z) = \tfrac{1}{2}\left(e^{iz} + e^{-iz}\right), \qquad z \in \mathbb{C}.$$

$$\sin(z) = \tfrac{1}{2i}\left(e^{iz} - e^{-iz}\right), \qquad z \in \mathbb{C}.$$

According to these applications, the trigonometric functions can be extracted from the complex exponential function. Both identities are conclusions from Euler's formula, since

$$e^{iz} = \cos(z) + i\,\sin(z) \qquad (1)$$

$$\begin{aligned} e^{-iz} &= \cos(-z) + i\,\sin(-z) \\ &= \cos(z) - i\,\sin(z) \qquad (2) \end{aligned}$$

If the equations (1) and (2) are added, then $e^{iz} + e^{-iz} = 2\cos(z)$.

If the equation (2) is subtracted from (1), then $e^{iz} - e^{-iz} = 2i\,\sin(z)$.

Dividing the factors gives the assertion in each case. $\qquad\square$

E5:
$$\cos^2(z) + \sin^2(z) = 1\,, \qquad z \in \mathbb{C}.$$

$(E5)$ is obtained from $(E4)$ by squaring and adding the two equations:

$$\begin{aligned} \cos^2(z) + \sin^2(z) &= \frac{1}{4}\left(e^{iz} + e^{-iz}\right)^2 - \frac{1}{4}\left(e^{iz} - e^{-iz}\right)^2 \\ &= \tfrac{1}{4}\left[\left(e^{iz}\right)^2 + 2\,e^{iz}\,e^{-iz} + \left(e^{-iz}\right)^2 \right. \\ &\qquad \left. - \left(e^{iz}\right)^2 + 2\,e^{iz}\,e^{-iz} - \left(e^{-iz}\right)^2\right] \\ &= e^{iz}\,e^{-iz} = e^{i(z-z)} = e^0 = 1. \qquad\square \end{aligned}$$

Application: Addition Theorems for Sine and Cosine

For $\alpha,\ \beta \in \mathbb{C}$ (or $\alpha,\ \beta \in \mathbb{R}$) the **addition theorems** apply

$$\cos(\alpha + \beta) = \cos\alpha\,\cos\beta - \sin\alpha\,\sin\beta \qquad (A1)$$

$$\sin(\alpha + \beta) = \sin\alpha\,\cos\beta + \cos\alpha\,\sin\beta \qquad (A2)$$

Proof of $(A1)$: Based on the representation of the cosine and sine functions by the complex exponential function $(E4)$ and the addition theorem $(E1)$, we conclude:

$$
\begin{aligned}
\cos\alpha\,&\cos\beta - \sin\alpha\,\sin\beta \\
&= \frac{1}{2}\left(e^{i\alpha}+e^{-i\alpha}\right)\frac{1}{2}\left(e^{i\beta}+e^{-i\beta}\right) \\
&\quad -\frac{1}{2i}\left(e^{i\alpha}-e^{-i\alpha}\right)\frac{1}{2i}\left(e^{i\beta}-e^{-i\beta}\right) \\
&= \frac{1}{4}\left(e^{i(\alpha+\beta)}+e^{-i\alpha+i\beta}+e^{i\alpha-i\beta}+e^{-i(\alpha+\beta)}\right) \\
&\quad +\frac{1}{4}\left(e^{i(\alpha+\beta)}-e^{-i\alpha+i\beta}-e^{i\alpha-i\beta}+e^{-i(\alpha+\beta)}\right) \\
&= \frac{1}{2}\left(e^{i(\alpha+\beta)}+e^{-i(\alpha+\beta)}\right) = \cos(\alpha+\beta).
\end{aligned}
$$

$(A2)$: Analogous to $(A1)$. $\square$

Conclusion: The addition theorems are used for converting a product of sine and cosine functions into a sum and a difference, respectively. For $\alpha,\ \beta \in \mathbb{C}$ (or $\alpha,\ \beta \in \mathbb{R}$) the formulas apply:

(1) $2\sin\alpha\,\sin\beta = \cos(\alpha-\beta) - \cos(\alpha+\beta)$

(2) $2\cos\alpha\,\cos\beta = \cos(\alpha-\beta) + \cos(\alpha+\beta)$

(3) $2\sin\alpha\,\cos\beta = \sin(\alpha-\beta) + \sin(\alpha+\beta)$

9.5.4 Complex Hyperbolic Functions

Definition: For $z \in \mathbb{C}$, the complex functions are defined

$$\cosh(z) := \tfrac{1}{2}\left(e^{z}+e^{-z}\right) \qquad \textbf{Hyperbolic Cosine}$$

$$\sinh(z) := \tfrac{1}{2}\left(e^{z}-e^{-z}\right) \qquad \textbf{Hyperbolic Sine}$$

We evaluate sine and cosine at $i\,z$ and obtain with property $(E4)$

$$\cos(i\,z) = \tfrac{1}{2}\left(e^{i(i\,z)}+e^{-i(i\,z)}\right) = \tfrac{1}{2}\left(e^{-z}+e^{z}\right) = \cosh(z),$$

$$\sin(i\,z) = \tfrac{1}{2i}\left(e^{i(i\,z)}-e^{-i(i\,z)}\right) = \tfrac{1}{2i}\left(e^{-z}-e^{z}\right) = i\,\sinh(z).$$

So the hyperbolic functions $\cosh(z)$ and $\sinh(z)$ are nothing more than the complex trigonometric functions cosine and sine evaluated at $i\,z$:

H1:
$$\cos(i\,z) = \cosh(z)\,, \qquad z \in \mathbb{C}.$$

H2:
$$\sin(i\,z) = i\,\sinh(z)\,, \qquad z \in \mathbb{C}.$$

We square $(H1)$ and $(H2)$, add the results and take equation $(E5)$ into account to obtain

H3:
$$\cosh^2(z) - \sinh^2(z) = 1\,, \qquad z \in \mathbb{C}.$$

9.5.5 Differentiation and Integration

Complex resistances have been introduced in Volume 1, Chapter 5 for the description of alternating current circuits. To obtain the corresponding rules, the function $e^{i\omega t}$ was differentiated with respect to t applying the formula

$$\left(e^{i\omega t}\right)' = i\omega\,e^{i\omega t}.$$

The imaginary unit i is considered as a constant factor and the function $e^{i\omega t}$ is differentiated with the chain rule with respect to t. The following theorem shows that this method is generally valid.

Differentiation of Complex Functions

Let $u, v : (a, b) \to \mathbb{R}$ be real, differentiable functions. Then the complex function

$$f : (a, b) \to \mathbb{C}\ , \quad x \mapsto f(x) := u(x) + i\,v(x)$$

can be differentiated with

$$f'(x) = u'(x) + i\,v'(x)\,.$$

This theorem states that a complex function is differentiated according to its real variable x by taking the usual derivative of the real part and the imaginary part. **When differentiating complex functions, all the differentiation rules can be used as for real-valued functions.** The formula for the derivative follows directly from the definition of the derivative, since

$$f'(x) = \lim_{h \to 0} \tfrac{1}{h} \left(f(x+h) - f(x) \right)$$
$$= \lim_{h \to 0} \tfrac{1}{h} \left(u(x+h) + i\,v(x+h) - (u(x) + i\,v(x)) \right)$$
$$= \lim_{h \to 0} \tfrac{1}{h} \left(u(x+h) - u(x) \right) + i \lim_{h \to 0} \tfrac{1}{h} \left(v(x+h) - v(x) \right)$$
$$= u'(x) + i\,v'(x).$$

Examples 9.39.

① Find the derivative of the function $f(t) = e^{i\omega t}$. Since

$$e^{i\omega t} = \cos(\omega t) + i\,\sin(\omega t)$$

we get

$$\left(e^{i\omega t} \right)' = \cos(\omega t)' + i\,\sin(\omega t)' = -\omega\,\sin(\omega t) + i\omega\,\cos(\omega t)$$
$$= i\omega\left(\cos(\omega t) + i\,\sin(\omega t) \right) = i\omega\,e^{i\omega t}.$$

② Find the derivative of the function $f(x) = x\,e^{i\,x}$. Using to the product rule,

$$f'(x) = e^{i\,x} + i\,x\,e^{i\,x} = (1 + i\,x)\,e^{i\,x}. \qquad \square$$

Integration of Complex Functions

Let $u,\,v : [a,\,b] \to \mathbb{R}$ be real, integrable functions. Then the complex function $f := u + i\,v$

$$f : [a,\,b] \to \mathbb{C}\ ,\qquad x \mapsto f(x) := u(x) + i\,v(x)$$

is integrated with

$$\int_a^b f(x)\,dx = \int_a^b u(x)\,dx + i \int_a^b v(x)\,dx.$$

When integrating a complex function $f(x) = u(x) + i\,v(x)$, the real and the imaginary parts are integrated and then the integral of f is formed from both parts. **When integrating complex functions, all the integration rules can be used as for real-valued functions.** The formula is checked in the same way as the differentiation formula.

Examples 9.40.

① Find the antiderivative function of $f(x) = e^{ix}$.

$$\int f(x)\, dx = \int (\cos x + i \sin x)\, dx = \int \cos x\, dx + i \int \sin x\, dx$$

$$= \sin x + i\,(-\cos x) + C = -i\,(\cos x + i \sin x) + C$$

$$= \frac{1}{i}\, e^{ix} + C.$$

② Find the indefinite integral $\int x\, e^{ix}\, dx$.
With partial integration ($u = x$, $v' = e^{ix} \hookrightarrow u' = 1$, $v = -i\,e^{ix}$) the following applies

$$\int x\, e^{ix}\, dx = -i\, x\, e^{ix} + i \int e^{ix}\, dx$$

$$= -i\, x\, e^{ix} + e^{ix} + C.$$

i is treated as a constant during integration. $\qquad\square$

Thumb rule: The complex function $e^{i\omega t}$ can be differentiated and integrated like the real exponential function by considering i as a constant factor.

9.6 Problems on Series and Taylor Series

9.1 Examine the following number series for convergence

$$\text{a) } \sum_{n=1}^{\infty} \frac{1}{\sqrt{n}} \qquad \text{b) } \sum_{n=1}^{\infty} n\,e^{-n^2} \qquad \text{c) } \sum_{n=1}^{\infty} \frac{\sin n}{n^2} \qquad \text{d) } \sum_{n=1}^{\infty} \frac{(-1)^n\,n}{2\,n+1}$$

$$\text{e) } \sum_{n=1}^{\infty} \frac{2^n}{n!} \qquad \text{f) } \sum_{n=1}^{\infty} n\left(\frac{1}{2}\right)^{n-1} \qquad \text{g) } \sum_{n=1}^{\infty} \frac{3^{2\,n}}{(2\,n)!} \qquad \text{h) } \sum_{n=1}^{\infty} \frac{(-1)^{n+1}\,n}{5^{2n-1}}$$

$$\text{i) } \sum_{n=1}^{\infty} \frac{(-1)^{n+1}}{2\,n+1} \qquad \text{j) } \sum_{n=1}^{\infty} \frac{1}{2^n\,n} \qquad \text{k) } \sum_{n=1}^{\infty} 2^n\,\frac{1}{n} \qquad \text{l) } \sum_{n=2}^{\infty} \frac{1}{(2n-1)(2n+1)}$$

9.2 Examine the convergence of the following series

$$\text{a) } \sum_{k=0}^{\infty} \frac{2^k+3}{4^k} \qquad\qquad \text{b) } \sum_{k=0}^{\infty} \frac{1}{k!} \qquad\qquad \text{c) } \sum_{n=1}^{\infty} \frac{n^2}{2^n}$$

9.3 Show the divergence of the series $\displaystyle\sum_{n=1}^{\infty} \frac{1\cdot 3\cdot 5\cdot\ldots\cdot(2\,n-1)}{2^n}\,(-1)^n$ and

the convergence of the series $\displaystyle\sum_{n=1}^{\infty} \frac{6^n}{\left(3^{n+1}-2^{n+1}\right)\left(3^n-2^n\right)}$.

9.4 Determine the convergence radius of the series

$$\text{a) } \sum_{n=1}^{\infty} (-1)^{n+1}\frac{1}{5^n\,n}\,x^n \qquad\qquad\qquad \text{b) } \sum_{n=1}^{\infty} \frac{\ln(n)}{n}\,x^n$$

$$\text{c) } \sum_{n=1}^{\infty} \left(1+\tfrac{1}{n}\right)^{n^2} x^n \qquad\qquad\qquad \text{d) } \sum_{n=1}^{\infty} \frac{(n!)^2}{(2n)!}\,x^{n+1}$$

9.5 Calculate the radius of convergence of

$$\text{a) } \sum_{n=1}^{\infty} \frac{n\,x^n}{2^n} \quad \text{b) } \sum_{n=1}^{\infty} \frac{x^n}{n^2+1} \quad \text{c) } \sum_{n=1}^{\infty} n\,x^n \quad \text{d) } \sum_{n=1}^{\infty} (-1)^n\,\frac{x^n}{n}$$

$$\text{e) } \sum_{n=0}^{\infty} \frac{x^n}{2^n} \quad \text{f) } \sum_{n=1}^{\infty} \frac{n}{n+1}\,x^{n+1} \quad \text{g) } \sum_{n=1}^{\infty} \frac{n+1}{n!}\,x^n \quad \text{h) } \sum_{n=1}^{\infty} 2^n\cdot\frac{1}{n}\,x^n$$

and discuss the convergence domain K.

9.6 Determine the convergence domain of the power series

$$\text{a) } \sum_{n=1}^{\infty} n\,e^{-n}\,(x-4)^n \qquad \text{b) } \sum_{i=0}^{\infty} \frac{i^3}{2^i+1}\,(x-1)^i$$

c) $\displaystyle\sum_{n=0}^{\infty} \frac{(-1)^n}{(2\,n)!}\, x^{2\,n}$ d) $\displaystyle\sum_{n=1}^{\infty} \frac{1}{n^n}\, (x-2)^n$

9.7 Show that the Taylor series of sine and cosine about the expansion point $x_0 = 0$ are given by

$$\sin x = \sum_{n=0}^{\infty} \frac{(-1)^n}{(2\,n+1)!}\, x^{2\,n+1} \quad , \quad \cos x = \sum_{n=0}^{\infty} \frac{(-1)^n}{(2\,n)!}\, x^{2\,n}.$$

Determine the convergence range of the power series.

9.8 Develop the function

$$f(x) = \frac{1}{x^2} - \frac{2}{x} \quad , \quad x > 0,$$

into a Taylor series about the expansion point $x_0 = 1$. Specify the corresponding convergence domain.

9.9 Calculate the Taylor series of the function $f(x) = \dfrac{1}{\sqrt{1+x}}$ about $x_0 = 0$ and determine the convergence domain.

9.10 Calculate the Taylor series of arc functions arcsin, arccos, arccot and determine the convergence domain.

9.11 Calculate the Taylor series of the Area functions $ar\sinh$, $ar\cosh$, $ar\coth$ and determine the area of domain.

9.12 Develop $f(x) = \cos x$ about $x_0 = \frac{\pi}{3}$ into a Taylor series and determine the convergence range.

9.14 The function $f(x) = x\,e^{-x}$ is to be approximated in the vicinity of zero by a third-order polynomial. Determine Taylor's series of this function using Taylor's series of the exponential.

9.15 Calculate the function value of $f(x) = \sqrt{1-x}$ at $x = 0.05$ up to six decimals, if a Taylor series approach with expansion point $x_0 = 0$ is selected as evaluation polynomial.

9.16 What is the maximum error in the interval $[0, \frac{1}{3}]$, if the function

$$f(x) = \frac{\sin x}{x}$$

is developed around the point $x_0 = 0$ up to order 2?

9.17 Calculate $\displaystyle\int_0^1 \frac{e^x - 1}{x}\, dx$ up to an accuracy of 3 digits.

9.18 Solve the indefinite integral $F(x) = \displaystyle\int_0^x \frac{1}{1+t^2}\, dt$, by first developing the integrand into a Taylor series about $x_0 = 0$ and then integrating it element by element.

9.19 If a body of the mass m falls into a liquid, the distance at t is

$$s\left(t\right) = \frac{m}{k} \ln\left(\cosh\left(\sqrt{\frac{k\,g}{m}}\,t\right)\right) \quad , t \geq 0,$$

where g is the acceleration due to gravity and k the friction factor.
a) Determine the velocity $v\left(t\right)$ and the acceleration $a\left(t\right)$.
b) Develop an approximation for small k.

9.20 Calculate the integral sine and the Gaussian error integral approximately by developing the integrand into a power series:

$$\operatorname{sinc}\left(x\right) = \int_0^x \frac{\sin\tilde{x}}{\tilde{x}}\,d\tilde{x} \quad , \quad \operatorname{erf}\left(x\right) = \frac{2}{\sqrt{\pi}} \int_0^x e^{-\tilde{x}^2}\,d\tilde{x}.$$

9.21 Split the complex functions into real and imaginary part by replacing z with $x + i\,y$:
a) $f\left(z\right) = z^3$ b) $f\left(z\right) = \frac{1}{1-z}$ c) $f\left(z\right) = e^{3\,z}$

9.22 Calculate $\left|e^{i\,z}\right|$ for $z = 6\,e^{i\frac{\pi}{3}}$.

9.23 Given are the complex functions $f : \mathbb{R} \to \mathbb{C}$ with:
i) $f\left(x\right) = \left(x + i\,x\right)^3$ ii) $f\left(x\right) = e^{3\,\left(x+i\,x\right)}$.
a) Differentiate these functions according to the real variable x.
b) Integrate these functions.

9.24 Prove the addition theorem

$$\sin\left(\alpha + \beta\right) = \sin\alpha\,\cos\beta + \cos\alpha\,\sin\beta,$$

by using the formulas

$$\sin(x) = \frac{1}{2i}\left(e^{i\,x} - e^{-i\,x}\right) \quad \text{and} \quad \cos(x) = \frac{1}{2}\left(e^{i\,x} + e^{-i\,x}\right).$$

Chapter 10

Differential Calculus for Multivariate Functions

10

The chapter on the functions of several variables consists of four sections. 10.1 introduces the multi-variable functions and gives the graphical representation of functions with two variables. The concept of continuity is generalized in 10.2. The mathematically most important section is 10.3, which uses the differential calculus to characterize and describe these functions in terms of their derivatives. The construction of the derivative is generalized such that partial derivative, total differentiability, gradient and directional derivative are introduced. Taylor's theorem provides the transition to the applications in 10.4, where the discussion of error calculation, local extrema, and the fitting curves are in the foreground.

10

10 Differential Calculus for Multivariate Functions

The chapter on the functions of several variables consists of four sections. 10.1 introduces the multi-variable functions and gives the graphical representation of functions with two variables. The concept of continuity is generalized in 10.2. The mathematically most important section is 10.3, which deals with the differential calculus to characterize and describe these functions in terms of their derivatives. The construction of the derivative is generalized such that partial derivative, total derivative, gradient and directional derivative are introduced. Taylor's theorem provides the transition to the applications in 10.4, where the discussion of error calculation, local extrema, and the fitting curves are in the foreground.

10.1 Multi-Variable Functions

This section introduces multi-variable functions based on physical problems. It focuses on the different representations of functions with two variables.

10.1.1 Introduction and Examples

A function f **of one** real variable x consists of the domain, the range and the function assignment:

$$f : \mathbb{D} \to \mathbb{R} \quad \text{with} \quad x \mapsto y = f(x).$$

The mapping is usually realized using a rule $y = f(x)$; the functions can then be displayed as a graph of the function.

Examples 10.1 (Functions of a variable):

① $f \colon \mathbb{R} \to \mathbb{R}$ with $f(x) = a\,x + b$ (Linear function).

② $f \colon \mathbb{R}_{>0} \to \mathbb{R}$ with $f(x) = \ln x \cdot \cos\left(x^2 - 1\right)$.

③ Potential in a plane plate capacitor

$$\Phi : [0,\, d] \to \mathbb{R} \quad \text{with} \quad \Phi(x) = \Delta\Phi \, \frac{x}{d}\,,$$

where d is the plate gap and $\Delta\Phi = \Phi_2 - \Phi_1$ is the potential difference. $\square$

But many applications in science are more complicated and cannot be expressed by a function with a **single** variable. Most physical laws represent relationships between **several** independent variables.

Examples 10.2 (Functions with two variables):

① For an ideal gas of one mole, the **equation of state** is

$$p = R \cdot \frac{T}{V}.$$

The pressure p depends on both the temperature T and volume V of the gas. R is the universal gas constant. Each pair of values (T, V) is assigned a pressure value $p\,(T, V)$ by the formula:

$$(T, V) \mapsto p\,(T, V) = R \cdot \frac{T}{V}.$$

In addition to the mapping rule, the domain must also be specified by a physical meaning $T > 0$, $V > 0$.

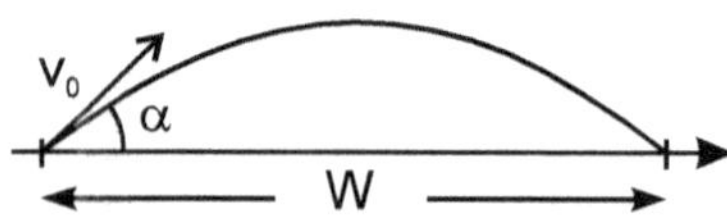

Figure 10.1. Throw distance W

② The **throw distance** W of a projectile is determined by the relationship

$$W = \frac{v_0^2}{g} \sin\left(2\,\alpha\right),$$

where v_0 is the initial velocity, α is the angle of the throw and g is the constant acceleration due to gravity. The throw distance thus depends on v_0 and α; each pair of numbers (v_0, α) is uniquely assigned a distance $W\,(v_0, \alpha)$:

$$(v_0, \alpha) \mapsto W\,(v_0, \alpha) = \frac{1}{g}\,v_0^2 \sin\left(2\,\alpha\right)$$

with $v_0 > 0$ and $0 < \alpha \leq 90°$.

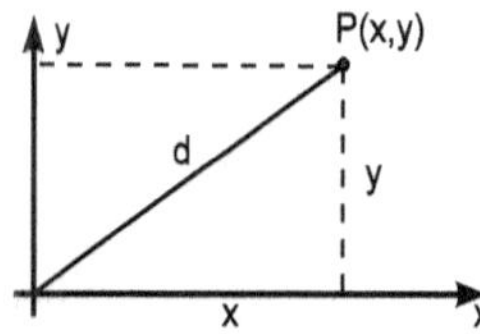

Figure 10.2. Distance d

③ The **distance** d of a point $P\,(x, y)$ in the plane from the origin is given by the Pythagorean theorem

$$d = \sqrt{x^2 + y^2}.$$

So each point (x, y) in the plane has exactly one functional value (its distance from the origin) $d\,(x, y)$

$$(x, y) \mapsto d\,(x, y) = \sqrt{x^2 + y^2}. \qquad \square$$

Examples 10.3 (Functions with more than two variables):

① In three-dimensional space, the **distance** d between two points $P_1(x_1, y_1, z_1)$ and $P_2(x_2, y_2, z_2)$ is

$$d = \left| \overrightarrow{P_1 P_2} \right| = \sqrt{(x_2 - x_1)^2 + (y_2 - y_1)^2 + (z_2 - z_1)^2}.$$

The distance d is a function of the 6 variables $x_1, x_2, y_1, y_2, z_1, z_2$:

$$(x_1, x_2, y_1, y_2, z_1, z_2) \mapsto d(x_1, \ldots, z_2) = \sqrt{(x_2 - x_1)^2 + \cdots + (z_2 - z_1)^2}.$$

② For a **series circuit** of n ohmic resistors $R_1, \ldots, R_n$, the total resistance is calculated by

$$R = R_1 + R_2 + \cdots + R_n.$$

The total resistance is a function of the n individual resistances

$$(R_1, R_2, \ldots, R_n) \mapsto R(R_1, \ldots, R_n) = R_1 + R_2 + \cdots + R_n. \qquad \square$$

Definition:
A real function f of n real variables $x_1, \ldots, x_n$ is a mapping that assigns each $(x_1, \ldots, x_n) \in \mathbb{D} \subset \mathbb{R}^n$ exactly one value in $\mathbb{R}$

$$f : \mathbb{D} \to \mathbb{R} \quad \text{with} \quad (x_1, \ldots, x_n) \mapsto y = f(x_1, \ldots, x_n).$$

The domain $\mathbb{D}$ is the set of n-tuples $(x_1, \ldots, x_n) \in \mathbb{R}^n$ of real numbers that can be inserted into the function expression.

Specifying both the domain and the range is quite tedious in practice, so the function $f(x_1, \ldots, x_n)$ is used instead. In this section, we will focus on functions with two variables. Many of the properties of multi-variable functions can already be explained here. Functions with two variables have the advantage that they can be represented graphically; functions with more than two variables no longer have this advantage!

$$z = f(x, y), \qquad (x, y) \in \mathbb{D}$$

is a real function of the two variables x and y defined on a domain $\mathbb{D} \subset \mathbb{R}^2$.

Example 10.4. In Fig. 10.3 some examples for two-dimensional domains are given. Domain (1) (*non-contiguous area*) is less important for our applications; domain (3) is called *contiguous*; areas (2) and (4) are called *simple contiguous*:

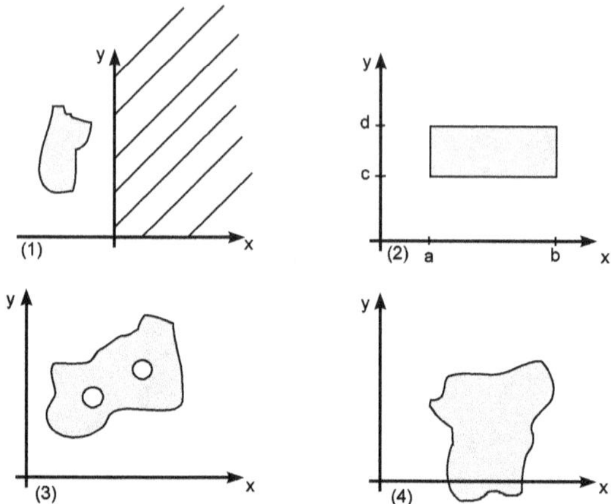

Figure 10.3. Examples of two-dimensional domains

10.1.2 Representing Functions with Two Variables

Three main display modes are used to illustrate functions with two variables:

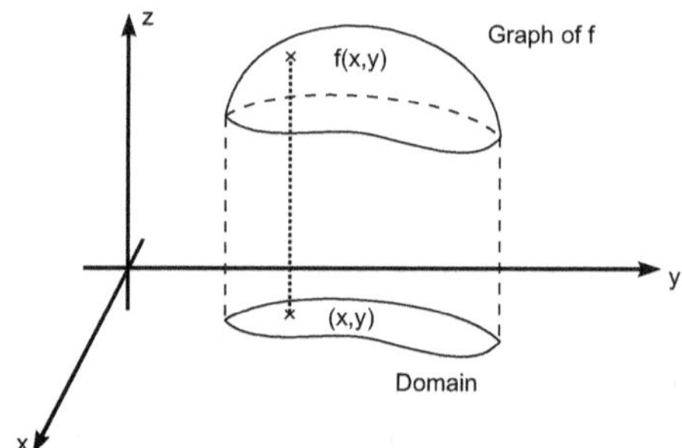

Figure 10.4. The graph of a two-dimensional function

(1) The Graph. The *Graph* of f is the set of points $(x, y, f(x, y))$ for which (x, y) are from the domain of f. In general, a graph is a curved surface in the three-dimensional space

$$\Gamma_f := \left\{ (x, y, z) \in \mathbb{R}^3 : \quad z = f(x, y) \quad \text{and} \quad (x, y) \in \mathbb{D} \right\}.$$

Example 10.5 (Mexican Hat, with MAPLE**-Worksheet).** Given is a function

$$f\left(r\right) = \frac{\sin\left(r\right)}{r}.$$

If we replace $r = \sqrt{x^2 + y^2}$ we get a function with two variables x and y:

$$f\left(x,y\right) = \frac{\sin\left(\sqrt{x^2 + y^2}\right)}{\sqrt{x^2 + y^2}}.$$

The representation of the function in the form of a three-dimensional graph is obtained with MAPLE as follows

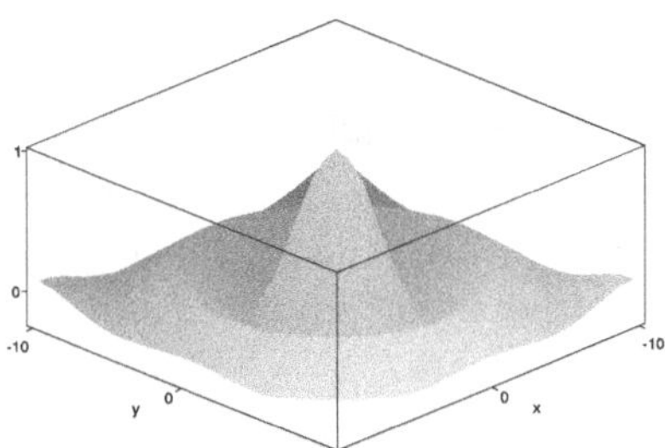

Figure 10.5. Mexican hat

(2) Contour Lines. A function f can also be represented graphically by its contour lines (level lines). The *contour line* of f at height c is the set of points $(x,\ y) \in \mathbb{D}$ which satisfy the implicit equation

$$f\left(x,\ y\right) = c.$$

Contour lines are intersections of the graph $(x,\ y,\ f\left(x,\ y\right))$ with planes parallel to $(x,\ y)$-plane with axis intercept c.

Example 10.6 (With MAPLE**-Worksheet).** Given is the function

$$f\left(x,\ y\right) = x^2 + y^2.$$

Fig. 10.6 (left) shows the graph of the function in the form of a three-dimensional representation. The contour lines are drawn as intersections of the graph with planes parallel to the (x, y) plane. If we choose $c_1 = 1$, $c_2 = 2$ and $c_3 = 3$ as function values (heights), the contour lines are concentric circles around the origin with radii $r_1 = 1$, $r_2 = \sqrt{2}$ and $r_3 = \sqrt{3}$. In Fig. 10.6 (right) the contour lines are projected into the (x, y) plane.

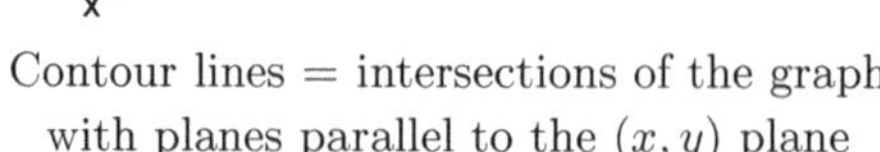

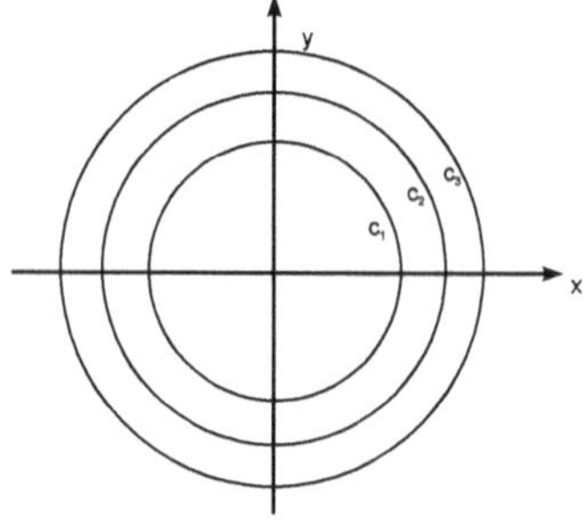

| Contour lines = intersections of the graph with planes parallel to the (x, y) plane | Contour lines projected in the (x, y) plane |

Figure 10.6. Graph of a function $f(x, y)$ and contour lines

To characterize the function graphically, we select several heights and draw the marked level lines in the (x, y)-plane. Examples of applications are equal pressure curves on a weather map (*isobars*), the contour lines on a map, or the lines of equal potential (*Equipotential Lines*) when describing electrostatic problems. $\qquad\square$

Application Example 10.7 (Electrostatic Potential).

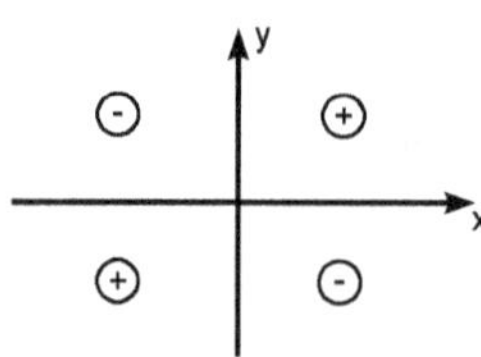

Figure 10.7. Quadrupole

The potential and contour lines of an electric *quadrupole* are searched for if the charges are placed in the corners of a square with edge length L (see Fig. 10.7). The electrostatic potential of an electric charge q located at the *origin* is given at point (x, y) by

$$\Phi(x, y) = \frac{1}{4\pi\,\varepsilon_0}\,\frac{q}{\sqrt{x^2 + y^2}}$$

with the dielectric constant $\varepsilon_0 = 8.8 \cdot 10^{-12}\,\frac{F}{m}$. A charge q localized at $\vec{r}_0 = (x_0, y_0)$ thus generates a potential at position $\vec{r} = (x, y)$ according to

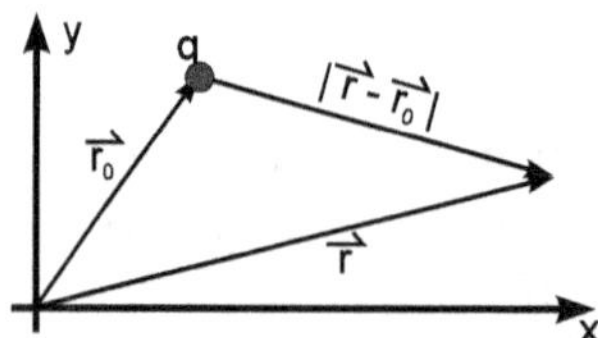

$$\Phi(x, y) = \frac{1}{4\pi\,\varepsilon_0}\,\frac{q}{\sqrt{(x - x_0)^2 + (y - y_0)^2}}.$$

For the potential of several point charges $q_1, \ldots, q_n$ the superposition principle applies

$$\Phi\left(x, y\right) = \Phi_1\left(x, y\right) + \cdots + \Phi_n\left(x, y\right).$$

So the potential of the electric quadrupole when the charges are at $\left(\frac{L}{2}, \frac{L}{2}\right)$, $\left(\frac{L}{2}, -\frac{L}{2}\right)$, $\left(-\frac{L}{2}, \frac{L}{2}\right)$, $\left(-\frac{L}{2}, -\frac{L}{2}\right)$ is given by

$$\Phi\left(x, y\right) = \frac{1}{4\pi\,\varepsilon_0}\left\{\frac{q}{\sqrt{\left(x - \frac{L}{2}\right)^2 + \left(y - \frac{L}{2}\right)^2}} + \frac{-q}{\sqrt{\left(x - \frac{L}{2}\right)^2 + \left(y + \frac{L}{2}\right)^2}}\right.$$
$$\left. + \frac{-q}{\sqrt{\left(x + \frac{L}{2}\right)^2 + \left(y - \frac{L}{2}\right)^2}} + \frac{q}{\sqrt{\left(x + \frac{L}{2}\right)^2 + \left(y + \frac{L}{2}\right)^2}}\right\}.$$

The graphical representation of the quadrupole is either in the form of a three-dimensional plot in Fig. 10.8 (a) or in the form of equipotential lines in Fig. 10.8 (b).

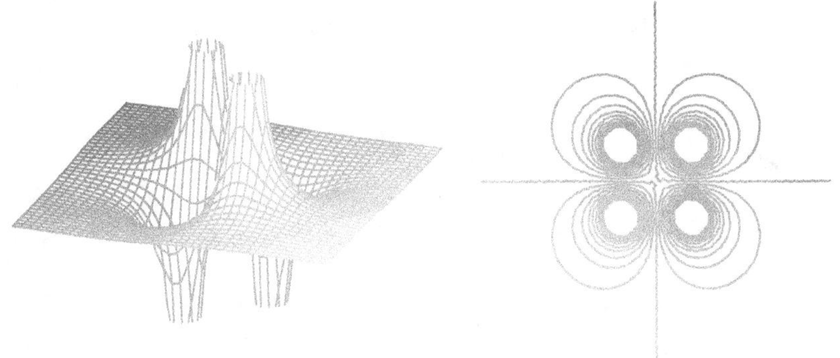

(a) 3D potential plot   (b) 2D equipotential lines

Figure 10.8. Potential plots for the electric quadrupole

(3) Intersection Curve Diagrams. Contour lines are the intersections of the graph of $z = f\left(x, y\right)$ with planes parallel to the (x, y) plane. If instead a plane parallel to the (x, z) or (y, z) plane is chosen, the so-called *intersection curve diagrams* are obtained

$$z = f\left(x = c, y\right) \quad \text{Intersection plane parallel to } (y, z),$$
$$z = f\left(x, y = c\right) \quad \text{Intersection plane parallel to } (x, z).$$

The following example illustrates this type of display in an application.

Application Example 10.8 (Equation of State for Ideal Gases).

The equation of state for a one-molar ideal gas is given by

$$p = R\,\frac{T}{V},$$

where p is the pressure, T is the temperature, V is the volume and R is the universal gas constant. Using constant values for the temperature T, $T_1 < T_2 < T_3 < T_4 < T_5$, gives the pressure as a function of the gas volume V. The curves at the same temperature are called *isotherms*.

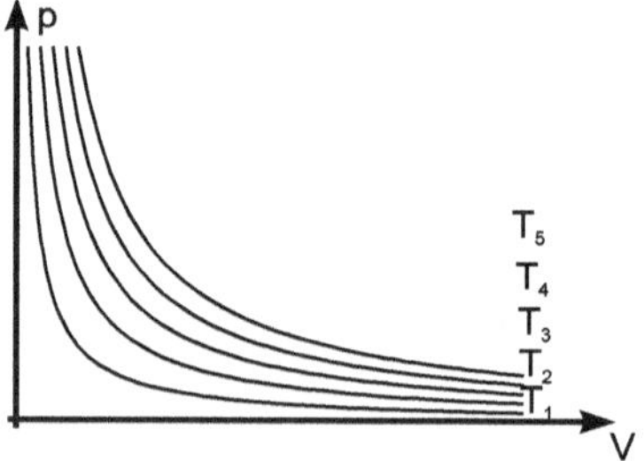

Figure 10.9. Representation of the pressure using isotherms

Visualization: The animation shows a function of two variables graphically from different angles. First a three-dimensional colored graph is shown, then the contour lines are faded in and only grey shades of the function are displayed.

10.2 Continuity

The *continuity* of a function $f(x)$ means, imprecisely expressed, that for any x_0 the graph of $f(x)$ converges to $f(x_0)$ as $x \longrightarrow x_0$. In this sense, a function in two variables is also called continuous.

The specification of the continuity term at the point x_0 is that the left and right function limits correspond to the value $f(x_0)$ of the function. Whether we approach x_0 from the left or from the right, the result is always the same function value $f(x_0)$: For any sequence $x_n \to x_0$ then $f(x_n) \to f(x_0)$ converges. This gives the following definition for functions with two variables.

Definition: (Continuity). *The function f is continuous at a point $(x_0,\, y_0) \in \mathbb{D}$ if for* **each** *sequence $(x_n,\, y_n) \in \mathbb{D}$ with $x_n \to x_0$ and $y_n \to y_0$*

$$f\,(x_n,\, y_n) \overset{n\to\infty}{\longrightarrow} f\,(x_0\, y_0)\,.$$

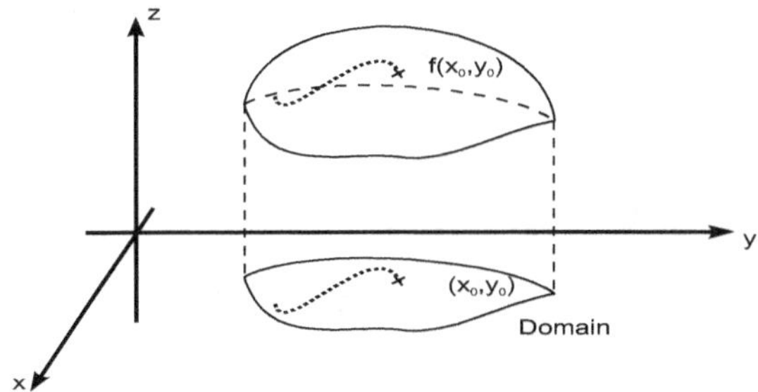

Figure 10.10. Continuity of a function

Remarks:

(1) Instead of $f\,(x_n,\, y_n) \overset{n\to\infty}{\longrightarrow} f\,(x_0,\, y_0)$ we also write $\lim\limits_{n\to\infty} f\,(x_n,\, y_n) = f\,(x_0,\, y_0)$.

(2) For $x_n \to x$ and $y_n \to y$ we use $(x_n,\, y_n) \to (x,\, y)$.

(3) ⚠ For the continuity of a function f at a point $(x_0,\, y_0) \in \mathbb{D}$ it is not sufficient that for **one** special sequence $(x_n,\, y_n) \to (x_0,\, y_0)$ the function sequence $f\,(x_n,\, y_n)$ converges, but the emphasis is on **each** sequence.

(4) If f is continuous at all points of the domain, f is *continuous in* $\mathbb{D}$.

(5) **Geometric interpretation:** f is continuous at a point $(x_0,\, y_0)$ if, for each sequence from the domain that converges to $(x_0,\, y_0)$, the function always tends to $f\,(x_0,\, y_0)$ (see Fig. 10.10).

Examples 10.9:

① For example, all polynomials in multiple variables are continuous

$$f\,(x,\, y) = 2 - x\,y + 3\,x^2\,y + 5\,x^9 - x^2\,y^3$$
$$g\,(x,\, y,\, z) = x^3 - 6\,x\,z - 3\,y\,z + 4\,x^2\,y^3\,z^4\,.$$

② The following functions are continuous for all $(x,\, y) \in \mathbb{R}^2$

$$e^{2x-3y}\,, \quad \ln\left(1 + x^4 + x^2\,y^2\right)\,, \quad \sin\left(x^2 + y^4\right)\,, \quad \sqrt{x^2 + y^2}\,.$$

③ Rational functions defined as the quotient of polynomials are also continuous functions at all points where the denominator polynomial does not disappear:

$$F\left(x,\,y\right) = \frac{x^3 - 2\,y^3 + x\,y}{x^2 - y^2}\ ,\quad G\left(x,\,y,\,z\right) = \frac{2\,x\,y^2 + 4\,y^2\,z}{x^2 + y^2 + z^2}\,.$$

The function F is continuous for all points of $\mathbb{R}^2$ that are not on the line $y = x$ or $y = -x$. G is continuous for all $(x,\,y,\,z) \in \mathbb{R}^3$ except for the origin.   □

Example 10.10. ⚠ **Caution:** The function

$$f\left(x,\,y\right) = \begin{cases} \dfrac{2\,x\,y}{x^2 + y^2} & \text{for}\quad (x,\,y) \neq (0,\,0) \\[2mm] 0 & \text{for}\quad (x,\,y) = (0,\,0) \end{cases}$$

is continuous for all points except at $(0,\,0)$: To see the discontinuity at $(0,\,0)$, we select the sequence $(x_n,\,y_n) = \left(\frac{1}{n},\,\frac{a}{n}\right) \rightarrow (0,\,0)$ from the domain and use the function f to obtain the sequence

$$f\left(x_n,\,y_n\right) = \frac{2 \cdot \frac{1}{n} \cdot \frac{a}{n}}{\frac{1}{n^2} + \frac{a^2}{n^2}} = \frac{2\,a}{1 + a^2}\,.$$

The function limit at $(0,\,0)$ **depends** on the selected sequence $(x_n,\,y_n) \rightarrow (0,\,0)$. So f is not continuous and cannot be continuously extended there. We have chosen $f(0,0) = 0$ because in the Example 10.16 we will show that with this definition the function can also be partially differentiated at $(0,0)$ although the function is not continuous at this point.   □

Remark: The continuity of a function f of two variables at a fixed point $(x_0,\,y_0)$ graphically means that the value of the function $f\left(x,\,y\right)$ is arbitrarily close to the value $f\left(x_0,\,y_0\right)$ if only the point $(x,\,y)$ is sufficiently close to the point $(x_0,\,y_0)$. This illustrative concept is made more precise by the following definition:

δ-ε-**continuity of a function:** *A function f is* **continuous** *at point* $(x_0,\,y_0) \in \mathbb{D}$, *if for any (arbitrarily small) number $\varepsilon > 0$ there exists a number $\delta > 0$ such that* $\boxed{\left|f\left(x,\,y\right) - f\left(x_0,\,y_0\right)\right| < \varepsilon}$ *for all points $(x,\,y) \in \mathbb{D}$ as $\left|x - x_0\right| < \delta$ and $\left|y - y_0\right| < \delta$.*

10.3 Differential Calculus

The most important term in the differential calculus of functions in several variables is the partial derivative. Partial derivatives are required for all further concepts such as total differentiability, gradient and directional derivatives or Taylor's theorem.

10.3.1 Partial Derivative

For functions of one variable, the derivative term plays a central role: The derivative of the function f at the point x_0 is defined as

$$f'(x_0) = \lim_{\triangle x \to 0} \frac{f(x_0 + \triangle x) - f(x_0)}{\triangle x}.$$

Geometrically, the derivative of the function f at x_0 is equal to the slope of the tangent.

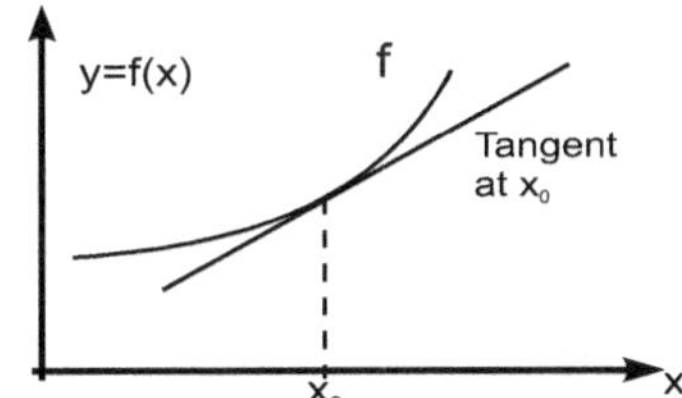

Figure 10.11. Derivative $=$ slope of the tangent line at x_0

This derivative term is extended to functions of two variables $f(x, y)$. If one variable is kept fixed (e.g. $y = y_0$), then

$$z = f(x, y = y_0)$$

is a function in the variable x. If this function can be differentiated in x_0, its derivative is called the *partial derivative with respect to x*. The partial derivative of f on y is defined analogously.

Definition: (Partial Derivative). *A function f is partially differentiable with respect to x at the point $(x_0, y_0) \in \mathbb{D}$ if the limit*

$$\frac{\partial f}{\partial x}(x_0, y_0) := \lim_{\triangle x \to 0} \frac{f(x_0 + \triangle x, y_0) - f(x_0, y_0)}{\triangle x}$$

*exists. It is called the **partial derivative of f with respect to x** at (x_0, y_0).*

Accordingly,

$$\frac{\partial f}{\partial y}(x_0,\,y_0) := \lim_{\triangle y \to 0} \frac{f(x_0,\,y_0 + \triangle y) - f(x_0,\,y_0)}{\triangle y}$$

is the **partial derivative of f with respect to** y *at the point* $(x_0,\,y_0)$ *if this limit exists.*

In vague terms, the definition can be summarized as follows:

> **Main Rule: The partial derivative with respect to a variable x_i is calculated by holding all other variables constant and taking the ordinary derivative with respect to x_i.** An important consequence of this is that all differentiation rules for functions of one variable apply directly to the partial differentiation.

Remarks:

(1) Note that, unlike ordinary derivatives, partial derivatives are not indicated by hyphens, but by indexing with the differentiation variable. Commonly used names are

$$f_x(x,\,y),\quad \frac{\partial f}{\partial x}(x,\,y),\quad \frac{\partial}{\partial x} f(x,\,y),\quad \partial_x f(x,\,y),$$

or short

$$f_x,\quad \frac{\partial f}{\partial x},\quad \frac{\partial}{\partial x} f,\quad \partial_x f.$$

To indicate that there are no ordinary derivatives, $\frac{d}{dx}$ is also replaced by $\frac{\partial}{\partial x}$. Analogous notations apply to the partial derivative with respect to y.

(2) f_x and f_y are called first-order partial derivatives.

(3) Alternative notations for the partial differential quotients are

$$\frac{\partial f}{\partial x}(x_0,\,y_0) = \lim_{x \to x_0} \frac{f(x,\,y_0) - f(x_0,\,y_0)}{x - x_0}$$
$$= \lim_{h \to 0} \frac{f(x_0 + h,\,y_0) - f(x_0,\,y_0)}{h}.$$

Analogous notations apply to the partial derivative with respect to y.

Examples 10.11 (Partial Derivatives):

① $f(x, y) = x^2 \cdot y^3 + x + y^2$:

$$\frac{\partial f}{\partial x}(x, y) = 2\,x \cdot y^3 + 1\,; \qquad \frac{\partial f}{\partial y}(x, y) = x^2 \cdot 3\,y^2 + 2\,y.$$

② $f(x, y) = \sin(x^2 - y)$:

$$\frac{\partial f}{\partial x}(x, y) = \cos(x^2 - y) \cdot 2\,x\,; \qquad \frac{\partial f}{\partial y}(x, y) = \cos(x^2 - y)\,(-1).$$

③ $f(x, y) = \ln\left(2\,x + 4 \cdot \frac{1}{y}\right)$:

$$\frac{\partial f}{\partial x}(x, y) = \frac{1}{2\,x + 4 \cdot \frac{1}{y}} \cdot 2\,; \qquad \frac{\partial f}{\partial y}(x, y) = \frac{1}{2\,x + 4 \cdot \frac{1}{y}} \cdot 4\,(-1)\ y^{-2}.$$

④ $U = R \cdot I$:

$$\frac{\partial U}{\partial R} = I\,; \qquad \frac{\partial U}{\partial I} = R.$$

⑤ $W = \frac{1}{g}\,v_0^2 \sin(2\alpha)$:

$$\frac{\partial W}{\partial v_0} = \frac{1}{g} \cdot 2\,v_0 \sin(2\alpha)\,; \qquad \frac{\partial W}{\partial \alpha} = \frac{1}{g}\,v_0^2 \cos(2\alpha) \cdot 2.$$

⑥ $p = R \cdot \dfrac{T}{V}$:

$$\frac{\partial p}{\partial V} = -R\,\frac{T}{V^2}\,; \qquad \frac{\partial p}{\partial T} = \frac{R}{V}. \qquad\qquad \square$$

Remark: Example 10.11 ⑥ shows the physical-chemical significance of the partial derivative: The pressure p of an ideal gas is proportional to the quotient $\frac{T}{V}$. So p is a function of the two variables T and V. $\frac{\partial p}{\partial V}$ then means the change in pressure as a function of the volume at constant temperature. In chemistry this is often symbolized by $\left(\frac{\partial p}{\partial V}\right)_{T=const}$. Similarly, $\frac{\partial p}{\partial T}$ is the change in pressure when the temperature changes at constant volume. $\square$

Geometric Interpretation: The vivid meaning of the partial derivatives is explained with the help of Fig. 10.12 and Fig. 10.13: The partial derivative $\frac{\partial}{\partial x} f(x_0, y_0)$ from f to x at the point (x_0, y_0) indicates the slope of the tangent at the point (x_0, y_0) parallel to the x-axis.

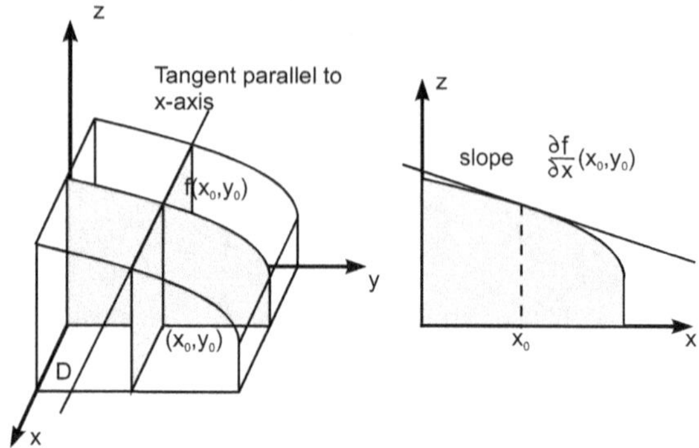

Figure 10.12. Partial derivative in x direction

Accordingly, the partial derivative $\frac{\partial}{\partial y} f\left(x_0,\, y_0\right)$ from f to y at points $\left(x_0,\, y_0\right)$ gives the slope of the tangent at point $\left(x_0,\, y_0\right)$ parallel to y-axis.

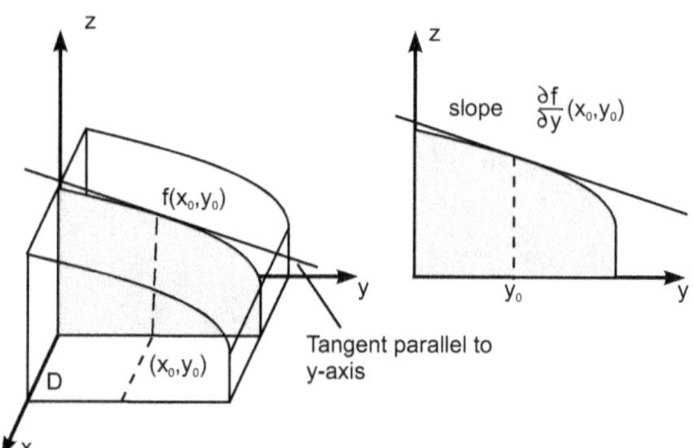

Figure 10.13. Partial derivative in y direction

⊙ Partial Differentiation of Multi-Variable Functions

If f is a function of the variables $(x_1, \ldots, x_n)$, then the partial derivative of f with respect to the variable x_i at a point in the domain $\left(x_1^0, \ldots, x_n^0\right)$ is defined as

$$
\frac{\partial f}{\partial x_i}\left(x_1^0, \ldots, x_n^0\right) = \lim_{h \to 0} \frac{f\left(x_1^0, \ldots, x_i^0 + h, \ldots, x_n^0\right) - f\left(x_1^0, \ldots, x_i^0, \ldots, x_n^0\right)}{h},
$$

if it exists. It is called the **partial derivative of f with respect to x_i** at the point $\left(x_1^0, \ldots, x_n^0\right)$ and is denoted by

$$
\frac{\partial f}{\partial x_i} \,, \quad \frac{\partial}{\partial x_i} f \,, \quad f_{x_i} \,, \quad \partial_{x_i} f.
$$

Example 10.12 (With MAPLE-Worksheet). Find all first-order partial derivatives of the function

$$f\left(x,\,y,\,z\right) = z \cdot e^{x^2+y^2} + \sqrt{1+x^2+z^4}.$$

Mainly using the chain rule, we get

$$f_x\left(x,\,y,\,z\right) = 2\,x\,z\,e^{x^2+y^2} + \frac{x}{\sqrt{1+x^2+z^4}}$$

$$f_y\left(x,\,y,\,z\right) = 2\,y\,z\,e^{x^2+y^2}$$

$$f_z\left(x,\,y,\,z\right) = e^{x^2+y^2} + \frac{2\,z^3}{\sqrt{1+x^2+z^4}}.$$

In addition, we calculate the partial derivatives at $(x_0, y_0, z_0) = (1, 2, 0)$ by inserting the point in the above formulas

$$f_x\left(1,\,2,\,0\right) = \tfrac{1}{2}\sqrt{2}; \quad f_y\left(1,\,2,\,0\right) = 0; \quad f_z\left(1,\,2,\,0\right) = e^5. \qquad \square$$

⊘ Higher Order Derivatives

If the partial derivatives $f_x\left(x,\,y\right)$ and $f_y\left(x,\,y\right)$ are again partially differentiable, then their partial derivatives $\frac{\partial}{\partial x} f_x\left(x,\,y\right)$, $\frac{\partial}{\partial y} f_x\left(x,\,y\right)$ and $\frac{\partial}{\partial x} f_y\left(x,\,y\right)$, $\frac{\partial}{\partial y} f_y\left(x,\,y\right)$ are partial derivatives of *second order*. The notation for the second order partial derivatives are

$$f_{xx} := \frac{\partial^2 f}{\partial x^2} := \frac{\partial}{\partial x}\left(\frac{\partial f}{\partial x}\right) \qquad \text{second partial derivative to } x.$$

$$f_{yy} := \frac{\partial^2 f}{\partial y^2} := \frac{\partial}{\partial y}\left(\frac{\partial f}{\partial y}\right) \qquad \text{second partial derivative to } y.$$

$$f_{xy} := \frac{\partial^2 f}{\partial y\,\partial x} := \frac{\partial}{\partial y}\left(\frac{\partial f}{\partial x}\right) \qquad \text{second partial derivative to } x \text{ and } y.$$

$$f_{yx} := \frac{\partial^2 f}{\partial x\,\partial y} := \frac{\partial}{\partial x}\left(\frac{\partial f}{\partial y}\right) \qquad \text{second partial derivative to } y \text{ and } x.$$

Its derivatives, if any, are the *third order partial derivatives* of f:

$$f_{xxx}, \ f_{xxy}, \ f_{xyx}, \ f_{yxx}, \ f_{xyy}, \ f_{yxy}, \ f_{yyx}, \ f_{yyy}.$$

Remarks:

(1) The order of differentiation is from inside to outside (from left to right): The derivative f_{xy} is calculated by taking the partial derivative with respect to y of the function f_x: $f_{xy} = (f_x)_y$.

(2) A derivative is called a *mixed* derivative when it has to be differentiated with respect to more than one variable.

(3) The *order* of the partial derivative corresponds to the total number of derivatives to be formed, i.e. the total number of indices:

$$f_{xyxx} \quad \text{is a 4th order derivative.}$$

Example 10.13. For the function

$$f(x, y) = x^3 y + y$$

the partial derivatives up to the 3rd order are given by

$$f_x = 3\,x^2\,y, \qquad\qquad\qquad f_y = x^3 + 1;$$

$$f_{xx} = 6\,x\,y, \quad f_{xy} = 3\,x^2 = f_{yx}, \qquad f_{yy} = 0;$$

$$f_{xxx} = 6\,y, \quad f_{xxy} = 6\,x = f_{xyx} = f_{yxx},$$
$$f_{xyy} = 0 = f_{yxy} = f_{yyx}, \qquad f_{yyy} = 0. \qquad\qquad \square$$

As we can see from this example, the sequence of the derivatives is not important, $f_{xy} = f_{yx}$, $f_{xxy} = f_{xyx} = f_{yxx}$ and so on. This property is confirmed for practically all functions occurring in applications. The following important statement applies

Schwarz's Theorem: Interchanging Mixed Derivatives

Given is a function $f(x, y)$ in two variables. If the mixed partial derivatives f_{xy} and f_{yx} are **continuous**, then the sequence of the derivatives to be formed does not matter:

$$f_{xy}(x, y) = f_{yx}(x, y).$$

Generalization: If for a function f all partial derivatives up to order k (≥ 2) are continuous, then for all partial derivatives up to order k the sequence of the derivatives is not important. Analogous to the higher partial derivatives for functions of two variables, they are formed for functions with more than two variables. Again, Schwarz's theorem applies.

According to the derivative scheme up to 2 for a function of three variables $f(x, y, z)$, the derivatives of f are calculated: To the left the derivative is with respect to x, in the middle to y and to the right the derivative is with respect to z.

$$f$$

$$f_x \qquad\qquad f_y \qquad\qquad f_z$$

$$f_{xx}\ \ \underline{f_{xy}}\ \ \underline{\underline{f_{xz}}} \qquad \underline{f_{yx}}\ \ f_{yy}\ \ \underline{\underline{\underline{f_{yz}}}} \qquad \underline{\underline{f_{zx}}}\ \ \underline{\underline{\underline{f_{zy}}}}\ \ f_{zz}$$

Example 10.14. For

$$f(x,\, y,\, z) = e^{x-y} \cos(5\,z)$$

all partial derivatives up to the order of 2 are searched:

First order derivatives:

$$f_x = e^{x-y} \cos(5\,z)$$
$$f_y = -e^{x-y} \cos(5\,z)$$
$$f_z = -5e^{x-y} \sin(5\,z)\,.$$

Second order derivatives (with respect to a variable):

$$f_{xx} = e^{x-y} \cos(5\,z)$$
$$f_{yy} = e^{x-y} \cos(5\,z)$$
$$f_{zz} = -25\,e^{x-y} \cos(5\,z)\,;$$

Second order derivatives (mixed):

$$f_{xy} = -e^{x-y} \cos(5\,z) = f_{yx}$$
$$f_{xz} = -5\,e^{x-y} \sin(5\,z) = f_{zx}$$
$$f_{yz} = 5\,e^{x-y} \sin(5\,z) = f_{zy}\,. \qquad\qquad \square$$

Examples 10.15:

① $\varphi(x, y) = \sqrt{x^2 + y^2}$:

$$\frac{\partial \varphi}{\partial x} = \frac{x}{\sqrt{x^2 + y^2}}\,; \qquad\qquad \frac{\partial \varphi}{\partial y} = \frac{y}{\sqrt{x^2 + y^2}}\,;$$

$$\frac{\partial^2 \varphi}{\partial x^2} = \frac{\sqrt{x^2 + y^2} - x\,\dfrac{x}{\sqrt{x^2+y^2}}}{x^2 + y^2} = \frac{x^2 + y^2 - x^2}{(x^2 + y^2)^{\frac{3}{2}}} = \frac{y^2}{(x^2 + y^2)^{\frac{3}{2}}}\,;$$

$$\frac{\partial^2 \varphi}{\partial y^2} = \frac{x^2}{(x^2 + y^2)^{\frac{3}{2}}}\,.$$

For the sum of the second derivatives $\varphi_{xx} + \varphi_{yy}$ it holds:

$$\varphi_{xx} + \varphi_{yy} = \frac{x^2 + y^2}{(x^2 + y^2)^{\frac{3}{2}}} = \frac{1}{(x^2 + y^2)^{\frac{1}{2}}} = \frac{1}{\varphi(x, y)}\,.$$

② $r(x, y, z) = \sqrt{x^2 + y^2 + z^2}$:

$$\frac{\partial r}{\partial x} = \frac{x}{\sqrt{x^2 + y^2 + z^2}} = \frac{x}{r}\,; \qquad \frac{\partial r}{\partial y} = \frac{y}{r}\,; \qquad \frac{\partial r}{\partial z} = \frac{z}{r}$$

$$\frac{\partial^2 r}{\partial x^2} = \frac{r - x\,\frac{x}{r}}{r^2} = \frac{r^2 - x^2}{r^3}\,; \qquad \frac{\partial^2 r}{\partial y^2} = \frac{r^2 - y^2}{r^3}\,; \qquad \frac{\partial^2 r}{\partial z^2} = \frac{r^2 - z^2}{r^3}$$

For the sum of the second derivatives $r_{xx} + r_{yy} + r_{zz}$ we conclude:

$$r_{xx} + r_{yy} + r_{zz} = \frac{3\,r^2 - (x^2 + y^2 + z^2)}{r^3} = \frac{2\,r^2}{r^3} = \frac{2}{r}\,.$$

③ $f(x, y, z) = \frac{1}{r} = \frac{1}{\sqrt{x^2 + y^2 + z^2}}$:

$$\frac{\partial f}{\partial x} = \frac{-x}{(x^2 + y^2 + z^2)^{\frac{3}{2}}} = \frac{-x}{r^3}\,; \qquad \frac{\partial f}{\partial y} = \frac{-y}{r^3}\,; \qquad \frac{\partial f}{\partial z} = \frac{-z}{r^3}$$

$$\frac{\partial^2 f}{\partial x^2} = -\frac{r^3 - x\,\frac{3}{2}\,r\,2x}{r^6} = -\frac{r^2 - 3\,x^2}{r^5}$$

$$\frac{\partial^2 f}{\partial y^2} = -\frac{r^2 - 3\,y^2}{r^5}\,; \qquad\qquad \frac{\partial^2 f}{\partial z^2} = -\frac{r^2 - 3\,z^2}{r^5}$$

For the sum of the second derivatives $f_{xx} + f_{yy} + f_{zz}$ we get:

$$f_{xx} + f_{yy} + f_{zz} = -\frac{3\,r^2 - 3\,(x^2 + y^2 + z^2)}{r^5} = 0.$$

$\square$

10.3.2 Total Differentiability

While differentiable functions of one variable are always continuous, this is not true for partially differentiable functions as the next example shows.

Example 10.16. We investigate the function

$$f(x, y) = \begin{cases} \dfrac{2\,x\,y}{x^2 + y^2} & \text{for} \quad (x, y) \neq (0, 0) \\ 0 & \text{for} \quad (x, y) = (0, 0) \end{cases}$$

which, according to the Example 10.10, is continuous for all points except at $(0, 0)$. Nevertheless, we can calculate the partial derivatives at **all** points including $(0, 0)$:

For $(x, y) \neq (0, 0)$ we calculate the partial derivatives with the quotient rule

$$f_x(x, y) = \frac{2\,y\,(y^2 - x^2)}{(x^2 + y^2)^2} \quad \text{and} \quad f_y(x, y) = \frac{2\,x\,(x^2 - y^2)}{(x^2 + y^2)^2}.$$

At the point $(x, y) = (0, 0)$ we take the definition of the partial derivative:

$$f_x(x_0, y_0) = \lim_{h \to 0} \frac{1}{h} \left(f(x_0 + h, y_0) - f(x_0, y_0) \right)$$

$$= \lim_{h \to 0} \frac{1}{h} \left(\frac{2\,h \cdot 0}{h^2 + 0^2} - 0 \right) = 0.$$

Similarly, $f_y(0, 0) = 0$ is calculated. So the partial derivatives of f exist, even though the function is not continuous at this point!   □

This is why we introduce the concept of *total* differentiability, which is different from *partial* differentiability. A function f of two variables is said to be totally differentiable at (x_0, y_0) if it can be approximated near this point by a plane called the *tangent plane*:

Definition: (Total Differentiability). *A function f is called **totally differentiable** at the point $(x_0, y_0) \in \mathbb{D}$ if there exist numbers $A, B \in \mathbb{R}$ and functions $\varepsilon_1(x, y), \varepsilon_2(x, y)$ such that for all (x, y) close to (x_0, y_0) the following holds*

$$f(x, y) = f(x_0, y_0) + A \cdot (x - x_0) + B \cdot (y - y_0)$$

$$+ \, \varepsilon_1(x, y)\,(x - x_0) + \varepsilon_2(x, y)\,(y - y_0), \quad (*)$$

when the functions ε_1 and ε_2 approach zero as $(x,\, y) \to (x_0,\, y_0)$:

$$\varepsilon_1\,(x,\, y) \to 0 \quad \text{for } (x,\, y) \to (x_0,\, y_0)$$
$$\varepsilon_2\,(x,\, y) \to 0 \quad \text{for } (x,\, y) \to (x_0,\, y_0).$$

The equation $(*)$ says that near the point $(x_0,\, y_0)$ the function values $f\,(x,\, y)$ are approximately described by the linear function

$$z\,(x,\, y) = f\,(x_0,\, y_0) + A\,(x - x_0) + B\,(y - y_0)\,.$$

The graph of z represents a plane in $\mathbb{R}^3$, the tangent plane, which passes through the point $(x_0,\, y_0,\, f\,(x_0,\, y_0))$.

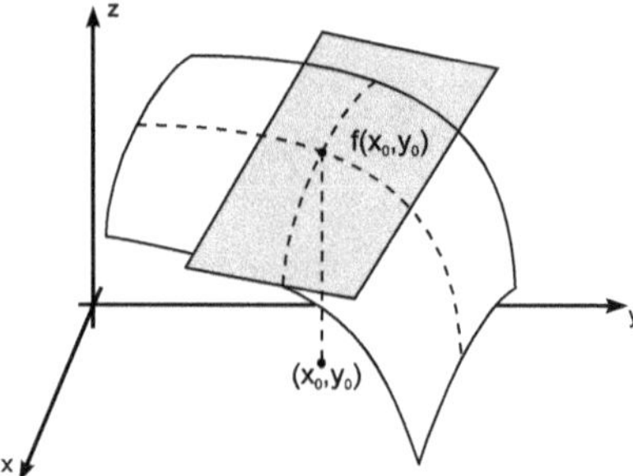

Figure 10.14. Function and tangent plane

From the total differentiability follows both the partial differentiability and the continuity of f:

Theorem:

If f is **totally differentiable** in $(x_0,\, y_0)$, then

(1) f is continuous at $(x_0,\, y_0)$.

(2) The partial derivatives exist $f_x\,(x_0,\, y_0)\,, \quad f_y\,(x_0,\, y_0)\,.$

(3) The numerical values A and B in the equation $(*)$ are calculated as follows $A = f_x\,(x_0,\, y_0)\,, \quad B = f_y\,(x_0,\, y_0)\,.$

(4) The function f can be approximated near the point by the **tangent plane** z_t

$$f\,(x,\, y) \approx z_t = f\,(x_0,\, y_0) + f_x\,(x_0,\, y_0)\,(x - x_0) + f_y\,(x_0,\, y_0)\,(y - y_0)$$

The definition of total differentiability is vividly illustrated, but difficult to verify in concrete cases. However, the partial derivatives can be used to decide whether a function f is totally differentiable:

Totally Differentiability

---

f is partially differentiable with respect to x and to y in an environment of $(x_0, y_0) \in \mathbb{D}$ and the **partial derivatives** f_x and f_y **are continuous at** (x_0, y_0). Then f is **totally differentiable** at (x_0, y_0).

Example 10.17 (Tangent Plane, with MAPLE-Worksheet). The tangent plane of the function

$$f(x, y) = e^{-\left(x^2 + y^2\right)} \quad \text{at point} \quad (x_0, y_0) = (0.15, 0.15).$$

The partial derivatives of the function

$$f_x(x, y) = -2\, x\, e^{-\left(x^2 + y^2\right)}$$
$$f_y(x, y) = -2\, y\, e^{-\left(x^2 + y^2\right)}$$

are continuous. Therefore, the function is totally differentiable and the tangent plane is given by

$$z = f(x_0, y_0) + f_x(x_0, y_0)(x - x_0) + f_y(x_0, y_0)(y - y_0)$$
$$= e^{-\frac{9}{200}} - \frac{3}{10} e^{-\frac{9}{200}} (x - 0.15) - \frac{3}{10} e^{-\frac{9}{200}} (y - 0.15).$$

Fig. 10.15 shows both the function and the tangent plane:

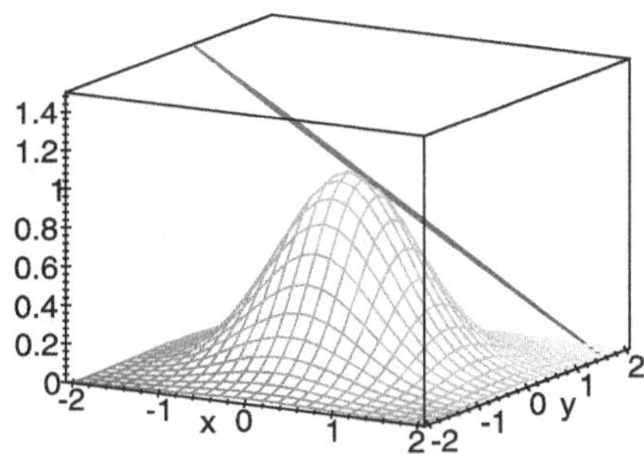

Figure 10.15. Function $f(x, y) = e^{-\left(x^2 + y^2\right)}$ with tangent plane

It can be seen that the tangent plane touches the graph of the function f at the point (x_0, y_0). $\qquad \square$

10.3.3 Gradient and Directional Derivative

In this section we assume a function $f(x, y)$ with two variables x and y, which is continuously partially differentiable in both x and y.

⊘ **The Gradient**

For functions of one variable, the derivative of the function at point x_0 gives the slope (= steepness) of the function. At points of high derivative the function changes rapidly, at points of low derivative it changes slightly. For functions of two variables, the partial derivative in both the x and y directions is taken into account. To characterize a function f in terms of its slope at a point (x_0, y_0), we introduce the vector $\operatorname{grad} f(x_0, y_0) := \begin{pmatrix} \frac{\partial f}{\partial x}(x_0, y_0) \\ \frac{\partial f}{\partial y}(x_0, y_0) \end{pmatrix}$ whose components are the partial derivatives with respect to x and y. It describes the slope of the tangent plane at the point (x_0, y_0) and it is called the *gradient*.

Definition: (Gradient)

The vector

$$\operatorname{grad}(f)\Big|_{(x_0, y_0)} := \begin{pmatrix} \frac{\partial f}{\partial x} \\ \frac{\partial f}{\partial y} \end{pmatrix} (x_0, y_0) := \begin{pmatrix} \frac{\partial f}{\partial x}(x_0, y_0) \\ \frac{\partial f}{\partial y}(x_0, y_0) \end{pmatrix}$$

*is called the **gradient of** f at the point (x_0, y_0). The Nabla operator* ∇ *is often used to represent the gradient:*

$$\nabla := \begin{pmatrix} \partial_x \\ \partial_y \end{pmatrix} \quad \Rightarrow \quad \operatorname{grad}(f) = \nabla f = \begin{pmatrix} \partial_x f \\ \partial_y f \end{pmatrix}.$$

Examples 10.18 (With Maple-Worksheet):

① Find the gradient of the function

$$f = \sqrt{x^2 + y^2 + 1}: \qquad \operatorname{grad}(f) = \begin{pmatrix} \frac{\partial f}{\partial x} \\ \frac{\partial f}{\partial y} \end{pmatrix} = \begin{pmatrix} \dfrac{x}{\sqrt{x^2 + y^2 + 1}} \\ \dfrac{y}{\sqrt{x^2 + y^2 + 1}} \end{pmatrix}.$$

If this result is represented in the form of a vector graphic, we get the two-dimensional representation:

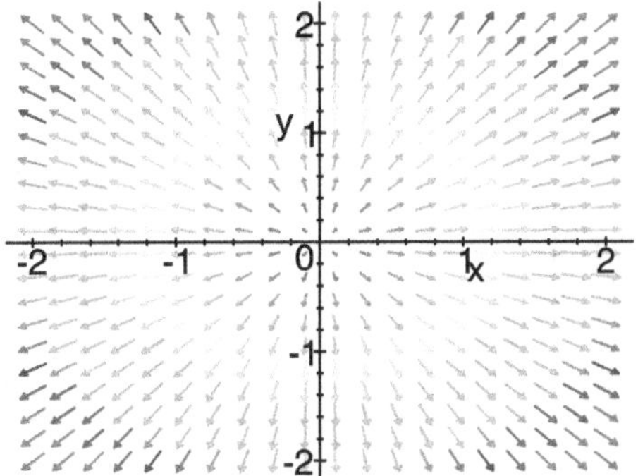

Figure 10.16. 2D representation of the gradient of a function $f(x, y)$

② Find the gradient of the function $f = \sqrt{x^2 + y^2 + z^2 + 1}$:

$$
\text{grad}\,(f) = \begin{pmatrix} \dfrac{\partial f}{\partial x} \\[2ex] \dfrac{\partial f}{\partial y} \\[2ex] \dfrac{\partial f}{\partial z} \end{pmatrix} = \begin{pmatrix} \dfrac{x}{\sqrt{x^2 + y^2 + z^2 + 1}} \\[2ex] \dfrac{y}{\sqrt{x^2 + y^2 + z^2 + 1}} \\[2ex] \dfrac{z}{\sqrt{x^2 + y^2 + z^2 + 1}} \end{pmatrix} .
$$

If this result is represented in the form of a vector graphic, the three-dimensional representation is obtained:

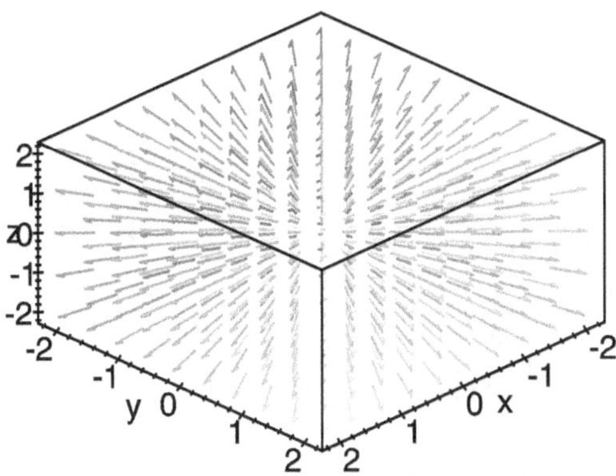

Figure 10.17. 3D representation of the gradient of a function $f(x, y, z)$

Application Example 10.19 (Electric Field of a Point Charge).

Given is a point charge q at the position $\vec{r}_0 = (x_0, y_0, z_0)$. The potential $\Phi(\vec{r})$ and the electric field $\vec{E}(\vec{r})$ at any point $\vec{r} = (x, y, z)$ is to be found. The potential induced by the point charge is

$$\Phi(\vec{r}) = \frac{1}{4\pi\,\varepsilon_0} \frac{q}{|\vec{r} - \vec{r}_0|} = \frac{1}{4\pi\,\varepsilon_0} \frac{q}{\sqrt{(x - x_0)^2 + (y - y_0)^2 + (z - z_0)^2}}$$

and the associated electric field is defined by

$$\vec{E}(\vec{r}) := -\,\mathrm{grad}\;\Phi(\vec{r}) = - \begin{pmatrix} \partial_x\,\Phi(x,y,z) \\ \partial_y\,\Phi(x,y,z) \\ \partial_z\,\Phi(x,y,z) \end{pmatrix}.$$

To avoid confusion with the partial derivatives of $\vec{E}$, the x component of the electric field is called E_1 instead of E_x, the y component is called E_2 instead of E_y, and the z component is called E_3 instead of E_z:

$$E_1(x, y, z) = -\partial_x\,\Phi(x, y, z)$$

$$= -\partial_x \frac{q}{4\pi\,\varepsilon_0} \left((x - x_0)^2 + (y - y_0)^2 + (z - z_0)^2 \right)^{-\frac{1}{2}}$$

$$= \frac{1}{4\pi\,\varepsilon_0} \frac{q}{\sqrt{(x - x_0)^2 + (y - y_0)^2 + (z - z_0)^2}^{\,3}} (x - x_0)$$

$$= \frac{1}{4\pi\,\varepsilon_0} \frac{q}{|\vec{r} - \vec{r}_0|^3} (x - x_0)$$

Analogous

$$E_2(x, y, z) = -\partial_y\,\Phi(x, y, z) = \frac{1}{4\pi\,\varepsilon_0} \frac{q}{|\vec{r} - \vec{r}_0|^3} (y - y_0)$$

$$E_3(x, y, z) = -\partial_z\,\Phi(x, y, z) = \frac{1}{4\pi\,\varepsilon_0} \frac{q}{|\vec{r} - \vec{r}_0|^3} (z - z_0)$$

Taking all three components together, we obtain the electric field

$$\vec{E}(\vec{r}) = \frac{1}{4\pi\,\varepsilon_0} \frac{q}{|\vec{r} - \vec{r}_0|^3} \begin{pmatrix} x - x_0 \\ y - y_0 \\ z - z_0 \end{pmatrix} = \frac{1}{4\pi\,\varepsilon_0} \frac{q}{|\vec{r} - \vec{r}_0|^3} (\vec{r} - \vec{r}_0).$$

$\vec{E}(\vec{r})$ is also called a *vector field*. $\square$

⊘ The Directional Derivative

In Fig. 10.18 the equipotential lines are schematically drawn for the two-electrode system of an electrostatic problem.

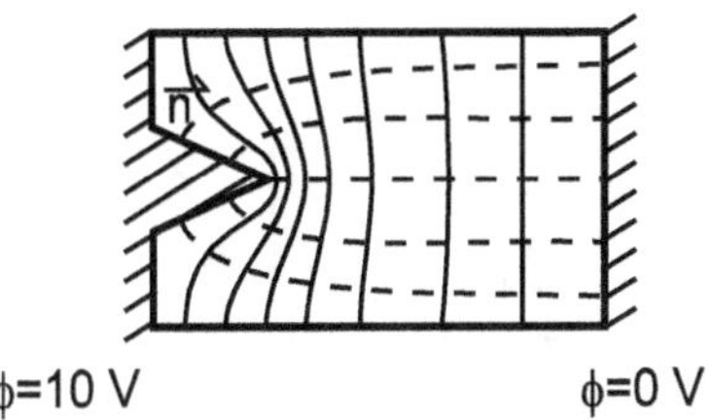

Figure 10.18. Qualitative course of the potential lines for a two-electrode system

The **gradient of Φ is perpendicular to the equipotential lines** and its magnitude is a measure of the density of the potential lines: At high density locations (near the edge) a large electric field is generated, at low density locations (right electrode) a small field is generated. For example, when looking for the electric field at the electrode surface, it is not the derivative of Φ in the direction x or y, but along a given direction $\vec{n}$.

Given is a function f, we look for the change of the function in the direction $\vec{n}$, where $\vec{n}$ is the directional unit vector. To find the directional derivative, we project the gradient in the direction of the vector $\vec{n}$. For the projection of a vector $\vec{b}$ in the direction $\vec{a}$ we use the formula $\vec{b}_a = \dfrac{\vec{a} \cdot \vec{b}}{|\vec{a}|^2} \cdot \vec{a}$, where $\vec{a} \cdot \vec{b}$ is the scalar product and $|\vec{a}|$ is the magnitude of the vector $\vec{a}$. In the case where $\vec{a}$ is a unit vector (i.e. $|\vec{a}| = 1$), the magnitude of $\vec{b}_a$ is

$$\left|\vec{b}_a\right| = b_a = \vec{a} \cdot \vec{b}.$$

Definition: (Directional Derivative). *The derivative of a function $f(x, y)$ in the direction of the unit vector $\vec{n}$ is defined by*

$$\frac{\partial f}{\partial \vec{n}} := \vec{n} \cdot \operatorname{grad}(f)$$

$$= \begin{pmatrix} n_1 \\ n_2 \end{pmatrix} \cdot \begin{pmatrix} \partial_x f(x, y) \\ \partial_y f(x, y) \end{pmatrix} = n_1\, \partial_x f(x, y) + n_2\, \partial_y f(x, y).$$

This is the **directional derivative** *of f in the direction $\vec{n}$.*

Notation: $\partial_{\vec{n}} f$, $\dfrac{\partial}{\partial \vec{n}} f$, $D_{\vec{n}} f$.

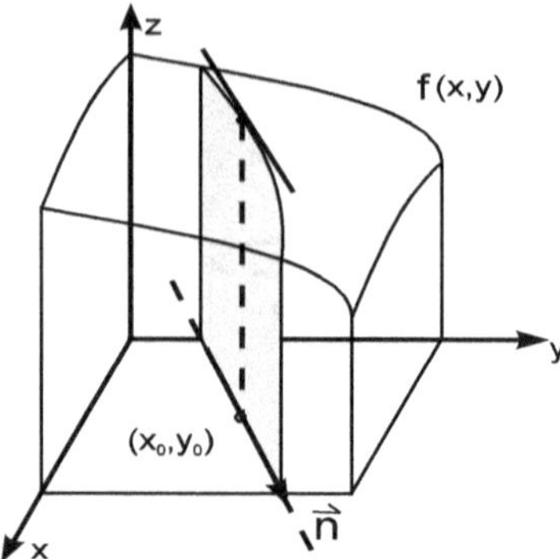

Figure 10.19. Directional derivative

⚠ **Caution:** Unlike the gradient, the directional derivative is not a vector but a scalar quantity. The partial derivatives with respect to x and y are special cases of the directional derivative.

Special Cases:

① We obtain the partial derivative with respect to x when $\vec{n} = \begin{pmatrix} 1 \\ 0 \end{pmatrix}$.

It is then $\partial_{\vec{n}} f = \frac{\partial f}{\partial x}$.

② We obtain the partial derivative with respect to y when $\vec{n} = \begin{pmatrix} 0 \\ 1 \end{pmatrix}$.

It is then $\partial_{\vec{n}} f = \frac{\partial f}{\partial y}$.

The directional derivative of f in direction $\vec{n}$ is the projection of the gradient $\mathrm{grad}\,(f)$ onto the line with direction $\vec{n}$. So we have:

Theorem:

(1) The directional derivative is greatest when the direction $\vec{n}$ is parallel to the gradient.

(2) The directional derivative is zero when $\vec{n}$ is perpendicular to the gradient $\mathrm{grad}\,(f)$.

(3) The gradient vector $\mathrm{grad}\,(f)$ points in the direction of the steepest slope of the function $f(x, y)$. Its magnitude indicates the size of the gradient.

Examples 10.20 (Directional Derivative, with MAPLE-Worksheet):

① $f(x, y) = \sqrt{x^2 + y^2}$. Find the derivative of f in the direction $\vec{n} = \frac{1}{\sqrt{2}} \begin{pmatrix} 1 \\ 1 \end{pmatrix}$: The partial derivatives of f with respect to x and y are

$$f_x = \frac{x}{\sqrt{x^2 + y^2}}, \quad f_y = \frac{y}{\sqrt{x^2 + y^2}}.$$

This determines the directional derivative in the direction of $\vec{n}$ by

$$\frac{\partial f}{\partial \vec{n}} = \vec{n} \cdot \operatorname{grad}(f) = \frac{1}{\sqrt{2}} \begin{pmatrix} 1 \\ 1 \end{pmatrix} \frac{1}{\sqrt{x^2 + y^2}} \begin{pmatrix} x \\ y \end{pmatrix} = \frac{1}{\sqrt{2}} \frac{x + y}{\sqrt{x^2 + y^2}}.$$

② $f(x, y) = x \cdot y$. Find the derivative of f in the direction $\vec{a} = \begin{pmatrix} 1 \\ 2 \end{pmatrix}$.

The partial derivatives are

$$f_x = y, \qquad f_y = x$$

and the direction vector for $\vec{a} = \begin{pmatrix} 1 \\ 2 \end{pmatrix}$ is $\vec{n} = \frac{1}{|\vec{a}|}\vec{a} = \frac{1}{\sqrt{5}} \begin{pmatrix} 1 \\ 2 \end{pmatrix}$.

$$\Rightarrow \frac{\partial f}{\partial \vec{n}} = \frac{1}{\sqrt{5}} \begin{pmatrix} 1 \\ 2 \end{pmatrix} \begin{pmatrix} y \\ x \end{pmatrix} = \frac{1}{\sqrt{5}} (2x + y). \qquad \qquad \square$$

Remark: The directional derivative of f in the direction of the unit vector $\vec{n} = \begin{pmatrix} n_1 \\ n_2 \end{pmatrix}$ at point (x_0, y_0) is often also expressed in the equivalent representation

$$\frac{\partial f}{\partial \vec{n}} = \lim_{h \to 0} \frac{f(x_0 + h\,n_1,\, y_0 + h\,n_2) - f(x_0, y_0)}{h}.$$

This limit value reflects the derivative along the line

$$\begin{pmatrix} x_0 + h\,n_1 \\ y_0 + h\,n_2 \end{pmatrix} = \begin{pmatrix} x_0 \\ y_0 \end{pmatrix} + h \begin{pmatrix} n_1 \\ n_2 \end{pmatrix}.$$

This line is defined in the point-direction representation by the point $\begin{pmatrix} x_0 \\ y_0 \end{pmatrix}$ and the direction $\vec{n} = \begin{pmatrix} n_1 \\ n_2 \end{pmatrix}$.

10.3.4 Taylor's Theorem

An essential property of differentiable functions in one variable is that they can be approximated by polynomials near a point. This is also possible in the multidimensional case. Instead of proving Taylor's theorem, we will introduce Taylor's formula for functions in two variables:

For a $(m+1)$-times continuously differentiable function $f(x)$, Taylor's formula from Section 9.3 holds at the expansion point $x_0 \in \mathbb{D}$

$$f(x) = f(x_0) + (x - x_0)\, f'(x_0) + \tfrac{1}{2}(x - x_0)^2\, f''(x_0) + \cdots +$$

$$+ \tfrac{1}{m!}(x - x_0)^m\, f^{(m)}(x_0) + R_m(x),$$

where the difference between the polynomial and the function is given by

$$R_m(x) = \tfrac{1}{(m+1)!}\, f^{(m+1)}(\xi)\, (x - x_0)^{m+1}.$$

The unknown value ξ is between x and x_0. In a modified notation, the expansion is

$$f(x) = f(x_0) + (x - x_0)\, \frac{d}{dx}\, f \Big|_{x_0}$$

$$+ \tfrac{1}{2!}(x - x_0)^2 \left(\frac{d}{dx}\right)^2 f \Big|_{x_0}$$

$$+ \cdots + \tfrac{1}{m!}(x - x_0)^m \left(\frac{d}{dx}\right)^m f \Big|_{x_0} + R_m(x)$$

$$f(x) = \sum_{n=0}^{m} \frac{1}{n!} \left[(x - x_0)\frac{d}{dx}\right]^n f \Big|_{x_0} + R_m(x). \qquad (*)$$

In this formula $\left[(x - x_0)\frac{d}{dx}\right]^n$ is replaced by the corresponding partial derivatives according to

$$(x - x_0)\frac{d}{dx} \quad \rightarrow \quad (x - x_0)\frac{\partial}{\partial x} + (y - y_0)\frac{\partial}{\partial y}$$

$$\left[(x - x_0)\frac{d}{dx}\right]^n \quad \rightarrow \quad \left[(x - x_0)\frac{\partial}{\partial x} + (y - y_0)\frac{\partial}{\partial y}\right]^n.$$

We use the binomial theorem $(a+b)^n = \sum_{k=0}^{n} \binom{n}{k} a^{n-k} b^k$, see Volume 1,

Section 1.2.5, so that with the binomial coefficients $\binom{n}{k} = \dfrac{n!}{(n-k)!\,k!}$ we

get

$$
\left[(x - x_0)\,\frac{d}{dx}\right]^n \rightarrow \sum_{k=0}^{n} \binom{n}{k} (x - x_0)^{n-k} \left(\frac{\partial}{\partial x}\right)^{n-k} (y - y_0)^k \left(\frac{\partial}{\partial y}\right)^k
$$

$$
\rightarrow \sum_{k=0}^{n} \binom{n}{k} \underbrace{(x - x_0)^{n-k} (y - y_0)^k}_{\text{polynomial of degree } n} \underbrace{\left(\frac{\partial}{\partial x}\right)^{n-k} \left(\frac{\partial}{\partial y}\right)^k}_{\substack{\text{partial derivative} \\ \text{of order } n}} .
$$

When inserted into the Taylor formula $(*)$, we get

Taylor's Theorem for Functions with Two Variables

Let $f(x, y)$ be a $(m + 1)$ times continuously partially differentiable function and $(x_0, y_0) \in \mathbb{D}$ the expansion point. For $(x, y) \in \mathbb{D}$ the following applies

$$
f(x, y) = \sum_{n=0}^{m} \frac{1}{n!} \left[(x - x_0)\,\frac{\partial}{\partial x} + (y - y_0)\,\frac{\partial}{\partial y}\right]^n f \Bigg|_{(x_0, y_0)} + R_m
$$

$$
= \sum_{n=0}^{m} \frac{1}{n!} \sum_{k=0}^{n} \binom{n}{k} (x - x_0)^{n-k} (y - y_0)^k \cdot
$$

$$
f \underbrace{x \ldots x}_{(n-k)\text{-times}} \underbrace{y \ldots y}_{k\text{-times}} (x_0, y_0) + R_m
$$

with the residual

$$
R_m = \frac{1}{(m+1)!} \sum_{k=0}^{m+1} \binom{m+1}{k} (x - x_0)^{m+1-k} (y - y_0)^k
$$

$$
f \underbrace{x \ldots x}_{(m+1-k)\text{-times}} \underbrace{y \ldots y}_{k\text{-times}} (\xi, \eta),
$$

where (ξ, η) is an unknown point on the straight line between (x_0, y_0) and (x, y), which must be completely inside $\mathbb{D}$.

Remarks:

(1) If (x, y) is sufficiently close to (x_0, y_0), then the line between these points is usually also contained in $\mathbb{D}$.

(2) The formula

$$\left[(x - x_0) \, \frac{\partial}{\partial x} + (y - y_0) \, \frac{\partial}{\partial y} \right]^n f \bigg|_{(x_0, y_0)}$$

can be interpreted as follows: Take the sum to the power of n, apply all derivatives to f, evaluate these derivatives at the (x_0, y_0) and multiply by the polynomial $(x - x_0)^i \, (y - y_0)^j$, where i is the order of the derivative $\frac{\partial}{\partial x}$, j is the order of the derivative $\frac{\partial}{\partial y}$, and $i + j = n$.

(3) **Taylor's theorem for functions with k variables:** Analogous to the derivation of Taylor's formula for a function with two variables, we get for a function f with k variables $(x_1, ..., x_k)$ the formula

$$f(x_1, ..., x_k) = \sum_{n=0}^{m} \frac{1}{n!} \left[\left(x_1 - x_1^{(0)} \right) \frac{\partial}{\partial x_1} + \cdots + \left(x_k - x_k^{(0)} \right) \frac{\partial}{\partial x_k} \right]^n$$
$$f \big|_{\left(x_1^{(0)}, ..., x_k^{(0)} \right)} + R_m \left(x_1, ..., x_k \right)$$

at the expansion point $(x_1^{(0)}, ..., x_k^{(0)})$ and the residual R_m, which is calculated in the same way as the residual of a function with two variables.

To illustrate the behavior of Taylor's formula for a function $f(x, y)$, we consider the special cases $n = 0, 1, 2$. $n = 1$ is often used in applications where functions are linearized. $n = 2$ becomes important when we will discuss the local extremes of a function. It will give us a criterion for deciding whether a point is an extreme value or not.

n = 0:

$$f(x, y) = \sum_{n=0}^{0} \frac{1}{n!} \left[(x - x_0) \, \frac{\partial}{\partial x} + (y - y_0) \, \frac{\partial}{\partial y} \right]^n f \bigg|_{(x_0, y_0)} + R_0(x, y)$$

$$= f(x_0, y_0) + R_0(x, y).$$

n = 1: Linearization

$$f(x, y) = \sum_{n=0}^{1} \frac{1}{n!} \left[(x - x_0) \, \frac{\partial}{\partial x} + (y - y_0) \, \frac{\partial}{\partial y} \right]^n f \bigg|_{(x_0, y_0)} + R_1(x, y)$$

$$f(x, y) = f(x_0, y_0) + (x - x_0)\, f_x(x_0, y_0) + (y - y_0)\, f_y(x_0, y_0) + R_1(x, y)$$

This formula is used to **linearize functions** by replacing $f(x, y)$ with the polynomial on the right side.

n = 2: Quadratic Approximation

$$
\begin{aligned}
f(x, y) &= \sum_{n=0}^{2} \frac{1}{n!} \left[(x - x_0)\, \frac{\partial}{\partial x} + (y - y_0)\, \frac{\partial}{\partial y} \right]^n f \bigg|_{(x_0, y_0)} + R_2(x, y) \\
&= f(x_0, y_0) + (x - x_0)\, f_x(x_0, y_0) + (y - y_0)\, f_y(x_0, y_0) \\
&\quad + \frac{1}{2!} \left[(x - x_0)^2\, \frac{\partial^2}{\partial x^2} + 2\,(x - x_0)(y - y_0)\, \frac{\partial}{\partial x}\, \frac{\partial}{\partial y} \right. \\
&\qquad\qquad \left. + (y - y_0)^2\, \frac{\partial^2}{\partial y^2} \right] f \bigg|_{(x_0, y_0)} + R_2(x, y)
\end{aligned}
$$

$$
\begin{aligned}
f(x, y) &= f(x_0, y_0) + (x - x_0)\, f_x(x_0, y_0) + (y - y_0)\, f_y(x_0, y_0) \\
&\quad + \frac{1}{2} \left((x - x_0)^2\, f_{xx}(x_0, y_0) + 2\,(x - x_0)(y - y_0)\, f_{xy}(x_0, y_0) \right. \\
&\qquad\qquad \left. + (y - y_0)^2\, f_{yy}(x_0, y_0) \right) + R_2(x, y)
\end{aligned}
$$

This formula is used in Section 10.4.3 to formulate a sufficient condition for a local extremum of a function in two variables. For functions with more than two variables, we need the following generalized formulation:

Generalization

We define the *Hesse matrix*

$$H(f) := \begin{pmatrix} f_{xx}(x, y) & f_{xy}(x, y) \\ f_{xy}(x, y) & f_{yy}(x, y) \end{pmatrix}.$$

With the vector $\vec{h} = \begin{pmatrix} x - x_0 \\ y - y_0 \end{pmatrix}$ we rewrite the formula for the quadratic approximation in the form of

$$f(x, y) - f(x_0, y_0) = \vec{h}^t \cdot \operatorname{grad}(f)\bigg|_{(x_0, y_0)} + \frac{1}{2}\, \vec{h}^t\, H(f)\bigg|_{(x_0, y_0)} \vec{h}$$

$$+ R_2(x, y)\,.$$

This formulation with the gradient and the Hesse matrix can be directly extended to functions with more than two variables (see Section 10.4.4).

Example 10.21 (Taylor Polynomial): (1) We are looking for the Taylor polynomial up to the order 2 for the function

$$f(x, y) = \sin\left(x^2 + 2y\right)$$

at the position $(x_0, y_0) = \left(0, \frac{\pi}{4}\right)$. (2) Using the Taylor approximation, we want to evaluate this polynomial at $(x, y) \neq \left(0, \frac{\pi}{4}\right)$ and obtain an estimate of the error. (3) Finally, we perform the numerical example for $(x, y) = \left(0.05, \frac{\pi}{4} + 0.05\right)$.

(1) The partial derivatives up to the 2nd order 2 are

$$f(x, y) = \sin\left(x^2 + 2y\right) \qquad\qquad f\left(0, \tfrac{\pi}{4}\right) = 1$$

$$f_x(x, y) = 2x \cos\left(x^2 + 2y\right) \qquad\qquad f_x\left(0, \tfrac{\pi}{4}\right) = 0$$
$$f_y(x, y) = 2 \cos\left(x^2 + 2y\right) \qquad\qquad f_y\left(0, \tfrac{\pi}{4}\right) = 0$$

$$f_{xx}(x, y) = -4x^2 \sin\left(x^2 + 2y\right) + 2 \cos\left(x^2 + 2y\right) \qquad f_{xx}\left(0, \tfrac{\pi}{4}\right) = 0$$
$$f_{yy}(x, y) = -4 \sin\left(x^2 + 2y\right) \qquad\qquad f_{yy}\left(0, \tfrac{\pi}{4}\right) = -4$$
$$f_{xy}(x, y) = -4x \sin\left(x^2 + 2y\right) \qquad\qquad f_{xy}\left(0, \tfrac{\pi}{4}\right) = 0.$$

Evaluation of the 2nd order Taylor polynomial

$$f(x, y) = 1 + \frac{1}{2}\left(y - y_0\right)^2 \cdot (-4) + R_2(x, y)$$
$$= 1 - 2\left(y - \frac{\pi}{4}\right)^2 + R_2(x, y).$$

(2) Calculation of the residual $R_2(x, y)$: The residual has to be estimated for an unknown point (ξ, η) on the line from (x, y) to $\left(0, \frac{\pi}{4}\right)$

$$R_2(x, y) = \frac{1}{3!}\left\{ f_{xxx}(\xi, \eta)(x - x_0)^3 + 3 f_{xxy}(\xi, \eta)(x - x_0)^2(y - y_0) \right.$$
$$\left. + 3 f_{xyy}(\xi, \eta)(x - x_0)(y - y_0)^2 + f_{yyy}(\xi, \eta)(y - y_0)^3 \right\}$$

To estimate $R_2(x, y)$ we determine the third-order partial derivatives
$$f_{xxx}(x, y) = -12x \sin\left(x^2 + 2y\right) - 8x^3 \cos\left(x^2 + 2y\right)$$
$$f_{xxy}(x, y) = -8x^2 \cos\left(x^2 + 2y\right) - 4 \sin\left(x^2 + 2y\right)$$
$$f_{xyy}(x, y) = -8x \cos\left(x^2 + 2y\right)$$
$$f_{yyy}(x, y) = -8 \cos\left(x^2 + 2y\right).$$

Let $\xi \in [0, x]$ and $\eta \in \left[\frac{\pi}{4}, y\right]$ be arbitrary. Assuming that $x > 0$ and $y > \frac{\pi}{4}$, the following estimates hold

$$
\begin{aligned}
|f_{xxx}(\xi, \eta)| &\leq 12\xi + 8\xi^3 &\leq 12x + 8x^3 \\
|f_{xxy}(\xi, \eta)| &\leq 8\xi^2 + 4 &\leq 8x^2 + 4 \\
|f_{xyy}(\xi, \eta)| &\leq 8\xi &\leq 8x \\
|f_{yyy}(\xi, \eta)| &\leq 8.
\end{aligned}
$$

Therefore, an upper bound for $R_2(x, y)$ is:

$$
\begin{aligned}
|R_2(x, y)| \leq\ & \tfrac{1}{3!}\left\{|f_{xxx}(\xi, \eta)|\, x^3 + 3\,|f_{xxy}(\xi, \eta)|\, x^2 \left(y - \tfrac{\pi}{4}\right)\right. \\
& \left. +3\,|f_{xyy}(\xi, \eta)|\, x \left(y - \tfrac{\pi}{4}\right)^2 + |f_{yyy}(\xi, \eta)| \left(y - \tfrac{\pi}{4}\right)^3\right\} \\[2mm]
\leq\ & \tfrac{1}{6}\left\{4x^4\left(2x^2 + 3\right) + 12x^2\left(y - \tfrac{\pi}{4}\right)\left(2x^2 + 1\right)\right. \\
& \left. +24x^2\left(y - \tfrac{\pi}{4}\right)^2 + 8\left(y - \tfrac{\pi}{4}\right)^3\right\}.
\end{aligned}
$$

(3) Numerical example: $(x, y) = \left(0.05,\ \frac{\pi}{4} + 0.05\right)$

$$
\begin{aligned}
f\left(0.05,\ \tfrac{\pi}{4} + 0.05\right) &= 0.99475 &&\text{(Exact value)} \\
f_T\left(0.05,\ \tfrac{\pi}{4} + 0.05\right) &= 0.995 &&\text{(Taylor polynomial)} \\
R_2\left(0.05,\ \tfrac{\pi}{4} + 0.05\right) &\leq 0.000455 &&\text{(Estimated error)} \qquad \square
\end{aligned}
$$

 **Visualization:** This visualization shows a function of two variables together with its Taylor expansion with increasing order in a three-dimensional animation.

Application: The expansion point (x_0, y_0) is held and (x, y) is any point near (x_0, y_0). The special cases for $n = 1$ and $n = 2$ lead to important *approximation expressions.*

(1) **Linearization:**

$$
f(x, y) \approx f(x_0, y_0) + (x - x_0)\, f_x(x_0, y_0) + (y - y_0)\, f_y(x_0, y_0).
$$

The right-hand side is a linear function in the variables x and y. The graph of this linear function represents a plane in three-dimensional space that has the common point $(x_0, y_0, f(x_0, y_0))$ with f and touches the graph of f there. This plane is called the *Tangent Plane* to the graph of f at the point (x_0, y_0) (see Section 10.3.2).

(2) Quadratic Approximation:

$$f(x, y) \approx f(x_0, y_0) + (x - x_0) f_x(x_0, y_0) + (y - y_0) f_y(x_0, y_0)$$
$$+ \tfrac{1}{2} (x - x_0)^2 f_{xx}(x_0, y_0) + (x - x_0)(y - y_0) f_{xy}(x_0, y_0)$$
$$+ \tfrac{1}{2} (y - y_0)^2 f_{yy}(x_0, y_0).$$

The right side is a quadratic function in the variables x and y. In general, the quadratic approximation for functions is better than the linear approximation: The graph of f is not approximated by a plane, but by a curved surface, which additionally has the same curvature as the function at the point (x_0, y_0).

Example 10.22. For the function

$$f(x, y) = x \cos(x + y) + (y - 1)^2 e^{-x^2}$$

the linear approximation at the point $(x_0, y_0) = (0, 0)$ is

$$f(x, y) \approx 1 + x - 2y$$

and the quadratic approximation is

$$f(x, y) \approx 1 + x - 2y - x^2 + y^2. \qquad \square$$

Application Example 10.23 (Oscillation Period of a Pendulum).

The oscillation period of a pendulum of length l is

$$T = 2\pi \sqrt{\frac{l}{g}}$$

($g = 9.81 \, \tfrac{m}{s^2}$). Find the linear expression in g and l which is a good approximation for T if l is only slightly different from $l_0 = 1\,m$ and g is only slightly different from $g_0 = 9.81 \, \tfrac{m}{s^2}$:

$$\frac{\partial T}{\partial l} = \frac{\pi}{\sqrt{l\,g}}; \qquad \frac{\partial T}{\partial g} = -\pi \sqrt{\frac{l}{g^3}}$$

$$T \approx T(l_0, g_0) + \frac{\partial T}{\partial l}(l_0, g_0) \cdot (l - l_0) + \frac{\partial T}{\partial g}(l_0, g_0) \cdot (g - g_0)$$
$$\approx \quad 2.006\,s \quad + \quad 1.0030\,(l - 1)\,s \quad - \quad 0.1022\,(g - 9.81)\,s. \qquad \square$$

10.4 Applications of Differential Calculus

This section discusses some important applications of the Taylor formula: The total differential as a linear approximation, error calculation, the theory of extremes for functions of two variables, and the determination of regression curves, especially of the regression line.

10.4.1 The Differential as Linear Approximation

The behavior of the function $z = f(x, y)$ in the immediate vicinity of a point (x_0, y_0) is studied by comparing the growth of the tangent plane dz with the growth of the function Δz.

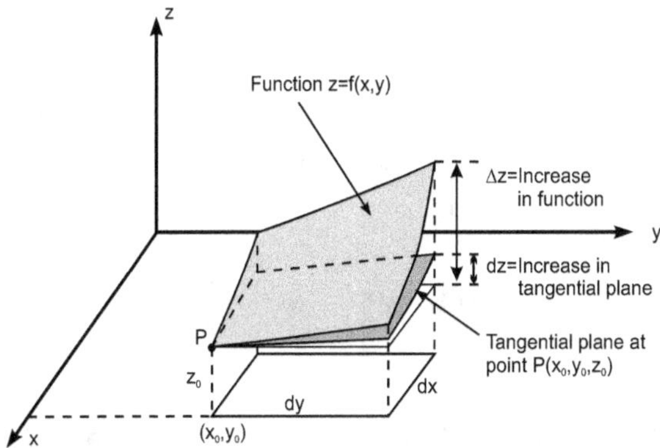

Figure 10.20. The concept of the complete differential

The tangent plane of the function $z = f(x, y)$ is defined at point (x_0, y_0) according to Section 10.3.2 by

$$z_t (x, y) = f (x_0, y_0) + (x - x_0)\, f_x (x_0, y_0) + (y - y_0)\, f_y (x_0, y_0) \ .$$

At the point $(x_0 + dx, y_0 + dy)$ the tangent plane has the value

$$z_t (x_0 + dx, y_0 + dy) = f (x_0, y_0) + dx\, f_x (x_0, y_0) + dy\, f_y (x_0, y_0) \ .$$

The change in the tangent plane dz is therefore

$$dz = z_t (x_0 + dx, y_0 + dy) - z_t (x_0, y_0) = dx\, f_x (x_0, y_0) + dy\, f_y (x_0, y_0) \ .$$

We call

dx, dy: *independent differential*

dz: *dependent differential* (= Change of the tangent plane).

Definition: (Total Differential of a Function with Two Variables). *The* **total differential** *of a function $z = f(x, y)$ at the point (x_0, y_0) is*

$$dz := f_x(x_0, y_0)\, dx + f_y(x_0, y_0)\, dy\,.$$

This describes the change in the tangent plane as we move from (x_0, y_0) towards $(x_0 + dx, y_0 + dy)$. It is also written df instead of dz.

Example 10.24. Find the total differential of the function

$$f(x, y) = x \ln(x + y)$$

at the point $(x_0, y_0) = \left(\frac{1}{2}, \frac{1}{2}\right)$.

$$\left.\begin{array}{l} \dfrac{\partial f}{\partial x} = \ln(x + y) + x \cdot \dfrac{1}{x + y} \\[4mm] \dfrac{\partial f}{\partial y} = x \cdot \dfrac{1}{x + y} \end{array}\right\} \quad df = \left(\ln(x + y) + \dfrac{x}{x + y}\right) dx + \dfrac{x}{x + y}\, dy\,.$$

$$\Rightarrow df(x_0, y_0) = \frac{1}{2}\, dx + \frac{1}{2}\, dy.$$
$\hfill \square$

We compare the **change in the tangent plane** dz with the **change in the function** $\triangle z$ by linearizing for the function $z = f(x, y)$:

$$f(x, y) - f(x_0, y_0) = (x - x_0)\, f_x(x_0, y_0) + (y - y_0)\, f_y(x_0, y_0) + R_1(x, y)\,.$$

The change of the function $\triangle z$ in (x_0, y_0) to $(x_0 + dx, y_0 + dy)$ is

$$\begin{aligned} \triangle z \;&=\; f(x, y) - f(x_0, y_0) = f(x_0 + dx, y_0 + dy) - f(x_0, y_0) \\[3mm] &=\; dx\, f_x(x_0, y_0) + dy\, f_y(x_0, y_0) + R_1(x_0 + dx, y_0 + dy) \\[3mm] &=\; dz + R_1(x_0 + dx, y_0 + dy). \end{aligned}$$

It corresponds to the change in the tangent plane dz

$$dz = dx\, f_x(x_0, y_0) + dy\, f_y(x_0, y_0)$$

except for $R_1(x_0 + dx, y_0 + dy)$. For $(dx, dy) \to (0, 0)$ we get
$R_1(x_0 + dx, y_0 + dy) \to R_1(x_0, y_0) = 0$, so that for small dx, dy we have

$$dz \approx \triangle z.$$
$\hfill \square$

⊘ Total Differential of Functions with n Variables

The concept of the total differential applies directly to functions with more than two variables:

Definition: (Total Differential of a Function with n Variables).
The **total differential of a function**

$$y = f(x_1, \ldots, x_n)$$

is the differential expression

$$dy = f_{x_1}\, dx_1 + f_{x_2}\, dx_2 + \cdots + f_{x_n}\, dx_n$$

$$= \frac{\partial f}{\partial x_1}\, dx_1 + \frac{\partial f}{\partial x_2}\, dx_2 + \cdots + \frac{\partial f}{\partial x_n}\, dx_n .$$

Notes:

(1) Instead of dy we now write df.

(2) The total differential describes approximately how the value of the function changes when the variables change slightly by dx_i $(i = 1, \ldots, n)$: $\triangle y \approx dy$.

Examples 10.25.

① Find the total differential of the function

$$f(x, y, z) = x\, e^{x\,y + 4\,z}$$

at point $(x_0, y_0, z_0) = (1, 0, 1)$:

We compute the partial derivatives

$$\frac{\partial f}{\partial x} = e^{x\,y + 4\,z} + x\, e^{x\,y + 4\,z}\, y \qquad \hookrightarrow f_x(1, 0, 1) = e^4$$

$$\frac{\partial f}{\partial y} = x^2\, e^{x\,y + 4\,z} \qquad \hookrightarrow f_y(1, 0, 1) = e^4$$

$$\frac{\partial f}{\partial z} = 4\,x\, e^{x\,y + 4\,z} \qquad \hookrightarrow f_z(1, 0, 1) = 4\,e^4$$

and evaluate the total differential

$$df = f_x \, dx + f_y \, dy + f_z \, dz.$$

So we obtain df:

$$df(x, y, z) = \left(e^{x\,y+4\,z} + x\,y\,e^{x\,y+4\,z}\right) dx + x^2\,e^{x\,y+4\,z}\,dy + 4\,x\,e^{x\,y+4\,z}\,dz$$
$$df(1, 0, 1) = e^4\,dx + e^4\,dy + 4\,e^4\,dz.$$

② An ideal gas satisfies the *equation of state*

$$p\,(V,\,T) = n\,R \cdot \frac{T}{V}\;.$$

The total differential of this function is

$$dp = \frac{\partial p}{\partial V}\,dV + \frac{\partial p}{\partial T}\,dT = -n\,R\,\frac{T}{V^2}\,dV + n\,\frac{R}{V}\,dT\;.$$

It approximately describes the change in gas pressure p with a small change in volume by dV and a simultaneous change in temperature by dT (see Example 10.11 ⑥). □

⊘ Linearization of Functions with Two Variables

For a function $z = f\,(x,\,y)$, for small dx and dy the following holds approximately

$$dz \approx \triangle z\;,$$

i.e. for small dx and dy the change in the function can be approximated by changing the tangent plane (total differential). This procedure is called the linearization of the function $z = f\,(x,\,y)$. We obtain

$$\triangle z = f\,(x,\,y) - f\,(x_0,\,y_0)\;\approx\;dz = f_x\,(x_0,\,y_0)\,dx + f_y\,(x_0,\,y_0)\,dy\;.$$

Using the relation $dx = x - x_0$ and $dy = y - y_0$ it follows

$$f\,(x,\,y)\;\approx\;f\,(x_0,\,y_0) + f_x\,(x_0,\,y_0)\,(x - x_0) + f_y\,(x_0,\,y_0)\,(y - y_0)$$

(Linearization of the function $z = f\,(x,\,y)$.)

Linearization is often used to approximate differences of the form

$$z\left(x_0 + \triangle x,\, y_0 + \triangle y\right) - z\left(x_0,\, y_0\right):$$

$$z\left(x_0 + \triangle x,\, y_0 + \triangle y\right) - z\left(x_0,\, y_0\right) \approx \left.\frac{\partial z}{\partial x}\right|_{(x_0,\,y_0)} \triangle x \; + \; \left.\frac{\partial z}{\partial y}\right|_{(x_0,\,y_0)} \triangle y$$

$$z\left(x,\, y\right) - z\left(x_0,\, y_0\right) \approx \left.\frac{\partial z}{\partial x}\right|_{(x_0,\,y_0)} \left(x - x_0\right) \; + \; \left.\frac{\partial z}{\partial y}\right|_{(x_0,\,y_0)} \left(y - y_0\right). \quad (1)$$

A generalized formula can be used to linearize functions of n variables:

Linearization of Functions with n Variables:

In the environment of a point $\left(x_1^0, \ldots, x_n^0\right)$, a non-linear function $f\left(x_1, \ldots, x_n\right)$ can be approximated by the *linear* function

$$y = f\left(x_1^0, \ldots, x_n^0\right) + \left(\frac{\partial f}{\partial x_1}\right)_0 \left(x_1 - x_1^0\right) + \ldots + \left(\frac{\partial f}{\partial x_n}\right)_0 \left(x_n - x_n^0\right).$$

The partial derivatives $\left(\dfrac{\partial f}{\partial x_i}\right)_0$ must be evaluated at the expansion point $\left(x_1^0, \ldots, x_n^0\right)$:

$$\left(\frac{\partial f}{\partial x_i}\right)_0 = \frac{\partial}{\partial x_i} f\left(x_1^0, \ldots, x_n^0\right) \; .$$

Example 10.26. Find the linearization of the function

$$f(x, y) = 5\, x\, e^{x - 4\, y^2}$$

at the point $\left(x_0,\, y_0\right) = \left(1,\, \tfrac{1}{2}\right)$:

With the partial derivatives

$$f_x\left(x,\, y\right) = 5\, e^{x - 4\, y^2} + 5\, x\, e^{x - 4\, y^2} \qquad \hookrightarrow \qquad f_x\left(1,\, \tfrac{1}{2}\right) = 5 + 5 = 10$$

$$f_y\left(x,\, y\right) = -40\, x\, y\, e^{x - 4\, y^2} \qquad \hookrightarrow \qquad f_y\left(1,\, \tfrac{1}{2}\right) = -20$$

we get the linear approximation

$$\Rightarrow z_l\left(x,\, y\right) = 5 + 10\left(x - 1\right) - 20\left(y - \tfrac{1}{2}\right). \qquad \square$$

10.4.2 Error Calculation

An application of the total differential occurs in error calculation, which plays a role wherever inaccurate measurements are investigated: A physical quantity y depends via a known law on n independent quantities $x_1, \ldots, x_n$:

$$y = f(x_1, \ldots, x_n) \ .$$

To evaluate y, the values of $x_1, \ldots, x_n$ must be measured, which is only possible with limited accuracy. The measured values are $x_1^0, \ldots, x_n^0$, the measurement errors (tolerances) $\triangle x_1, \ldots, \triangle x_n$. The question arises: How much can the value $f\left(x_1^0 + \triangle x_1, \ldots, x_n^0 + \triangle x_n\right)$, calculated from the measured values with tolerances, differ at most from the value $f\left(x_1^0, \ldots, x_n^0\right)$?

The difference

$$\triangle f = f\left(x_1^0 + \triangle x_1, \ldots, x_n^0 + \triangle x_n\right) - f\left(x_1^0, \ldots, x_n^0\right)$$

is called the **Absolute Error** of f. An estimate of $\triangle f$ is sought if we know the maximum size of the measurement errors $\triangle x_i$.

For small $dx_i = \triangle x_i$, the total differential df is approximately equal to the change in the function $\triangle f$:

$$\triangle f \approx df = \left(\frac{\partial f}{\partial x_1}\right)_0 \cdot dx_1 + \cdots + \left(\frac{\partial f}{\partial x_n}\right)_0 \cdot dx_n \ .$$

This formula is used to calculate the influence of **small** errors on the result. Since a tolerance $\pm \triangle x_i$ is usually given for the measurement errors, an upper limit for the error in the linear approximation is given by

$$|\triangle f| \approx |df| \leq \left|\left(\frac{\partial f}{\partial x_1}\right)_0\right| |\triangle x_1| + \cdots + \left|\left(\frac{\partial f}{\partial x_n}\right)_0\right| |\triangle x_n| = \overline{df} \ .$$

The term $\overline{df}$ is called **Absolute Error in Linear Approximation**. The partial derivatives $\left(\frac{\partial f}{\partial x_i}\right)_0$ for the measured values $\left(x_1^0, \ldots, x_n^0\right)$ are to be evaluated and $|\triangle x_i|$ gives the maximum error of these measured values. Often the **Relative Error** is also of interest

$$\left|\frac{\overline{df}}{f_0}\right| , \qquad \left|\frac{\triangle x_i}{x_i^0}\right| \qquad (i = 1, \ldots, n) \ ,$$

which is usually expressed as a percentage: $100 \left|\frac{\overline{df}}{f_0}\right| \%$.

Summary: Error Propagation According to Gauss

The quantity y depends on the independent variables $x_1, \ldots, x_n$ according to the law

$$y = f(x_1, \ldots, x_n).$$

When the individual values $x_1, \ldots, x_n$ are measured, they are expressed in the form

$$x_i^0 \pm \triangle x_i,$$

where x_i^0 is the mean value of the sizes x_i; $\triangle x_i$ is the error tolerance. Then the error in y is approximately given by

$$\overline{dy} = \left| \left(\frac{\partial f}{\partial x_1} \right)_0 \right| \cdot |\triangle x_1| + \cdots + \left| \left(\frac{\partial f}{\partial x_n} \right)_0 \right| \cdot |\triangle x_n|,$$

if the partial derivatives $\left(\frac{\partial f}{\partial x_i} \right)_0$ are evaluated at $(x_1^0, \ldots, x_n^0)$. The size y has the value

$$y = f(x_1^0, \ldots, x_n^0) \pm \overline{dy}.$$

$\overline{dy}$ is the **Absolute Maximum Error** in linear approximation and

$$\frac{\overline{dy}}{f(x_1^0, \ldots, x_n^0)}$$

is the **Relative Maximum Error**.

⚠ **Caution:** Since the error is given in linear approximation $|df|$ instead of $\triangle f$, this procedure is only useful for small deviations $\triangle x_i$. For large error tolerances, the change in the function $\triangle f$ is not comparable to the change in the tangent plane!

Application Example 10.27

The voltage drop U across an electrical resistance is related to the direct current I by Ohm's law

$$R = \frac{U}{I}.$$

If U is incorrectly measured with dU and I with dI, the upper bound for the error in R is

$$\triangle R \approx dR = \frac{\partial R}{\partial U} \cdot dU + \frac{\partial R}{\partial I} \cdot dI$$

$$\Rightarrow \quad |\triangle R| \leq \left| \frac{1}{I} \, dU \right| + \left| -\frac{U}{I^2} \, dI \right| = \overline{dR}.$$

Numerical example: $U = (110 \pm 2)\, V$, $I = (20 \pm 0.5)\, A$ results in

$$\overline{dR} = \left(\tfrac{1}{20} \cdot 2 + \tfrac{110}{400} \cdot 0.5 \right)\, \Omega = 0.2375\, \Omega \Rightarrow R = (5.5 \pm 0.24)\, \Omega \ .$$

The relative error is $\frac{\overline{dR}}{R} = \frac{0.24\,\Omega}{5.5\,\Omega} = 0.044 = 4.4\,\%$. $\square$

Application Example 10.28   Two resistors of size R_0 with a tolerance of $10\,\%$ and $5\,R_0$ with a tolerance of $2\,\%$ are connected a) in series b) in parallel. In both cases the total resistance and the relative error are required.

a)
$$R_{tot} = R_1 + R_2 \,(= 6\,R_0)$$

$$\Rightarrow \quad \frac{\partial R_{tot}}{\partial R_1} = 1; \qquad \frac{\partial R_{tot}}{\partial R_2} = 1.$$

$$\overline{dR} = \left(\frac{\partial R_{tot}}{\partial R_1} \right)_0 \cdot |\triangle R_1| + \left(\frac{\partial R_{tot}}{\partial R_2} \right)_0 \cdot |\triangle R_2|$$
$$= 10\,\%\, R_0 + 10\,\%\, R_0 = 20\,\%\, R_0.$$

This gives in the relative error to $\dfrac{\overline{dR}}{R_{tot}} = \dfrac{20\,\%\, R_0}{6\,R_0} \approx 3.3\,\%$.

b)
$$\frac{1}{R_{tot}} = \frac{1}{R_1} + \frac{1}{R_2}$$

$$\Rightarrow \quad R_{tot} = \frac{R_1 \cdot R_2}{R_1 + R_2} \left(= \frac{5\,R_0^2}{6\,R_0} = \frac{5}{6}\,R_0 \right).$$

$$\frac{\partial R_{tot}}{\partial R_1} = \frac{R_2^2}{(R_1 + R_2)^2}; \qquad \frac{\partial R_{tot}}{\partial R_2} = \frac{R_1^2}{(R_1 + R_2)^2} \ .$$

$$\overline{dR} = \frac{R_2^2}{(R_1 + R_2)^2}\, |\triangle R_1| + \frac{R_1^2}{(R_1 + R_2)^2}\, |\triangle R_2|$$
$$= \frac{25}{36}\, 10\,\%\, R_0 + \frac{1}{36}\, 10\,\%\, R_0 = \frac{260}{36}\,\%\, R_0 = 7.2\,\%\, R_0.$$

The relative error is $\dfrac{\overline{dR}}{R_{tot}} = \dfrac{7.2\,\%\, R_0}{\frac{5}{6}\,R_0} = 8.66\,\%$. $\square$

10.4.3 Local Extrema for Functions with Two Variables

We now will consider a function $z = f(x, y)$ with two variables x and y, whose partial derivatives are continuous up to the second order. To characterize the function we look for the points at which the function f becomes largest or smallest, respectively. In this discussion, we limit ourselves to the *local* maxima and minima:

> **Definition: (Local Extremes).** *A function $z = f(x, y)$ has a **relative maximum** at $(x_0, y_0) \in \mathbb{D}$, if in an area surrounding the point (x_0, y_0) for all $(x, y) \neq (x_0, y_0)$ it is*
>
> $$f(x_0, y_0) > f(x, y) .$$
>
> *A function $z = f(x, y)$ has a **relative minimum** at $(x_0, y_0) \in \mathbb{D}$, if in an area surrounding the point (x_0, y_0) for all $(x, y) \neq (x_0, y_0)$ it holds*
>
> $$f(x_0, y_0) < f(x, y) .$$

The relative maxima and minima are called *relative extremes*. Relative extremes are sometimes also called local extremes, because the extreme location need only be in the immediate vicinity of (x_0, y_0), but not globally.

Examples 10.29 (With MAPLE-Worksheet).

① If the function

$$f(x, y) = x^2 + y^2 + 4$$

is drawn (e.g. with MAPLE) it can be seen that it has a local (and also a global) minimum at the point $(x_0, y_0) = (0, 0)$, see Fig. 10.21.

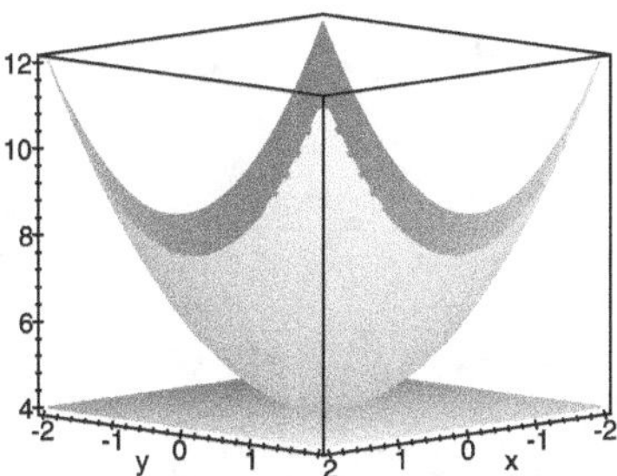

Figure 10.21. Local minimum of $f(x, y) = x^2 + y^2 + 4$ at $(0, 0)$

In addition to the function f, the tangent plane is also drawn at the point $(0, 0)$. We can see that it is parallel to the (x, y)-plane.

② The function

$$f(x, y) = e^{-(x^2 + y^2)}$$

has a local (even global) maximum at the point $(x_0, y_0) = (0, 0)$, as can be seen from the graph in Fig. 10.22:

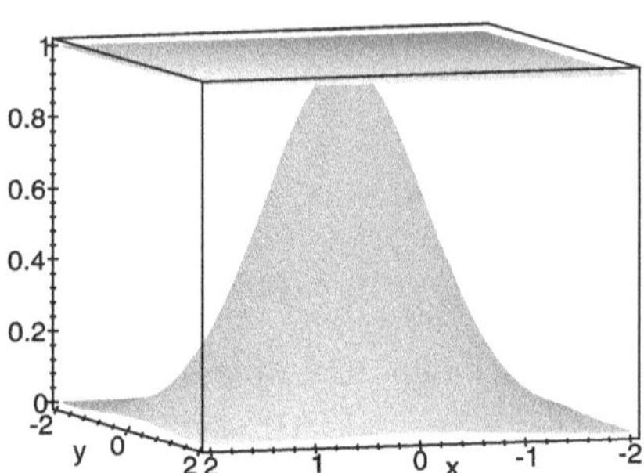

Figure 10.22. Local maximum of the function $f(x, y) = e^{-(x^2 + y^2)}$

Again the tangent plane is parallel to the (x, y) plane.   □

For a continuously differentiable function f in *one* variable x, the necessary condition for a local extreme $x_0 \in \mathbb{D}$ is that the tangent at the point $(x_0, f(x_0))$ is parallel to the x-axis: $f'(x_0) = 0$. If a function f of *two* variables in (x_0, y_0) has a local extreme, then the tangent plane is parallel to the (x, y) plane. The tangent plane z_t is defined in Section 10.3.2

$$z_t(x, y) = f(x_0, y_0) + \frac{\partial f}{\partial x}(x_0, y_0)(x - x_0) + \frac{\partial f}{\partial y}(x_0, y_0)(y - y_0) \ .$$

It is parallel to the (x, y) plane if both partial derivatives to x and to y disappear at the point (x_0, y_0).

Condition for a Relative Extreme

In a *relative extreme* $(x_0, y_0) \in \mathbb{D}$ the function $f(x, y)$ has a tangent plane parallel to the (x, y) plane:

$$(x_0, y_0) \text{ relative extreme} \Rightarrow \frac{\partial f}{\partial x}(x_0, y_0) = 0 \text{ and } \frac{\partial f}{\partial y}(x_0, y_0) = 0.$$

Also: If (x_0, y_0) is a relative extreme, then the gradient of f in (x_0, y_0) disappears: $\operatorname{grad} f(x_0, y_0) = \vec{0}$.   But beware, this statement is not reversible, as the following example shows:

Example 10.30 (With MAPLE-Worksheet). The function

$$f(x, y) = x \cdot y$$

has a horizontal tangent plane at the point $(x_0, y_0) = (0, 0)$:

$$\left.\begin{aligned}
\frac{\partial f}{\partial x}(x, y) = y \Rightarrow \frac{\partial f}{\partial x}(0, 0) = 0 \\[2mm]
\frac{\partial f}{\partial y}(x, y) = x \Rightarrow \frac{\partial f}{\partial y}(0, 0) = 0
\end{aligned}\right\} \quad \Rightarrow \ \operatorname{grad}\ f(0, 0) = \vec{0}.$$

But in a region around the point $(0, 0)$ there are positive **and** negative function values, as shown in the function graph in Fig. 10.23:

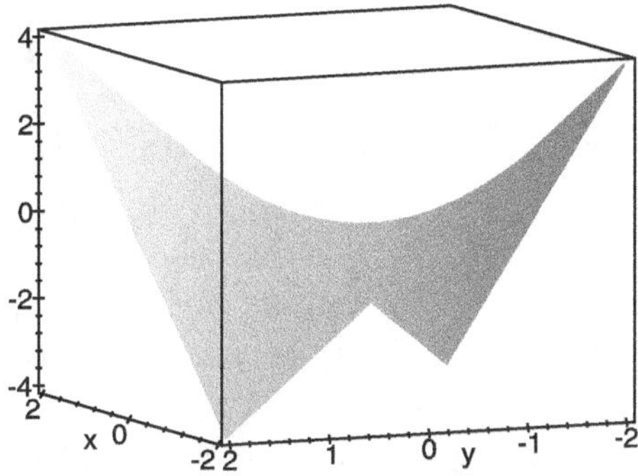

Figure 10.23. Function with saddle point

So f does not have a local extreme at $(0, 0)$, but a *saddle point*.   □

Definition: (Stationary Point). *A point (x_0, y_0) is called a* **stationary point** *of f, if*

$$\operatorname{grad} f(x_0, y_0) = \vec{0}.$$

Relative extremes are therefore stationary points, but not all stationary points are relative extremes according to Example 10.30. A stationary point P is characterized by the fact that the tangent plane of f in P is parallel to the (x, y) plane. To decide whether there is a local extreme in a stationary point (x_0, y_0), consider f according to Taylor's theorem in the vicinity of (x_0, y_0) up to order 2

$$f(x, y) = f(x_0, y_0) + f_x(x_0, y_0)(x - x_0) + f_y(x_0, y_0)(y - y_0)$$
$$+\frac{1}{2!}\left(f_{xx}(x_0, y_0)(x - x_0)^2 + 2 f_{xy}(x_0, y_0)(x - x_0)(y - y_0)\right.$$
$$\left.+f_{yy}(x_0, y_0)(y - y_0)^2\right) + R_2(x, y)$$

Since $f_x(x_0, y_0) = f_y(x_0, y_0) = 0$, the first-order derivatives are omitted; for the second-order derivatives, the following is rewritten

$$f(x, y) = f(x_0, y_0)$$
$$+\frac{1}{2} f_{xx}(x_0, y_0)\left\{(x - x_0) + \frac{f_{xy}(x_0, y_0)}{f_{xx}(x_0, y_0)}(y - y_0)\right\}^2$$
$$+\frac{1}{2}\frac{1}{f_{xx}(x_0, y_0)}\left\{f_{xx}(x_0, y_0) f_{yy}(x_0, y_0) - f_{xy}^2(x_0, y_0)\right\}(y - y_0)^2$$
$$+ R_2(x, y).$$

From this representation of the function we can read:

i) $f_{xx}(x_0, y_0) > 0$ and $f_{xx}(x_0, y_0) f_{yy}(x_0, y_0) - f_{xy}^2(x_0, y_0) > 0$

$$\Rightarrow f(x, y) > f(x_0, y_0) + R_2(x, y) \ .$$

i.e. in an area around the point (x_0, y_0) the function values are **greater** than $f(x_0, y_0)$. So there is a local minimum at (x_0, y_0).

ii) $f_{xx}(x_0, y_0) < 0$ and $f_{xx}(x_0, y_0) f_{yy}(x_0, y_0) - f_{xy}^2(x_0, y_0) > 0$

$$\Rightarrow f(x, y) < f(x_0, y_0) + R_2(x, y) \ .$$

This means that in an area around the point (x_0, y_0) the function values are **smaller** than $f(x_0, y_0)$. So there is a local maximum at (x_0, y_0).

The expression

$$f_{xx}(x_0, y_0) f_{yy}(x_0, y_0) - f_{xy}^2(x_0, y_0) > 0,$$

determines whether there is an extremum at a stationary point:

> ### Sufficient Condition for a Local Extreme Value
>
> $f(x, y)$ is twice continuously partially differentiable at $(x_0, y_0) \in \mathbb{D}$.
> The point (x_0, y_0) is a local extremum if
>
> (1) $f_x(x_0, y_0) = 0, \quad f_y(x_0, y_0) = 0.$
>
> (2) $\Delta := f_{xx}(x_0, y_0) \cdot f_{yy}(x_0, y_0) - f_{xy}^2(x_0, y_0) > 0.$
>
> > For $f_{xx}(x_0, y_0) < 0$ there is a **relative maximum**.
> >
> > For $f_{xx}(x_0, y_0) > 0$ there is a **relative minimum**.
>
> However, if
>
> $$f_{xx}(x_0, y_0)\, f_{yy}(x_0, y_0) - f_{xy}^2(x_0, y_0) < 0 \,,$$
>
> then (x_0, y_0) is a stationary point without being an extreme value,
> but a **saddle point**.

Notes:

(1) As with functions of one variable, the second derivative is used to decide whether there is a local extreme value or not.

(2) Using the *Hesse matrix*

$$H(f) = \begin{pmatrix} f_{xx} & f_{xy} \\ f_{yx} & f_{yy} \end{pmatrix},$$

the determinant of the Hesse matrix at the point (x_0, y_0) decides whether there is an extreme value or not. The following applies:

$\det(H(f)) < 0$ not an extreme value, but a saddle point.

$\det(H(f)) = 0$ no decision possible, whether
    there is an extremum at the position (x_0, y_0)

$\det(H(f)) > 0$ a local extremum is present.

In this section we will use the gradient and the Δ formula to identify the extreme values of a function in two variables. However, in the next section 10.4.4, when working with functions of more than two variables, we will need to work with the Hessian matrix in addition to the gradient. To confirm that a stationary point is an extreme, we use properties of symmetric matrices, summarized in Section 11.5.

Example 10.31. The function

$$f\left(x,\,y\right) = x^2 + y^2 + c$$

has a local minimum at point $(x_0,\,y_0) = (0,\,0)$:

(i) The stationary points follow from $\mathrm{grad}\,(f) = \overrightarrow{0}$:

$$\left.\begin{array}{l} f_x\left(x,\,y\right) = 2\,x = 0 \Rightarrow x = 0 \\ f_y\left(x,\,y\right) = 2\,y = 0 \Rightarrow y = 0 \end{array}\right\} \Rightarrow (x,\,y) = (0,\,0) \quad \text{is a stationary point.}$$

(ii) Insert the stationary point in the $\triangle$ formula:

$$\left.\begin{array}{l} f_{xx}\left(0,\,0\right) = 2 \\ f_{yy}\left(0,\,0\right) = 2 \\ f_{xy}\left(0,\,0\right) = 0 \end{array}\right\} \Rightarrow \triangle = f_{xx}\left(0,\,0\right) \cdot f_{yy}\left(0,\,0\right) - f_{xy}^2\left(0,\,0\right) = 4 > 0\,.$$

$\Rightarrow (0,\,0)$ is relative extreme value.
Since $f_{xx}\left(0,\,0\right) > 0$, it is a relative minimum. $\qquad\qquad\square$

Example 10.32. The function

$$f\left(x,\,y\right) = c + x^2 - y^2$$

has a saddle point at $(0,\,0)$:

If the partial derivatives are set to zero $f_x\left(x,\,y\right) = 2\,x = 0$ and $f_y\left(x,\,y\right) = 2\,y = 0$, then $(x,\,y) = (0,\,0)$ is the only stationary point of the function. Since

$$f_{xx}\left(0,\,0\right) \cdot f_{yy}\left(0,\,0\right) - f_{xy}^2\left(0,\,0\right) = -4 < 0$$

The function has **no** local extrema but a saddle point at $(0,\,0)$. $\qquad\square$

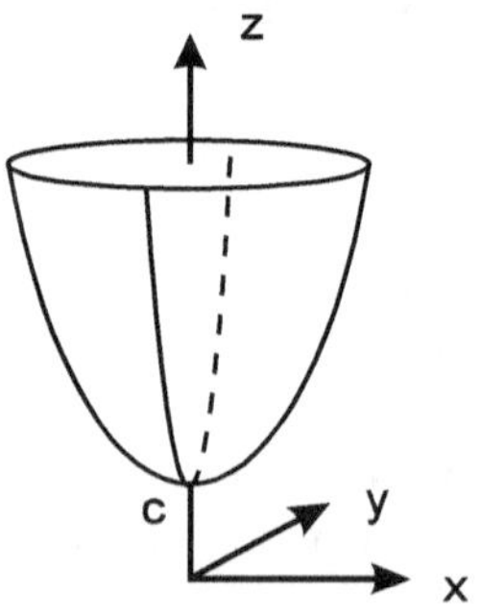

Ex. 10.31 $f(x,y) = c + x^2 + y^2$

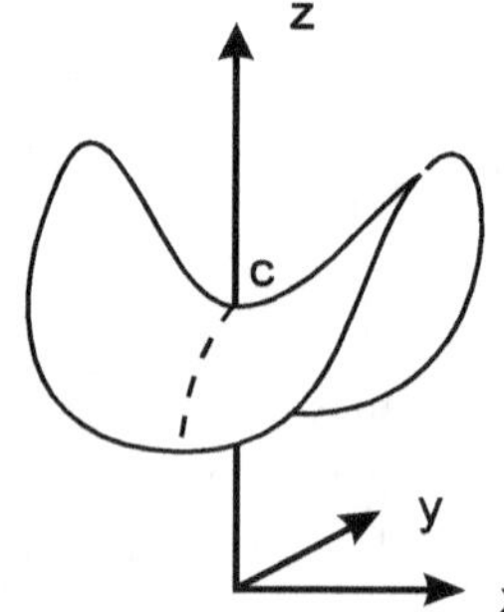

Ex. 10.32 $f(x,y) = c + x^2 - y^2$

Prime Example 10.33 . Find the local extrema of the function

$$f\left(x,\, y\right) = \left(x^2 + y^2\right)^2 - 2\left(x^2 - y^2\right)\ .$$

(i) The stationary points follow from $\operatorname{grad}\left(f\right) = \vec{0}$:

$$f_x\left(x,\, y\right) = 4\,x\left(x^2 + y^2\right) - 4\,x = 0$$
$$f_y\left(x,\, y\right) = 4\,y\left(x^2 + y^2\right) + 4\,y = 0.$$

This is a coupled system of non-linear equations for the unknowns x and y. By simplifying each equation we obtain

$$4\,x\left(x^2 + y^2 - 1\right) = 0 \tag{1}$$
$$4\,y\left(x^2 + y^2 + 1\right) = 0. \tag{2}$$

Equation (1) gives $x = 0$ or $x^2 + y^2 - 1 = 0$.
  We insert $x = 0$ in equation (2): So $4y(y^2 + 1) = 0$ or $y = 0$.
  $\Rightarrow (x, y) = (0, 0)$ is a stationary point.

  We substitute $x^2 + y^2 - 1 = 0$ in equation (2): $4y \cdot 2 = 0$. Inserting $y = 0$ into $x^2 + y^2 - 1 = 0$ gives $x^2 - 1 = 0$ or $x = \pm 1$.
  $\Rightarrow (x, y) = (1, 0)$ and $(x, y) = (-1, 0)$ are also stationary points.

So the function has three stationary points: $(0,\,0)$, $(1,\,0)$ and $(-1,\,0)$. It is only at these points that f can be extreme.

(ii) Insert the stationary points in the $\triangle$ formula: We calculate the second-order partial derivatives

$$f_{xx}\left(x,\, y\right) = 12\,x^2 + 4\,y^2 - 4$$
$$f_{yy}\left(x,\, y\right) = 12\,y^2 + 4\,x^2 + 4$$
$$f_{xy}\left(x,\, y\right) = 8\,xy$$

and insert the stationary points:
$\triangle\left(0,\,0\right) = f_{xx}\left(0,\,0\right) \cdot f_{yy}\left(0,\,0\right) - f_{xy}^2\left(0,\,0\right) = -4 \cdot 4 - 0 = -16 < 0.$

  $\Rightarrow$ At $(0,\,0)$ there are **no** extremes, but a saddle point.

$\triangle\left(1,\,0\right) = f_{xx}\left(1,\,0\right) \cdot f_{yy}\left(1,\,0\right) - f_{xy}^2\left(1,\,0\right) = 8 \cdot 8 - 0 = 64 > 0.$

  $\Rightarrow$ At $(1,\,0)$ there is a local minimum, since $f_{xx}\left(1,\,0\right) = 8 > 0$.

Due to the symmetry $f\left(-x,\, y\right) = f\left(x,\, y\right)$ there is also a local minimum at the point $(-1,\,0)$. $\qquad\square$

10.4.4 Relative Extremes for Functions with Multiple Variables

The concept of relative extremes is directly transferred to functions with more than two variables $y = f(x_1, \ldots, x_n)$: A **necessary condition** for an extreme at $\left(x_1^0, \ldots, x_n^0\right)$ is that the gradient of f disappears at this point:

Extreme Values (Necessary Condition)

The function $y = f(x_1, \ldots, x_n)$ has a **local extreme** at point $\left(x_1^0, \ldots, x_n^0\right)$, then

$$\operatorname{grad} f\left(x_1^0, \ldots, x_n^0\right) = \vec{0} \, .$$

This means that at the point $\left(x_1^0, \ldots, x_n^0\right)$ all partial derivatives disappear

$$\frac{\partial}{\partial x_1} f\left(x_1^0, \ldots, x_n^0\right) = 0, \; \ldots, \; \frac{\partial}{\partial x_n} f\left(x_1^0, \ldots, x_n^0\right) = 0.$$

To obtain a **sufficient condition**, we go to the corresponding Hesse matrix

$$H(f) := \begin{pmatrix} f_{x_1 x_1} & f_{x_1 x_2} & \cdots & f_{x_1 x_n} \\ f_{x_2 x_1} & f_{x_2 x_2} & \cdots & f_{x_2 x_n} \\ \vdots & & & \\ f_{x_n x_1} & f_{x_n x_2} & \cdots & f_{x_n x_n} \end{pmatrix}$$

of all partial derivatives of order 2. According to Taylor's theorem, see Section 10.3.4 on page 140, we rewrite the formula for the quadratic approximation in the form

$$f(\vec{x}) - f(\vec{x}_0) = \vec{h}^{\,t} \cdot \operatorname{grad}(f)\big|_{(\vec{x}_0)} + \tfrac{1}{2} \vec{h}^{\,t} H(f)\big|_{(\vec{x}_0)} \vec{h} + R_2(\vec{x}) \, .$$

For a stationary point $(\vec{x}_0) = (x_1^0, \ldots, x_n^0)$ the gradient disappears so that

$$f(\vec{x}) - f(\vec{x}_0) = \tfrac{1}{2} \vec{h}^{\,t} H(f)\big|_{(\vec{x}_0)} \vec{h} + R_2(\vec{x}) \, .$$

If the expression $\vec{h}^{\,t} \, H(f)\big|_{(\vec{x}_0)} \, \vec{h}$ is always positive independently of h, then the difference $f(\vec{x}) - f(\vec{x}_0)$ is positive. So the value of the function increases near $f(\vec{x}_0)$ and there is a local minimum at $\vec{x}_0$. This is the case if the Hessian $H(f)$ at $\vec{x}_0$ is positive definite.

However, if the expression $\vec{h}^{\,t} \, H(f)\big|_{(\vec{x}_0)} \, \vec{h}$ is always negative independently of h, then the difference $f(\vec{x}) - f(\vec{x}_0)$ is negative. So the value of the function decreases near $f(\vec{x}_0)$ and there is a local maximum at $\vec{x}_0$. This is the case if the Hessian $H(f)$ at $\vec{x}_0$ is negative definite.

Since we assume in this section that the functions are twice continuously partially differentiable, Schwarz's theorem states that the mixed second-order derivatives are the same: $f_{x_i\,x_j} = f_{x_j\,x_i}$. The Hessian matrix is therefore symmetric. To characterize the extreme values, we introduce the Sylvester criterion for symmetric matrices, which will be discussed in more detail in the next chapter in Section 11.5 on positive definite matrices:

Applying Sylvester's Criterion

Let $A = \begin{pmatrix} a_{11} & \cdots & a_{1n} \\ \vdots & & \vdots \\ a_{n1} & \cdots & a_{nn} \end{pmatrix}$ be a symmetric $(n \times n)$ matrix.

A is **positive definite**, if all sub-determinants (**principal minors or leading minors**) $\Delta_k := \det \begin{pmatrix} a_{11} & \cdots & a_{1k} \\ \vdots & & \vdots \\ a_{k1} & \cdots & a_{kk} \end{pmatrix}$ are greater than zero: $\Delta_k > 0$ for all $k = 1, \ldots, n$.

A is positive semi-definite, if all principal minors are greater than or equal to zero: $\Delta_k \geq 0$ for all $k = 0, \ldots, n$.

A is **negative definite**, if the matrix $(-A)$ is positive definite.

A is negative semi-definite, if the matrix $(-A)$ is positive semi-definite.

Otherwise A is indefinite.

Examples 10.34:

① $A = \begin{pmatrix} 1 & 2 \\ 2 & 5 \end{pmatrix}$ is positive definite:

$\Delta_1 = \det(1) = 1 > 0$ and $\Delta_2 = \det(A) = 1 > 0$.

② $A = \begin{pmatrix} -3 & -3 & 1 \\ -3 & -5 & 0 \\ 1 & 0 & -4 \end{pmatrix}$ is negative definite:

Since $\Delta_1 = \det(-3)$ is negative, A is not positive definite. So we check

$$-A = \begin{pmatrix} 3 & 3 & -1 \\ 3 & 5 & 0 \\ -1 & 0 & 4 \end{pmatrix}. \text{ With this matrix we calculate}$$

$$\Delta_1 = \det(3) = 3 > 0, \quad \Delta_2 = \det\begin{pmatrix} 3 & 3 \\ 3 & 5 \end{pmatrix} = 15 - 9 = 6 > 0 \text{ and}$$

$$\Delta_3 = \det\begin{pmatrix} 3 & 3 & -1 \\ 3 & 5 & 0 \\ -1 & 0 & 4 \end{pmatrix} = 19 > 0$$

All three principal minors are positive. So $-A$ is positive definite and A is negative definite. $\qquad\square$

We will use the Sylvester criterion to check whether a stationary point $\vec{x}_0$ of a function f is an extreme or not. Summarizing the Section 11.5, we get the important result:

> **Sufficient Condition for Local Extremes**
>
> Let $y = f(x_1, \ldots, x_n)$ be a function in n variables. At a stationary point $(x_1^0, \ldots, x_n^0)$ it applies:
>
> If the Hessian $H(f)$ is negative definite at the point $(x_1^0, \ldots, x_n^0)$, then a **local maximum** exists.
>
> If the Hessian $H(f)$ is positive definite at the point $(x_1^0, \ldots, x_n^0)$, then a **local minimum** exists.

Example 10.35. Given is the function

$$f(x_1, x_2, x_3) = 5\,x_1^2 + 6\,x_2^2 + 7\,x_3^2 - 4\,x_1 x_2 + 4\,x_2 x_3 - 10\,x_1 + 8\,x_2 + 14\,x_3 - 6.$$

We look for the extremes and their type.

(1) **Stationary Points:**

First we determine the stationary points by solving $\operatorname{grad}(f) = \vec{0}$.

$$\begin{aligned} f_{x_1} &= 10\,x_1 - 4\,x_2 - 10 = 0 \\ f_{x_2} &= -4\,x_1 + 12\,x_2 + 4\,x_3 + 8 = 0 \\ f_{x_3} &= 4\,x_2 + 14\,x_3 + 14 = 0 \end{aligned} \qquad\Longrightarrow\qquad \left(\begin{array}{ccc|c} 10 & -4 & 0 & 10 \\ -4 & 12 & 4 & -8 \\ 0 & 4 & 14 & -14 \end{array}\right).$$

This system has the unique solution $\vec{x}_0 = \begin{pmatrix} 1 \\ 0 \\ -1 \end{pmatrix}$.

(2) **Hesse Matrix:** With the second derivatives we set up the Hesse matrix
and we evaluate the Hesse matrix at this stationary point $\vec{x}_0$.

$$
H(f) = \begin{pmatrix} f_{x_1 x_1} & f_{x_1 x_2} & f_{x_1 x_3} \\ f_{x_2 x_1} & f_{x_2 x_2} & f_{x_2 x_3} \\ f_{x_3 x_1} & f_{x_3 x_2} & f_{x_3 x_3} \end{pmatrix} = \begin{pmatrix} 10 & -4 & 0 \\ -4 & 12 & 4 \\ 0 & 4 & 14 \end{pmatrix}
$$

The matrix $H(f)|_{\vec{x}_0}$ is positive definite since all minors $\Delta_1 = 10$, $\Delta_2 = 104$ and $\Delta_3 = 1296$ are positive. So $\vec{x}_0$ is a local minimum.

Example 10.36. Given is the function

$$
f(x, y) = 3\,x^2 y + 4\,y^3 - 3\,x^2 - 12\,y^2 + 1.
$$

Show that $P_1(0,0)$, $P_2(0,2)$, $P_3(2,1)$ and $P_4(-2,1)$ are stationary points. Which of the stationary points are extreme values? Which are saddle points?

(1) **Stationary Points:**
We compute the the first-order derivatives of f and check whether the evaluation at the given points is zero:

$$
\begin{aligned}
f_x &= 6\,x\,y - 6\,x \\
f_y &= 3\,x^2 + 12\,y^2 - 24\,y.
\end{aligned}
$$

Inserting $P_1(0,0)$, $P_2(0,2)$, $P_3(2,1)$ and $P_4(-2,1)$ into these derivatives gives zero, confirming that all the points are stationary points.

(2) **Hesse Matrix:** With the second derivatives we set up the Hesse matrix

$$
H = \begin{pmatrix} f_{x,x} & f_{x,y} \\ f_{y,x} & f_{y,y} \end{pmatrix} = \begin{pmatrix} 6y - 6 & 6x \\ 6x & 24y - 24 \end{pmatrix}
$$

which must be evaluated at the four points.

$P_1(0,0)$: $H(0,0) = \begin{pmatrix} -6 & 0 \\ 0 & -24 \end{pmatrix}$. Since $\Delta_1 = -6$ we go over to $-H(0,0) = \begin{pmatrix} 6 & 0 \\ 0 & 24 \end{pmatrix}$. The principal minors are both positive: $\Delta_1 = 6$ and $\Delta_2 = 144$. So $-H(0,0)$ is positive definite and $H(0,0)$ is negative definite. P_1 is a local maximum.

$P_2(0,2)$: $H(0,2) = \begin{pmatrix} 6 & 0 \\ 0 & 24 \end{pmatrix}$. $H(0,2)$ is positive definite, since both principal minors are positive: $\Delta_1 = 6$ and $\Delta_2 = 144$. So P_2 is a local minimum.

$P_3(2,1)$: $H(2,1) = \begin{pmatrix} 0 & 12 \\ 12 & 0 \end{pmatrix}$. $H(2,1)$ is indefinite: $\Delta_1 = 0$ and $\Delta_2 = -144$. So P_3 is not a local extreme. It is a saddle point.

$P_4(-2,1)$: $H(-2,1) = \begin{pmatrix} 0 & -12 \\ -12 & 0 \end{pmatrix}$. $H(-2,1)$ is indefinite: $\Delta_1 = 0$ and $\Delta_2 = -144$. So P_4 is not a local extreme. It is a saddle point. □

Example 10.37. Given is the function

$$f(x, y, z) = 2\,x^2 - xy + 2\,xz - y - y^3 + z^2.$$

We look for the extremes and their type.

(1) **Stationary Points:** From the condition $\mathrm{grad}\,(f) = 0$, we obtain the first-order derivatives

$$f_x = 4x - y + 2z = 0$$
$$f_y = 3y^2 - x - 1 = 0$$
$$f_z = 2x + 2z = 0.$$

Solving the three equations for the unknowns, we get

$$x = \frac{1}{3}, \; y = \frac{2}{3}, \; z = -\frac{1}{3}$$

and

$$x = -\frac{1}{4}, \; y = -\frac{1}{2}, \; z = \frac{1}{4}$$

as the solution. So we identify two stationary points $P_1(\frac{1}{3}, \frac{2}{3}, -\frac{1}{3})$ and $P_2(-\frac{1}{4}, -\frac{1}{2}, \frac{1}{4})$.

(2) **Hesse Matrix:** With the second-order derivatives we set up the Hesse matrix

$$H = \begin{pmatrix} 4 & -1 & 2 \\ -1 & 6y & 0 \\ 2 & 0 & 2 \end{pmatrix}.$$

The Hesse matrix for point P_1 evaluates to

$$H(\tfrac{1}{3}, \tfrac{2}{3}, -\tfrac{1}{3}) = \begin{pmatrix} 4 & -1 & 2 \\ -1 & 4 & 0 \\ 2 & 0 & 2 \end{pmatrix}.$$

H is positive definite, since all principal minors are positive:

$$\det\begin{pmatrix} 4 \end{pmatrix} = 4 > 0, \quad \det\begin{pmatrix} 4 & -1 \\ -1 & 4 \end{pmatrix} = 15 > 0,$$

$$\det\begin{pmatrix} 4 & -1 & 2 \\ -1 & 4 & 0 \\ 2 & 0 & 2 \end{pmatrix} = 14 > 0.$$

So P_1 is a local minimum. The Hesse matrix for point P_2

$$H(-\tfrac{1}{4}, -\tfrac{1}{2}, \tfrac{1}{4}) = \begin{pmatrix} 4 & -1 & 2 \\ -1 & -3 & 0 \\ 2 & 0 & 2 \end{pmatrix}$$

is indefinite, since for the principal minors

$$\Delta_1 = 4 > 0, \ \Delta_2 = -13 < 0, \ \Delta_3 = -14 < 0.$$

So P_2 is a saddle point. $\qquad\qquad\qquad\qquad\qquad\qquad\qquad\qquad\square$

Example 10.38. Given is the function

$$f(x, y, z) = e^x \left(x^2 + y^2 + z^2 + \tfrac{3}{4}\right).$$

The stationary points are $P_1(-\tfrac{1}{2}, 0, 0)$ and $P_2(-\tfrac{3}{2}, 0, 0)$. And the Hesse matrix of f is

$$H = e^x \begin{pmatrix} x^2 + 4x + y^2 + z^2 + \tfrac{11}{4} & 2y & 2z \\ 2y & 2 & 0 \\ 2 & & 0 & 2 \end{pmatrix}.$$

For P_1 the Hesse matrix is positive definite, since all principal minors

$$\Delta_1 = e^{-\frac{1}{2}} \cdot 1 > 0, \ \Delta_2 = e^{-\frac{1}{2}} \cdot 2 > 0, \ \Delta_3 = e^{-\frac{1}{2}} \cdot 4 > 0$$

are positive. For P_2 the Hesse matrix is indefinite, since for H the principal minors are all negative. For $-H$ they are positive and twice negative. $\qquad\square$

10.4.5 Compensation of Measurement Errors; Regression Line

An application of the theory of extreme values is the compensation calculation: Measurements have been recorded that determine the dependence of a quantity y on another quantity x using n pairs of values $(x_1,\, y_1)$, ..., $(x_n,\, y_n)$. The task of the compensation calculation is to find a function f that fits the existing data points as well as possible and has a smooth slope.

Problem: The temperature dependence of an ohmic resistance is measured. A straight line is required that adequately represents the measurement points.

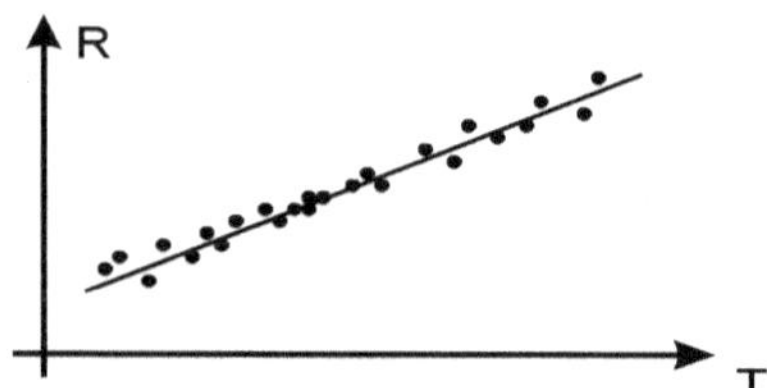

Figure 10.24. Temperature dependence of an ohmic resistance

⊘ **Least Squares Method**

The searched function f should have the property that the sum of the squares of the distances of the measuring points to the compensation function becomes minimal:

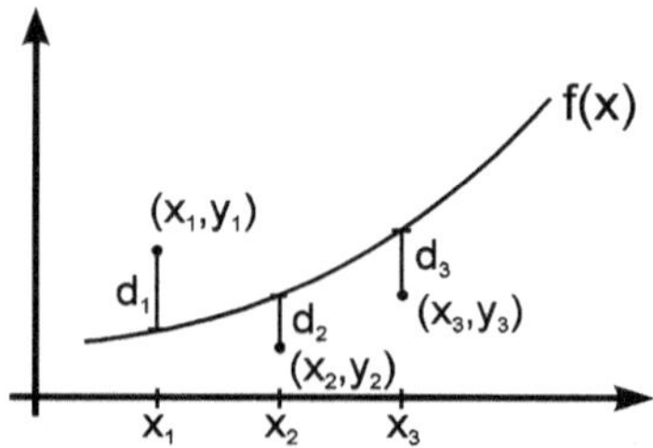

Figure 10.25. Distances d_i from the compensation function

The distance of the point y_i from the compensating function f at point x_i is $d_i = y_i - f(x_i)$. The sum of all squares of the distances is therefore

$$\sum_{i=1}^{n} d_i^2 = \sum_{i=1}^{n} (y_i - f(x_i))^2 \,.$$

Depending on the assumed functional relationship, a functional approach is chosen. Frequently used compensation functions are shown in Table 10.1.

Table 10.1: Compensation Functions

Compensation function		Parameter
Linear Function	$f(x) = a\,x + b$	a, b
Quadratic Function	$f(x) = a\,x^2 + b\,x + c$	a, b, c
Polynomial	$f(x) = a_n\,x^n + \cdots + a_1\,x + a_0$	$a_n, \ldots, a_0$
Power Function	$f(x) = a\,x^b$	a, b
Exponential	$f(x) = a\,e^{bx}$	a, b

Each fitting function $f(x)$ contains parameters that must be determined in such a way that the sum of the squares of the distances becomes minimal (*principle of least squares*):

$$F(a,\,b,\,c,\ldots) := \sum_{i=1}^{n} (y_i - f(x_i))^2 \to \text{ minimal.}$$

This sum F is again a function of the parameters $a,\,b,\,c,\ldots$. A minimum of this function is required. From the theory of extremes in 10.4.3 there is a necessary condition for a local extreme value: All partial derivatives must vanish at the stationary point

$$\Rightarrow \frac{\partial F}{\partial a} = 0\,, \ \frac{\partial F}{\partial b} = 0\,, \ \frac{\partial F}{\partial c} = 0, \ldots \quad .$$

This gives as many equations as there are parameters in the approach.

Least Squares Method

For n given measurement points $(x_i,\,y_i)_{i=1,\ldots,n}$, a fitting curve can be determined as follows:

(1) Select an approximation function (straight line, parabola, …). This fitting function contains undefined parameters $a,\,b,\,c,\ldots$.

(2) The sum of the squares of the distances is calculated, which is a function of the free parameters

$$F(a,\,b,\,c,\ldots) = \sum_{i=1}^{n} (y_i - f(x_i))^2\,.$$

(3) The parameters are then determined so that the function $F(a,\,b,\,c,\ldots)$ becomes minimal, i.e.

$$\frac{\partial F}{\partial a} = 0\,, \quad \frac{\partial F}{\partial b} = 0\,, \quad \frac{\partial F}{\partial c} = 0, \ldots \quad .$$

⊘ Regression Line

Only the most important special case of the regression line for the applications will be discussed. Assume that for n different x-values $x_1, \ldots, x_n$ there are corresponding y-values $y_1, \ldots, y_n$ are present, so that n measuring points

$$(x_1,\, y_1)\,, \ (x_2,\, y_2)\,, \ \ldots, \ (x_n,\, y_n)$$

are given. The parameters a and b are to be found in the regression line

$$f(x) = a\,x + b\,,$$

so that the sum of the squares of the distances becomes minimal. The distance d_i of the measuring points from the compensation line is

$$d_i = y_i - a\,x_i - b\,.$$

The sum of the squares of the distances

$$F(a,\, b) = \sum_{i=1}^{n} d_i^2 = \sum_{i=1}^{n} (y_i - a\,x_i - b)^2$$

is therefore a function of the two parameters a and b. From this function $F(a,\, b)$ the minimum has to be found. To find the stationary points, the partial derivatives of F with respect to a and b are set to zero:

$$\frac{\partial F}{\partial a} = 2 \sum_{i=1}^{n} (y_i - a\,x_i - b)\,(-x_i)$$

$$= -2 \sum_{i=1}^{n} y_i\, x_i + 2\,a \sum_{i=1}^{n} x_i^2 + 2\,b \sum_{i=1}^{n} x_i = 0.$$

$$\frac{\partial F}{\partial b} = 2 \sum_{i=1}^{n} (y_i - a\,x_i - b)\,(-1)$$

$$= -2 \sum_{i=1}^{n} y_i + 2\,a \sum_{i=1}^{n} x_i + 2 \sum_{i=1}^{n} b = 0.$$

These necessary conditions form a system of linear equations for a, b:

$$\left(\sum_{i=1}^{n} x_i^2 \right) a + \left(\sum_{i=1}^{n} x_i \right) b = \sum_{i=1}^{n} x_i\, y_i$$

$$\left(\sum_{i=1}^{n} x_i \right) a + \quad n \quad b = \sum_{i=1}^{n} y_i\,.$$

We introduce the coefficient matrix

$$D = \begin{vmatrix} \sum x_i^2 & \sum x_i \\ \sum x_i & n \end{vmatrix} = n \sum_{i=1}^{n} x_i^2 - \left(\sum_{i=1}^{n} x_i \right)^2 = \frac{1}{2} \sum_{j=1}^{n}\sum_{i=1}^{n} (x_i - x_j)^2 > 0.$$

Using Cramer's rule, for example, we obtain the unique solution to the system of linear equations

$$a = \hat{a} := \frac{1}{D} \left(n \cdot \sum_{i=1}^{n} x_i y_i - \left(\sum_{i=1}^{n} x_i \right) \left(\sum_{i=1}^{n} y_i \right) \right)$$

$$b = \hat{b} := \frac{1}{D} \left(\left(\sum_{i=1}^{n} x_i^2 \right) \left(\sum_{i=1}^{n} y_i \right) - \left(\sum_{i=1}^{n} x_i \right) \left(\sum_{i=1}^{n} x_i y_i \right) \right).$$

To check that F has a local minimum at the point $(\hat{a}, \hat{b})$, we compute the second partial derivatives

$$F_{aa} = 2 \sum_{i=1}^{n} x_i^2; \qquad F_{bb} = 2\,n; \qquad F_{ab} = 2 \sum_{i=1}^{n} x_i$$

and substitute them into the $\triangle$-formula, $\triangle = F_{aa}(\hat{a}, \hat{b}) \cdot F_{bb}(\hat{a}, \hat{b}) - F_{ab}^2(\hat{a}, \hat{b})$:

$$\triangle = F_{aa} \cdot F_{bb} - F_{ab}^2 = 4\,n \sum_{i=1}^{n} x_i^2 - 4 \left(\sum_{i=1}^{n} x_i \right)^2 = 4\,D > 0.$$

Since $\triangle > 0$ and $F_{aa} > 0$, this is a local minimum.

Summary: Regression Line

Given are n data points $(x_i, y_i)_{i=1,\dots,n}$. The **regression line** or **compensation line** $y = \hat{a}\,x + \hat{b}$ has the coefficients

$$\hat{a} = = \frac{1}{D} \left(n \cdot \sum_{i=1}^{n} x_i y_i - \left(\sum_{i=1}^{n} x_i \right) \left(\sum_{i=1}^{n} y_i \right) \right)$$

$$\hat{b} = \frac{1}{D} \left(\left(\sum_{i=1}^{n} x_i^2 \right) \left(\sum_{i=1}^{n} y_i \right) - \left(\sum_{i=1}^{n} x_i \right) \left(\sum_{i=1}^{n} x_i y_i \right) \right),$$

where

$$D = n \sum_{i=1}^{n} x_i^2 - \left(\sum_{i=1}^{n} x_i \right)^2.$$

Example 10.39 (With MAPLE**-Worksheet).** A compensation line for the following measured values is required:

x	0	1	2	3	4	5
y	3	5	7	8	10	10

With these values we calculate D, $\hat{a}$ and $\hat{b}$:

$$D = 6 \cdot \left(1 + 2^2 + 3^2 + 4^2 + 5^2\right) - \left(1 + 2 + 3 + 4 + 5\right)^2 = 105$$
$$\hat{a} = \frac{1}{105} \left[6 \cdot \left(1 \cdot 5 + 2 \cdot 7 + 3 \cdot 8 + 4 \cdot 10 + 5 \cdot 10\right)\right.$$
$$\left. - \left(1 + 2 + 3 + 4 + 5\right)\left(3 + 5 + 7 + 8 + 10 + 10\right)\right] = \frac{51}{35}.$$

Similarly, $\hat{b} = \frac{74}{21}$. So the regression line has the form

$$y = \frac{51}{35}x + \frac{74}{21}.$$

In Fig. 10.26 the compensation line is graphically displayed together with the measured values:

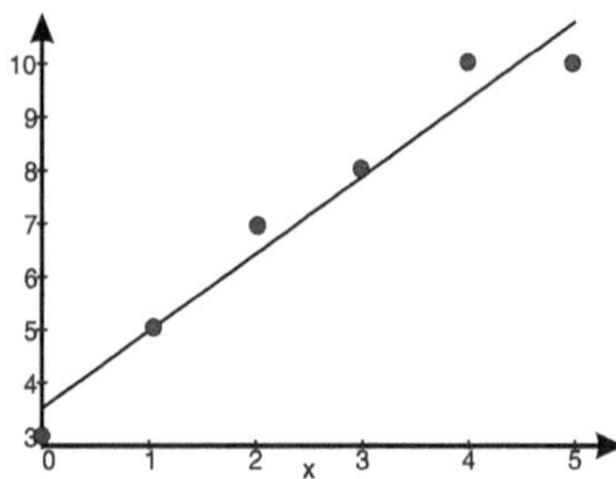

Figure 10.26. Data values with regression line

For a larger number of measuring points, the determination of the regression line must be carried out on a computer.   □

The case of the compensation line has been discussed in detail, as it also includes logarithmic, exponential and power fitting by modifying the measured values:

Remarks:

(1) If a **logarithmic fit** is to be found

$$f(x) = a \ln x + b$$

the auxiliary variable $z = \ln x$ is introduced. The logarithm is formed from the x-values of the measurement and is passed from the measured values (x_i, y_i), $i = 1, \ldots, n$ to the value pairs $(\ln(x_i), y_i)$, $i = 1, \ldots, n$. These data are used to determine the compensation line.

(2) If we want to find an **exponential fit**

$$f(x) = a\,e^{b\,x}$$

to the measured values, we form from the equation

$$y = a\,e^{b\,x}$$

the logarithm

$$\ln y = \ln a + b \cdot x = A\,x + B\,.$$

From the measured data (x_i, y_i) we go to the value pairs $(x_i, \ln(y_i))$ and determine the parameters A and B of the compensating line. Then $b = A$ and $a = e^{B}$.

(3) If we want to find a **power fit**

$$f(x) = a\,x^{b}$$

we derive from the equation

$$y = a\,x^{b}$$

the logarithm

$$\ln y = \ln a + b \ln x\,.$$

From the measured data (x_i, y_i) we go to the value pairs $(\ln(x_i), \ln(y_i))$ and determine the compensating line

$$y = A\,x + B\,.$$

Then $b = A$ and $a = e^{B}$.

(4) The logarithmic, exponential and power fits correspond to the representation of the measured values in a logarithmic or double-logarithmic plot.

10.5 Problems on Differential Calculus

10.1 Use the **plot3d** command to graphically display the functions in MAPLE
a) $z = x \cdot y$ b) $z = x + y$ c) $z = x^2 - y^2$
d) $z = x^2 + y^2$ e) $z = (x - y)^2$ f) $z = e^{-(x^2 + y^2)}$

10.2 Use the option **style = contour** to insert 20 contour lines into the charts and vary interactively the viewing angle.

10.3 Calculate all 1st order partial derivatives of the functions
a) $f(x, y) = x^3 + x \cdot y - y^{-2}$ b) $f(a, t) = 3 \cdot a \cdot x + y \cdot \ln(t^2)$
c) $f(u, v) = \frac{u+w}{u+v}$ d) $f(x, y, z) = \operatorname{arcsinh}(x^2 + z^2)$
e) $f(x_1, x_2, x_3) = x_2$ f) $f(a, b) = (a x + b x^2)^{-1} + y \cdot e^{a b}$

10.4 Calculate the 1st and 2nd order partial derivatives of the functions
a) $f(x, y) = 3 x^2 + 4 x y - 2 y^2$ b) $f(x, y) = 2 \cos(3 x y)$
c) $f(x, y) = (3 x - 5 y)^4$ d) $f(x, y) = \frac{x^2 - y^2}{x + y}$
e) $f(x, y) = 3 x \cdot e^{x y}$ f) $f(x, y) = \sqrt{x^2 - 2 x y}$

10.5 Given is the function $f(x, y) = \sin(x^2 + 2 y)$.
Confirm Schwarz's theorem that $f_{xy} = f_{yx}$.

10.6 Calculate all 2nd order partial derivatives for the function
$$f(x_1, x_2, x_3) = x_1 \cdot \ln(x_2^2 + x_3^2).$$

10.7 Find the second-order partial derivatives of the function
a) $f(x, y) = (3 x - 5 y)^4$ b) $f(x, y, z) = e^{x-y} \cdot \cos(5 z)$

10.8 Show that the function
$$f(x, y, z) = \frac{a}{\sqrt{x^2 + y^2 + z^2}}$$

is a solution of Laplace's equation $f_{xx} + f_{yy} + f_{zz} = 0$.

10.9 Show that $z = x \cdot e^{y/x}$
satisfies the partial differential equation $x \frac{\partial z}{\partial x} + y \frac{\partial z}{\partial y} = z$.

10.10 Show that the function
$$f(x, y) = \frac{1}{2} \cdot \ln(x^2 + y^2)$$

satisfies the partial differential equation $f_{xx} + f_{yy} = 0$.

10.11 Determine the equation of the tangent plane in point $P(1, 0)$ to the surface $z = (3 x + x \cdot y)^2$.

10.12 At point $P(1, 2, 0)$ calculate the total differential of
$$f(x, y, z) = y \cdot \cos(z) + \frac{\ln(1 + x^2)}{y}.$$

10.13   Calculate gradient and directional derivative in direction $\vec{a} = \begin{pmatrix} 2 \\ 4 \end{pmatrix}$ for the function
$$f(x, y) = (3x + x \cdot y)^2 .$$

10.14   Calculate gradient and directional derivative in direction $\vec{a} = \begin{pmatrix} 3 \\ -1 \\ 2 \end{pmatrix}$ for function
$$f(x, y, z) = y \cdot \cos(z) + \frac{\ln(1 + x^2)}{y} .$$

10.15   Determines the total differential of the functions
a) $z(x, y) = 4x^3 y - 3x \cdot e^y$
b) $z(x, y) = \dfrac{x^2 + y^2}{x - y}$
c) $f(x, y, z) = \ln \sqrt{x^2 + y^2 + z^2}.$

10.16   Consider the differentiable functions $f_1, f_2 : \mathbb{R} \to \mathbb{R}$ and $g : \mathbb{R}^2 \to \mathbb{R}$ and form the concatenation $h(x_1, x_2) = g(f_1(x_1), f_2(x_2))$. Calculate h and the first partial derivatives of h in the following cases
a) $f_1(x_1) = a_0 + a_1 x_1$; $f_2(x_2) = b_0 + b_1 x_2$; $g(u_1, u_2) = c_0 + c_1 u_1 + c_2 u_2$.
b) $f_1(x_1) = \sin x_1$; $f_2(x_2) = \cos x_2$; $g(u_1, u_2) = u_1^2 + u_1 u_2$.

10.17   Calculate Taylor's series of the function f about point (x_0, y_0) up to order 2 for
a) $f(x, y) = \dfrac{(x - y)}{(x + y)}$, $(x_0, y_0) = (1, 1)$
b) $f(x, y) = e^{x^2 + y^2}$, $(x_0, y_0) = (1, 0)$

10.18   Calculate the total differential of
a) $f(x, y) = \sin(x^2 + 2y)$ b) $f(x, y) = 3x^2 + 4xy - 2y^2$
c) $f(x, y) = y \cdot \cos(x - 2y)$ d) $f(x, y, z) = x^2 z - y z^3 + x^4$

10.19   The diameter $(6.0 \pm 0.003)\, m$ and the height $(4.0 \pm 0.02)\, m$ of a straight circular cylinder are measured. What is the absolute and relative error of the cylinder volume?

10.20   To calculate an electrical resistance $R = \frac{U}{I}$ the current $I = (15 \pm 0.3)\, A$ and the voltage $U = (110 \pm 2)\, V$ are measured. What is the relative maximum error of R.

10.21   The modulus of elasticity E of a cylindrical wire (r: radius of the wire's cross-section, l: length of the wire) is determined by measuring the increase in length z of the wire under the influence of the force k:
$$E = \frac{l \cdot k}{\pi r^2 \cdot z} \qquad \text{(E-module)}.$$

How large and with what accuracy is E determined if the measured values are $l = (2000 \pm 3)\, mm$, $r = (0.2 \pm 0.002)\, mm$, $k = (200 \pm 0.05)\, N$ and $z = (15 \pm 0.1)\, mm$?

10.22 The density ρ of a piece of brass is to be determined by the buoyancy method: If m is the weight in air, $\bar{m}$ the weight in water, then

$$\rho = \frac{m}{m - \bar{m}} = \frac{\text{Weight in air}}{\text{Volume}} \; .$$

How big is the relative error of ρ if $m = \left(100 \pm 5 \cdot 10^{-3}\right)$ g and $\bar{m} = \left(88 \pm 8 \cdot 10^{-3}\right)$ g?

10.23 Linearize the function $f\left(x,\, y,\, z\right) = y \cdot \cos\left(z\right) + \frac{\ln\left(1+x^2\right)}{y}$ at the point $\left(x_0,\, y_0,\, z_0\right) = \left(1,\, 2,\, 0\right)$.

10.24 First determine the critical points for the following functions and decide whether (and if so, which) are local extreme points

a) $f\left(x,\, y\right) = x^2 + \cos\left(y\right)$
b) $f\left(x,\, y\right) = 3\,x^2 + 3\,x\,y - 18\,y^2$
c) $f\left(x,\, y\right) = \left(x - y\right)^3 + 12\,x\,y$

10.25 Which point of the area $z = \sqrt{1 + (x - 2\,y)^2}$ has the smallest distance from the point $\left(1,\, -2,\, 0\right)$?

10.26 Show that the function $f\left(x,\, y\right) = c - x^2 - y^2$ has a local maximum at the point $\left(0,\, 0\right)$.

10.27 Determine the relative extremes of the function

$$f\left(x,\, y\right) = x^3 + y^3 - 3\,x - 12\,y + 20.$$

10.28 Find the relative extremes of the functions

a) $f\left(x,\, y\right) = 3\,x\,y - x^3 - y^3$
b) $f\left(x,\, y\right) = x^2 + y^2 + x - y$
c) $f\left(x,\, y\right) = 1 - x + y - 2\,x\,y + x^2 - y^2$
d) $f\left(x,\, y\right) = e^{x^2 + y^2} - 2\,x^4 - 2\,y^2$

10.29 Determine all stationary points of the function

$$f\left(x,\, y,\, z\right) = e^{-\left(x^2 + y^2 + z^2\right)} \cdot \left(x^2 - z^2\right)$$

and check whether they are local extremes.

10.30 Determine the compensation line for the following measurement series

a)

x_i	0	1	2	3	4	5	6
y_i	2.1	0.81	−0.5	−2.1	−3.4	−4.3	−5.8

b)

x_i	1.5	1.8	2.4	3.0	3.5	4.0	4.5	6.0
y_i	1.9	2.1	2.8	3.4	4.0	4.1	5.1	6.1

Enter the points together with the compensation line in a diagram!

11

Chapter 11
Positive Definite Matrices

In this chapter we summarize important results that we need in order to understand the generalization of calculus to extreme values for functions with more than two variables. Fundamental is the notion of eigenvectors and eigenvalues for a quadratic matrix. Symmetric matrices have enough eigenvectors to form a basis of $\mathbb{R}^n$. And among the symmetric matrices, we have the positive definite matrices with very special properties, which we used when discussing the Hessian matrix in the last chapter.

11

11 Positive Definite Matrices

In this chapter we summarize important results that we need in order to understand the generalization of calculus to extreme values for functions with more than two variables. Fundamental is the notion of eigenvectors and eigenvalues for a quadratic matrix. Symmetric matrices have enough eigenvectors to form a basis of $\mathbb{R}^n$. And among the symmetric matrices, we have the positive definite matrices with very special properties, which we used when discussing the Hessian matrix in Section 10.4.4.

11.1 Eigenvalues and Eigenvectors

Definition: *Let A be an $(n \times n)$ matrix and $\vec{x}$ a vector $\vec{x} \neq \vec{0}$. Then we call $\vec{x}$ an* **Eigenvector** *of A, if there exists a complex number λ with*

$$A\vec{x} = \lambda\vec{x}.$$

λ is then called **Eigenvalue** *from A to the eigenvector $\vec{x}$.*

 **Visualization:** On the homepage there is an animation to graphically determine eigenvalues and eigenvectors for a (2×2) matrix. With the corresponding MAPLE procedure you can select your own (2×2) matrices.

Example 11.1. Given is the (3×3) matrix $A = \begin{pmatrix} 4 & 7 & -5 \\ 0 & 3 & -1 \\ 2 & 8 & -4 \end{pmatrix}$ and the

vectors $\vec{x}_1 = \begin{pmatrix} 2 \\ 1 \\ 3 \end{pmatrix}$, $\vec{x}_2 = \begin{pmatrix} 1 \\ 1 \\ 2 \end{pmatrix}$, $\vec{x}_3 = \begin{pmatrix} -1 \\ 1 \\ 1 \end{pmatrix}$. Then

$$A\vec{x}_1 = \begin{pmatrix} 4 & 7 & -5 \\ 0 & 3 & -1 \\ 2 & 8 & -4 \end{pmatrix} \begin{pmatrix} 2 \\ 1 \\ 3 \end{pmatrix} = \begin{pmatrix} 8+7-15 \\ 0+3-3 \\ 4+8-12 \end{pmatrix} = \begin{pmatrix} 0 \\ 0 \\ 0 \end{pmatrix} = 0 \cdot \begin{pmatrix} 2 \\ 1 \\ 3 \end{pmatrix}$$

$\hookrightarrow \vec{x}_1$ is an eigenvector to the eigenvalue $\lambda_1 = 0$.

$$A\,\vec{x}_2 = \begin{pmatrix} 4 & 7 & -5 \\ 0 & 3 & -1 \\ 2 & 8 & -4 \end{pmatrix} \begin{pmatrix} 1 \\ 1 \\ 2 \end{pmatrix} = \begin{pmatrix} 4+7-10 \\ 0+3-2 \\ 2+8-8 \end{pmatrix} = \begin{pmatrix} 1 \\ 1 \\ 2 \end{pmatrix} = 1 \cdot \begin{pmatrix} 1 \\ 1 \\ 2 \end{pmatrix}$$

$\hookrightarrow \vec{x}_2$ is an eigenvector to the eigenvalue $\lambda_2 = 1$.

$$A\,\vec{x}_3 = \begin{pmatrix} 4 & 7 & -5 \\ 0 & 3 & -1 \\ 2 & 8 & -4 \end{pmatrix} \begin{pmatrix} -1 \\ 1 \\ 1 \end{pmatrix} = \begin{pmatrix} -4+7-5 \\ 0+3-1 \\ -2+8-4 \end{pmatrix} = \begin{pmatrix} -2 \\ 2 \\ 2 \end{pmatrix} = 2 \cdot \begin{pmatrix} -1 \\ 1 \\ 1 \end{pmatrix}$$

$\hookrightarrow \vec{x}_3$ is an eigenvector to the eigenvalue $\lambda_3 = 2$.

In Example 11.1 we only checked that the given vectors are eigenvectors. The next step is to calculate the eigenvectors and eigenvalues of a matrix A. We first compute all the eigenvalues of A and then for each eigenvalue we compute the corresponding eigenvectors:

11.2 Calculation of the Eigenvalues of a Matrix

Let A be an $(n \times n)$ matrix and $\lambda \in \mathbb{C}$ be an eigenvalue of A. Then there exists a vector $\vec{x} \neq 0$ such that

$$A\vec{x} = \lambda\,\vec{x}. \tag{3}$$

Let us rewrite the matrix equation with the identity matrix

$$I_n = \begin{pmatrix} 1 & & 0 \\ & \ddots & \\ 0 & & 1 \end{pmatrix}$$

and the vector

$$\vec{x} = \begin{pmatrix} x_1 \\ \vdots \\ x_n \end{pmatrix} = \begin{pmatrix} 1 & & 0 \\ & \ddots & \\ 0 & & 1 \end{pmatrix} \begin{pmatrix} x_1 \\ \vdots \\ x_n \end{pmatrix} = I_n\,\vec{x}.$$

Then the equivalent of equation (3) is

$$A\vec{x} = \lambda\,I_n\,\vec{x} \;\Leftrightarrow\; A\vec{x} - \lambda\,I_n\,\vec{x} = \vec{0} \;\Leftrightarrow\; (A - \lambda\,I_n)\,\vec{x} = \vec{0}.$$

This is a system of linear equations with the matrix $B = (A - \lambda\,I_n)$. A linear system of equations is uniquely solvable according to the fundamental theorem of LEq (see Volume 1, Section 3.3.1) if $\det(B) \neq 0$. Then $\vec{x} = \vec{0}$

is the only solution of $B\,\vec{x} = \vec{0}$. However, this solution does not give an eigenvector, because the eigenvectors must be non-zero. So in order to get an eigenvector, $\vec{x} = \vec{0}$ cannot be the only solution, i.e. the LEq must not be uniquely solvable! So it must be $\det(B) = 0$, which means

$$\det\left(A - \lambda\, I_n\right) = 0.$$

Eigenvalues of a Matrix

λ is an eigenvalue of the matrix A $\Leftrightarrow$ $\det\left(A - \lambda\, I_n\right) = 0.$

Example 11.2. Given is the (3×3) matrix $A = \begin{pmatrix} 5 & 7 & -5 \\ 0 & 4 & -1 \\ 2 & 8 & -3 \end{pmatrix}$. Find all the eigenvalues of this matrix.

To find the eigenvalues, we calculate $A - \lambda\, I_3$

$$A - \lambda\, I_3 = \begin{pmatrix} 5 & 7 & -5 \\ 0 & 4 & -1 \\ 2 & 8 & -3 \end{pmatrix} - \lambda \begin{pmatrix} 1 & 0 & 0 \\ 0 & 1 & 0 \\ 0 & 0 & 1 \end{pmatrix} = \begin{pmatrix} 5 - \lambda & 7 & -5 \\ 0 & 4 - \lambda & -1 \\ 2 & 8 & -3 - \lambda \end{pmatrix}$$

and compute the determinant

$$\det\left(A - \lambda\, I_3\right) = \begin{vmatrix} 5 - \lambda & 7 & -5 \\ 0 & 4 - \lambda & -1 \\ 2 & 8 & -3 - \lambda \end{vmatrix}$$

$$= (5 - \lambda) \begin{vmatrix} 4 - \lambda & -1 \\ 8 & -3 - \lambda \end{vmatrix} + 2 \begin{vmatrix} 7 & -5 \\ 4 - \lambda & -1 \end{vmatrix}$$

$$= -\lambda^3 + 6\lambda^2 - 11\lambda + 6 = -\left(\lambda - 1\right)\left(\lambda - 2\right)\left(\lambda - 3\right).$$

From the condition $\det\left(A - \lambda\, I_3\right) = 0$, the eigenvalues are identified to be: $\lambda_1 = 1$, $\lambda_2 = 2$ and $\lambda_3 = 3$. $\qquad\qquad\square$

We observe that for the (3×3) matrix A, the determinant $\det\left(A - \lambda\, I_3\right) = -\lambda^3 + 6\lambda^2 - 11\lambda + 6$ is a polynomial in λ of degree 3. In general, it is

> **Characteristic Polynomial**
>
> ---
>
> If A is an $(n \times n)$ matrix, then
>
> $$P(\lambda) := \det(A - \lambda I_n)$$
>
> is a polynomial in λ of degree n. $P(\lambda)$ is called the **Characteristic Polynomial. The zeros of the characteristic polynomial are the eigenvalues of the matrix A.**

According to the Fundamental Theorem of Algebra (see Volume 1, Section 5.2.7) every complex polynomial of degree n has exactly n zeros. So the characteristic polynomial in the complex always splits into linear factors. In practice, however, for systems with more than 3 equations, numerical methods such as Newton's method (Volume 1, Section 7.8) are required to determine the eigenvalues.

11.3 Calculation of the Eigenvectors

Once the eigenvalues of a matrix have been determined, the corresponding eigenvectors must be calculated for each eigenvalue by solving the system of linear equations $(A - \lambda I_n)\,\vec{x} = \vec{0}$:

> **Eigenvectors**
>
> ---
>
> If λ is an eigenvalue of the matrix A, then all eigenvectors to the eigenvalue λ are defined as the solutions of the system of linear equations
>
> $$(A - \lambda I_n)\,\vec{x} = \vec{0}.$$

Example 11.3 (Eigenvectors). Given is the matrix $A = \begin{pmatrix} 5 & 7 & -5 \\ 0 & 4 & -1 \\ 2 & 8 & -3 \end{pmatrix}$ from Example 11.2 with the eigenvalues $\lambda_1 = 1$, $\lambda_2 = 2$, $\lambda_3 = 3$. Find the eigenvectors associated with these eigenvalues.

i) Calculation of an eigenvector for the eigenvalue $\lambda_1 = 1$: We search for vectors $\vec{x} \neq \vec{0}$, so that $(A - \lambda_1 I_3)\,\vec{x} = \vec{0}$. We solve the LEq

$$\begin{pmatrix} 5-1 & 7 & -5 \\ 0 & 4-1 & -1 \\ 2 & 8 & -3-1 \end{pmatrix} \begin{pmatrix} x_1 \\ x_2 \\ x_3 \end{pmatrix} = \begin{pmatrix} 0 \\ 0 \\ 0 \end{pmatrix} :$$

$$\hookrightarrow \left(\begin{array}{ccc|c} 4 & 7 & -5 & 0 \\ 0 & 3 & -1 & 0 \\ 2 & 8 & -4 & 0 \end{array} \right) \hookrightarrow \left(\begin{array}{ccc|c} 4 & 7 & -5 & 0 \\ 0 & 3 & -1 & 0 \\ 0 & -9 & 3 & 0 \end{array} \right) \hookrightarrow \left(\begin{array}{ccc|c} 4 & 7 & -5 & 0 \\ 0 & 3 & -1 & 0 \\ 0 & 0 & 0 & 0 \end{array} \right).$$

That is $x_3 = t$; $3\,x_2 - t = 0 \hookrightarrow x_2 = \frac{1}{3}\,t$; $4\,x_1 + \frac{7}{3}\,t - 5\,t = 0 \hookrightarrow x_1 = \frac{2}{3}\,t$.

$$\Rightarrow \mathbb{L}_1 = \left\{ \vec{x} \in \mathbb{R}^3 : \vec{x} = t \begin{pmatrix} \frac{2}{3} \\ \frac{1}{3} \\ 1 \end{pmatrix}, \quad t \in \mathbb{R} \right\}$$

is the solution set. For example, if we choose $t = 3$, then $\vec{x}_1 = \begin{pmatrix} 2 \\ 1 \\ 3 \end{pmatrix}$ is an eigenvector for the eigenvalue $\lambda_1 = 1$.

ii) Calculation of an eigenvector for the eigenvalue $\lambda_2 = 2$: The linear system of equations is to be solved $(A - \lambda_2\, I_3)\, \vec{x} = \vec{0}$:

$$\left(\begin{array}{ccc|c} 5-2 & 7 & -5 & 0 \\ 0 & 4-2 & -1 & 0 \\ 2 & 8 & -3-2 & 0 \end{array} \right) \hookrightarrow \left(\begin{array}{ccc|c} 3 & 7 & -5 & 0 \\ 0 & 2 & -1 & 0 \\ 0 & 0 & 0 & 0 \end{array} \right).$$

That is $x_3 = t$; $2\,x_2 - t = 0 \hookrightarrow x_2 = \frac{1}{2}\,t$; $3\,x_1 + \frac{7}{2}\,t - 5\,t = 0 \hookrightarrow x_1 = \frac{1}{2}\,t$.

$$\Rightarrow \mathbb{L}_2 = \left\{ \vec{x} \in \mathbb{R}^3 : \vec{x} = t \begin{pmatrix} \frac{1}{2} \\ \frac{1}{2} \\ 1 \end{pmatrix}, \quad t \in \mathbb{R} \right\}$$

is the solution set of the linear system of equations. An eigenvector for the eigenvalue $\lambda_2 = 2$ is obtained by choosing e.g. $t = 2$: $\vec{x}_2 = \begin{pmatrix} 1 \\ 1 \\ 2 \end{pmatrix}$.

iii) Calculation of an eigenvector for the eigenvalue $\lambda_3 = 3$: By solving the linear system of equations $(A - \lambda_3\, I_3)\, \vec{x} = \vec{0}$ we get e.g. $\vec{x}_3 = \begin{pmatrix} -1 \\ 1 \\ 1 \end{pmatrix}$ as an eigenvector for the eigenvalue $\lambda_3 = 3$. $\qquad \square$

If $\vec{x}_1$ and $\vec{x}_2$ are two eigenvectors for the **same** eigenvalue λ, then any linear combination of $\vec{x}_1$ and $\vec{x}_2$ is also an eigenvector for the same eigenvalue:

$$\begin{aligned} A\,(c_1\,\vec{x}_1 + c_2\,\vec{x}_2) &= c_1\,A\,\vec{x}_1 + c_2\,A\,\vec{x}_2 = c_1\,\lambda\,\vec{x}_1 + c_2\,\lambda\,\vec{x}_2 \\ &= \lambda\,(c_1\,\vec{x}_1 + c_2\,\vec{x}_2). \end{aligned}$$

Consequently, $c_1\,\vec{x}_1 + c_2\,\vec{x}_2$ is also an eigenvector for the eigenvalue λ for every $c_1,\,c_2 \in \mathbb{C}$. With the UVR criterion (see Volume 1, Section 2.4.2) we conclude

---

Eigenspace

Let A be an $(n \times n)$ matrix and λ an eigenvalue of A. Then the set of eigenvectors for the eigenvalue λ forms a vector space.

$$\begin{aligned} \mathrm{Eig}\,(A,\,\lambda) &= \text{Set of all eigenvectors for the eigenvalue } \lambda \\ &= \text{Eigenspace of } A \text{ for the eigenvalue } \lambda. \end{aligned}$$

The dimension of this vector space is called the **geometric order** of the eigenvalue λ.

If $\vec{x}_1$ and $\vec{x}_2$ are two eigenvectors for **different** eigenvalues λ_1 and λ_2, then $\vec{x}_1$ and $\vec{x}_2$ are linearly independent:

Check: Let $\lambda_1 \neq \lambda_2$ be different eigenvalues for the eigenvectors $\vec{x}_1$ and $\vec{x}_2$, e.g. $A\,\vec{x}_1 = \lambda_1\,\vec{x}_1$ and $A\,\vec{x}_2 = \lambda_2\,\vec{x}_2$. Then we apply the matrix A to

$$c_1\,\vec{x}_1 + c_2\,\vec{x}_2 = \vec{0}$$

$$\Rightarrow c_1\,A\vec{x}_1 + c_2\,A\vec{x}_2 = A\,\vec{0} \quad \Rightarrow \quad c_1\,\lambda_1\vec{x}_1 + c_2\,\lambda_2\vec{x}_2 = \vec{0}$$

In the last identity we replace $c_1\,\vec{x}_1 = -c_2\,\vec{x}_2$, so

$$-c_2\lambda_1\vec{x}_2 + c_2\,\lambda_2\vec{x}_2 = \vec{0} \quad \Rightarrow \quad (\lambda_2 - \lambda_1)\,c_2\,\vec{x}_2 = \vec{0}$$

Since $\lambda_1 \neq \lambda_2$, it must be $c_2 = 0$ and hence also $c_1 = 0$. $\qquad\square$

Different eigenvalues

Eigenvectors for different eigenvalues are linearly independent.

If we have n different eigenvalues, we get a basis of eigenvectors. If we have an eigenvalue of algebraic order k and the geometric order is also k, then we get a basis of eigenvectors. This situation applies, for example, to the important application of symmetric matrices. Only for those matrices where the algebraic order is greater than the geometric order will we not find enough eigenvectors to form a basis of $\mathbb{R}^n$.

Basis of eigenvectors

Let A be an $(n \times n)$ matrix and let $\vec{x}_1, \ldots, \vec{x}_n$ be a basis of eigenvectors for the eigenvalues $\lambda_1, \ldots, \lambda_n$. Then there exists an invertible matrix S and a diagonal matrix D such that

$$A = S\,D\,S^{-1} \quad \text{or} \quad D = S^{-1} A\,S.$$

Proof: The diagonal matrix D is defined by the eigenvalues $\lambda_1, \ldots, \lambda_n$

$$D = \begin{pmatrix} \lambda_1 & & 0 \\ & \ddots & \\ 0 & & \lambda_n \end{pmatrix}$$

and the columns of the matrix S are defined by the eigenvectors

$$S = (\vec{x}_1 \ \ldots \ \vec{x}_n).$$

So $S\,\vec{e}_j = \vec{x}_j$ and for the inverse matrix S^{-1} it follows $S^{-1}\vec{x}_j = \vec{e}_j$. For each eigenvector $\vec{x}_j$ we compute

$$S\,D\,S^{-1}\vec{x}_j = S\,D\,(S^{-1}\vec{x}_j) = S\,D\,\vec{e}_j = S\,\lambda_j\vec{e}_j$$
$$= \lambda_j\,(S\,\vec{e}_j) = \lambda_j\,\vec{x}_j = A\,\vec{x}_j.$$

Since $\vec{x}_1, \ldots, \vec{x}_n$ is a basis, we write each vector $\vec{x}$ as $\vec{x} = c_1\,\vec{x}_1 + \cdots + c_n\,\vec{x}_n$ and conclude that $S\,D\,S^{-1}(\vec{x}) = A(\vec{x})$ for all $\vec{x} \in \mathbb{R}^n$

$$S\,D\,S^{-1}\vec{x} = S\,D\,S^{-1}\,(c_1\,\vec{x}_1 + \cdots + c_n\,\vec{x}_n)$$
$$= (c_1\,S\,D\,S^{-1}\,\vec{x}_1 + \cdots + c_n\,S\,D\,S^{-1}\,\vec{x}_n)$$
$$= (c_1\,A\,\vec{x}_1 + \cdots + c_n\,A\,\vec{x}_n)$$
$$= A\,(c_1\,\vec{x}_1 + \cdots + c_n\,\vec{x}_n) = A\,\vec{x}.$$

So the matrix on the left side is identical to the matrix on the right side. $\qquad\square$

Example 11.4 (Diagonalization). Given is the matrix

$$A = \begin{pmatrix} 5 & 7 & -5 \\ 0 & 4 & -1 \\ 2 & 8 & -3 \end{pmatrix}$$

from Example 11.2 with the eigenvalues $\lambda_1 = 1$, $\lambda_2 = 2$, $\lambda_3 = 3$. We are looking for the decomposition of the matrix A using the diagonal matrix with the eigenvalues.

The eigenvectors associated with these eigenvalues are, according to Example 11.3,

$$\vec{x}_1 = \begin{pmatrix} 2 \\ 1 \\ 3 \end{pmatrix}, \ \vec{x}_2 = \begin{pmatrix} 1 \\ 1 \\ 2 \end{pmatrix}, \ \vec{x}_3 = \begin{pmatrix} -1 \\ 1 \\ 1 \end{pmatrix}.$$

With the eigenvectors we define the transfer matrix S and the diagonal matrix D consisting of the eigenvalues of A:

$$S = \begin{pmatrix} 2 & 1 & -1 \\ 1 & 1 & 1 \\ 3 & 2 & 1 \end{pmatrix}, \ D = \begin{pmatrix} 1 & 0 & 0 \\ 0 & 2 & 0 \\ 0 & 0 & 3 \end{pmatrix}.$$

With the inverse of S

$$S^{-1} = \begin{pmatrix} -1 & -3 & 2 \\ 2 & 5 & -3 \\ -1 & -1 & 1 \end{pmatrix}$$

we can verify that

$$S^{-1} A S = \begin{pmatrix} -1 & -3 & 2 \\ 2 & 5 & -3 \\ -1 & -1 & 1 \end{pmatrix} \cdot \begin{pmatrix} 5 & 7 & -5 \\ 0 & 4 & -1 \\ 2 & 8 & -3 \end{pmatrix} \cdot \begin{pmatrix} 2 & 1 & -1 \\ 1 & 1 & 1 \\ 3 & 2 & 1 \end{pmatrix}$$

$$= \begin{pmatrix} -1 & -3 & 2 \\ 2 & 5 & -3 \\ -1 & -1 & 1 \end{pmatrix} \cdot \begin{pmatrix} 2 & 2 & -3 \\ 1 & 2 & 3 \\ 3 & 4 & 3 \end{pmatrix}$$

$$= \begin{pmatrix} 1 & 0 & 0 \\ 0 & 2 & 0 \\ 0 & 0 & 3 \end{pmatrix} = D.$$

Furthermore, if we compute the determinant, we get

$$\det(D) = \det(S^{-1} A S) = \det(S^{-1}) \cdot \det(A) \cdot \det(S)$$
$$= \frac{1}{\det(S)} \cdot \det(A) \cdot \det(S) = \det(A)$$

So

$$\det(A) = 1 \cdot 2 \cdot 3 = 6. \qquad \square$$

11.4 Symmetric Matrices

It is fundamental for real symmetric matrices that they have only real eigenvalues with eigenvectors that form a basis of $\mathbb{R}^n$, which we will prove below. Thus, if we define the columns of the transformation matrix S by the eigenvectors, then $D = S^{-1} \cdot A \cdot S$, where D is the diagonal matrix with the corresponding eigenvalues. Using the Gram-Schmidt orthogonalization, we generate an orthonormal basis of eigenvectors and $S^{-1} = S^t$.

Gram-Schmidt Orthogonalization Method

Let $\vec{v}_1, \ldots, \vec{v}_k$ be k linearly independent vectors. Then there are k orthogonal vectors $\vec{s}_1, \ldots, \vec{s}_k$ such that $\mathrm{span}(\vec{v}_1, \ldots, \vec{v}_k) = \mathrm{span}(\vec{s}_1, \ldots, \vec{s}_k)$.

The Gram-Schmidt algorithm guarantees that we can always construct from k linearly independent vectors another k orthogonal vectors, which generate the same vector space with dimension k. After normalization we then have k orthonormal vectors.

Proof:

If we have k linearly independent vectors $\vec{v}_1, \ldots, \vec{v}_k$, then we start with $\vec{s}_1 = \vec{v}_1$ and continue with the recursion $(j = 2, \ldots, k)$

$$\vec{s}_j = \vec{v}_j - \sum_{i=1}^{j-1} \frac{\vec{v}_j^{\,t} \vec{s}_i}{|\vec{s}_i|^2} \vec{s}_i$$

to generate k orthogonal eigenvectors. We check directly that $\vec{s}_k$ is orthogonal to any $\vec{s}_j$ for $j = 1, \ldots, k-1$ by taking the scalar product of any $\vec{s}_j$ with $\vec{s}_k$:

$$\vec{s}_j^{\,t} \cdot \vec{s}_k = \vec{s}_j^{\,t} \cdot \left(\vec{v}_k - \sum_{i=1}^{k-1} \frac{\vec{v}_k^{\,t} \vec{s}_i}{|\vec{s}_i|^2} \, \vec{s}_i \right) = \vec{s}_j^{\,t} \vec{v}_k - \vec{s}_j^{\,t} \cdot \sum_{i=1}^{k-1} \frac{\vec{v}_k^{\,t} \vec{s}_i}{|\vec{s}_i|^2} \, \vec{s}_i$$

$$= \vec{s}_j^{\,t} \vec{v}_k - \sum_{i=1}^{k-1} \frac{\vec{v}_k^{\,t} \vec{s}_i}{|\vec{s}_i|^2} \, \vec{s}_j^{\,t} \cdot \vec{s}_i.$$

By recursive construction it is $\vec{s}_j^{\,t} \cdot \vec{s}_i = 0$ for $i \neq j$ and $\vec{s}_j^{\,t} \cdot \vec{s}_i \neq 0$ for $i = j$. So

$$\vec{s}_j^{\,t} \cdot \vec{s}_k = \vec{s}_j^{\,t} \vec{v}_k - \frac{\vec{v}_k^{\,t} \vec{s}_j}{|\vec{s}_j|^2} \, \vec{s}_j^{\,t} \vec{s}_j = \vec{s}_j^{\,t} \cdot \vec{v}_k - \vec{v}_k^{\,t} \vec{s}_j = 0. \qquad \square$$

Note

Let A be a real symmetric $(n \times n)$ matrix. Let S be an invertible matrix with $S^{-1} = S^t$.

(1) $S^t \cdot A \cdot S$ is symmetric.

(2) If λ is an eigenvalue of A, then λ is also an eigenvalue of $S^t \cdot A \cdot S$ and vice verse.

(3) The characteristic polynomials of A and $S^t \cdot A \cdot S$ have the same roots.

Proof:

(1) $(S^t \cdot A \cdot S)^t = S^t \cdot A^t \cdot (S^t)^t = S^t \cdot A \cdot S.$

(2) If λ is an eigenvalue of A, then there exists an eigenvector $\vec{x}$ such that $A\vec{x} = \lambda\vec{x}$. Then $\vec{y} = S^t \vec{x}$ is an eigenvector of $S^t \cdot A \cdot S$:

$$S^t A S (S^t \vec{x}) = S^t A (S S^t) \vec{x} = S^t A \vec{x} = S^t \lambda \vec{x} = \lambda \vec{y},$$

since $(S S^t) = (S S^{-1}) = I_n$. And similarly, if λ is an eigenvalue of $S^t \cdot A \cdot S$ with eigenvector $\vec{y}$, then $\vec{x} = S \vec{y}$ is an eigenvector of A.

(3) Consequence of Note (2).

With these preliminary remarks, we can prove the following properties of real symmetric matrices.

Properties of Real Symmetric Matrices

Let A be a real symmetric $(n \times n)$ matrix.

(1) A has only real eigenvalues.

(2) The geometric order of an eigenvalue is equal to the algebraic order.

(3) The eigenvectors for different eigenvalues are orthogonal.

(4) Orthogonal eigenvectors for the same eigenvalue can be constructed.

(5) If $(\vec{s}_1, \ldots \vec{s}_n)$ is an orthonormal basis, then the transformation matrix $S = (\vec{s}_1, \ldots \vec{s}_n)$ has the inverse S^t.

Note: Note (1) says that the characteristic polynomial of degree n has only real zeros. They can be single zeros or zeros of order k. This number is called the algebraic order. If the algebraic order is k, we need as many linearly independent eigenvectors as the algebraic order to diagonalize the matrix. The number of linearly independent eigenvectors for an eigenvalue is the geometric order. From point (2) we conclude that any real symmetric matrix can be diagonalized.

From points (3) and (4) we can conclude that there is a basis of eigenvectors that stay orthogonal and have the lengths of 1. So we have an orthonormal basis of eigenvectors. According to point (5), the transformation matrix S consisting of the orthonormal eigenvectors has S^t as its inverse.

Proof of the properties: For A to be real means $\bar{A} = A$. For A to be symmetric means $A^t = A$. An important rule for the transposition of two matrices is $(A \cdot B)^t = B^t \cdot A^t$. Consequently, for the matrix-vector multiplication we have $(A \cdot \vec{x})^t = \vec{x}^{\,t} \cdot A^t$.

(1) If λ is an eigenvalue, then there exists an eigenvector $\vec{v} \neq 0$ such that $A\vec{v} = \lambda\vec{v}$. We take the complex conjugate of both sides and obtain

$$\overline{A\vec{v}} = \overline{\lambda\vec{v}} \quad \Rightarrow \quad \overline{A}\,\overline{\vec{v}} = \overline{\lambda}\,\overline{\vec{v}} \quad \Rightarrow \quad A\overline{\vec{v}} = \overline{\lambda}\,\overline{\vec{v}}.$$

So $\overline{\lambda}$ is an eigenvalue of the eigenvector $\overline{\vec{v}}$. Taking the scalar product of the last identity with $\vec{v}$, we get

$$\vec{v}^{\,t} A \overline{\vec{v}} = \vec{v}^{\,t} \overline{\lambda}\, \overline{\vec{v}} \quad \Rightarrow \quad (A\,\vec{v})^t\, \overline{\vec{v}} = \overline{\lambda}\, \vec{v}^{\,t}\, \overline{\vec{v}}$$

$$\Rightarrow \quad (\lambda\,\vec{v})^t\, \overline{\vec{v}} = \overline{\lambda}\, \vec{v}^{\,t}\, \overline{\vec{v}} \quad \Rightarrow \quad \lambda\,\vec{v}^{\,t}\, \overline{\vec{v}} = \overline{\lambda}\, \vec{v}^{\,t}\, \overline{\vec{v}}.$$

Since $\vec{v}^{\,t}\, \overline{\vec{v}} = |\vec{v}|^2 \neq 0$, we get $\lambda = \overline{\lambda}$ and λ must be a real number.

(2) The geometric order of an eigenvalue is equal to the algebraic order: Let $\mu \in \mathbb{R}$ be an eigenvalue of A with the geometric order l. Then there exist l linearly independent eigenvectors $(\vec{x}_1, \ldots, \vec{x}_l)$ of A.

Using the Gram-Schmidt orthogonalization procedure, we can assume that these are orthonormal vectors. We extend this list of vectors by $n - l$ vectors $(\vec{x}_{l+1}, \ldots, \vec{x}_n)$ such, that $(\vec{x}_1, \ldots, \vec{x}_l;\ \vec{x}_{l+1}, \ldots, \vec{x}_n)$ is an orthonormal basis of $\mathbb{R}^n$. Again using the Gram-Schmidt algorithm. With these vectors we define the matrix $S = (\vec{x}_1, \ldots, \vec{x}_l;\ \vec{x}_{l+1}, \ldots, \vec{x}_n)$. Then we have

$$S^t \cdot A \cdot S = \left(
\begin{array}{ccccc|ccc}
\mu & 0 & \cdots & & 0 & 0 & \cdots \cdots & 0 \\
0 & \mu & \ddots & & \vdots & \vdots & & \vdots \\
\vdots & \ddots & \ddots & & 0 & 0 & \cdots \cdots & 0 \\
0 & \cdots & 0 & \mu & & & & \\
\hline
& 0 \cdots \cdots 0 & & & & & & \\
& \vdots & \vdots & & & & Y & \\
& \vdots & \vdots & & & & & \\
& 0 \cdots \cdots 0 & & & & & &
\end{array}
\right).$$

The left upper block is the diagonal $(l \times l)$ sub-matrix with the eigenvalue μ, the left lower block is the zero $((n - l) \times l)$ sub-matrix, since the first l columns are the images of the eigenvectors. For A, $S^t \cdot A \cdot S$ is also a symmetric matrix, so the right upper block must be the $(l \times (n - l))$ zero sub-matrix. According to Note (3), the characteristic polynomials of A and $S^t \cdot A \cdot S$ have the same roots. Up to a constant factor c they are identical:

$$p_A(\lambda) = c\, p_{S^t A S}(\lambda) = c\, (\mu - \lambda)^l \det(Y - \lambda\, I_{n-l}).$$

If μ were a zero of the characteristic polynomial of Y, it would have an eigenvector $\vec{y}_{n-l} \in \mathbb{R}^{n-l}$. Then

$$\vec{y} = \begin{pmatrix} 0 \\ \vdots \\ 0 \\ \vec{y}_{n-l} \end{pmatrix}$$

is an eigenvector of $S^t \cdot A \cdot S$ for the eigenvalue μ. So μ would also be an eigenvalue of $S^t \cdot A \cdot S$ and

$$S\,\vec{y} = (\vec{x}_1, \ldots, \vec{x}_l \,;\, \vec{x}_{l+1}, \ldots, \vec{x}_n) \begin{pmatrix} 0 \\ \vdots \\ 0 \\ \vec{y}_{n-l} \end{pmatrix}$$
$$= \vec{x}_1 \cdot 0 + \cdots + \vec{x}_l \cdot 0 + \vec{x}_{l+1} \cdot y_1 + \cdots + \vec{x}_n \cdot y_l \notin \text{span}\,(\vec{x}_1, \ldots, \vec{x}_l).$$

This contradicts the assumption that only $\vec{x}_1, \ldots, \vec{x}_l$ are linearly independent eigenvectors of A. Therefore, μ cannot be an eigenvalue of Y and the algebraic order of the eigenvalue μ is also l.

(3) Let $\vec{v}_1$ be an eigenvector to the eigenvalue λ_1 (i.e. $A\,\vec{v}_1 = \lambda_1\,\vec{v}_1$) and $\vec{v}_2$ be an eigenvector to the eigenvalue λ_2 (i.e. $A\,\vec{v}_2 = \lambda_2\,\vec{v}_2$) and $\lambda_1 \neq \lambda_2$. Using the symmetry property ($A^t = A$), we obtain

$$(\lambda_1\,\vec{v}_1)^t\,\vec{v}_2 = (A\,\vec{v}_1)^t\,\vec{v}_2 = \vec{v}_1^{\,t}\,A^t\,\vec{v}_2 = \vec{v}_1^{\,t}\,A\,\vec{v}_2 = \vec{v}_1^{\,t}\,\lambda_2\,\vec{v}_2\,.$$

Therefore,

$$\lambda_1\,\vec{v}_1^{\,t}\,\vec{v}_2 - \lambda_2\,\vec{v}_1^{\,t}\,\vec{v}_2 = 0 \quad \Rightarrow \quad (\lambda_1 - \lambda_2)\,\vec{v}_1^{\,t}\,\vec{v}_2 = 0.$$

Since the eigenvalues are different, it must be $\vec{v}_1^{\,t}\,\vec{v}_2 = 0$. This is the scalar product of $\vec{v}_1$ and $\vec{v}_2$. So they are orthogonal.

(4) According to point (2), eigenvectors to different eigenvalues stay orthogonal. If the geometric order of an eigenvalue is greater than 1, we can construct orthogonal eigenvectors to the same eigenvalue using the Gram-Schmidt orthogonalization procedure.

(5) If $(\vec{s}_1, \ldots, \vec{s}_n)$ is an orthonormal basis, then the transformation matrix $S = (\vec{s}_1, \ldots, \vec{s}_n)$ has the inverse S^t. According to items (3) and (4), an orthonormal basis of eigenvectors $\vec{s}_i$ $(i = 1, \ldots, n)$ can be constructed by finally normalizing them: They remain orthogonal to each other and

have lengths of 1, which can be simplified in the property

$$\vec{s}_i^{\,t} \cdot \vec{s}_j = \delta_{ij} = \begin{cases} 1 & \text{for } i = j \\ 0 & \text{for } i \neq j \end{cases}.$$

If we now define the transformation matrix $S = (\vec{s}_1, \ldots, \vec{s}_n)$ with the eigenvectors as the columns, then $S^t = \begin{pmatrix} \vec{s}_1^{\,t} \\ \vdots \\ \vec{s}_n^{\,t} \end{pmatrix}$ consists of the rows as the eigenvectors and the product of S^t with S gives the identity matrix $S^t \cdot S = (\delta_{ij}) = I_n$. So S^t is the inverse of S. $\qquad\square$

As a consequence of Note (5), for any given real symmetric matrix A we find an invertible matrix S with $S^{-1} = S^t$ such that

$$D = S^t \cdot A \cdot S,$$

where D is the diagonal matrix consisting of all eigenvalues of A. Depending on these eigenvalues, we define positive or negative definite matrices:

11.5 Positive Definite Matrices

Definition: *Let A be a real symmetric $(n \times n)$ matrix.*

(1) *A is called **positive definite** if all eigenvalues are greater than zero: $\lambda_k > 0$ for all $k = 1, \ldots, n$.*

(2) *A is called **positive semi-definite** if all eigenvalues are greater than or equal to zero: $\lambda_k \geq 0$ for all $k = 1, \ldots, n$.*

(3) *A is called **negative definite**, if all eigenvalues are less than zero: $\lambda_k < 0$ for all $k = 1, \ldots, n$.*

(4) *A is called **negative semi-definite**, if all eigenvalues are less than or equal to zero: $\lambda_k \leq 0$ for all $k = 1, \ldots, n$.*

(5) *Otherwise, A is called indefinite.*

Note: A is negative definite, if $-A$ is positive definite. A is negative semi-definite, if $-A$ is positive semi-definite.

Just looking at the definition of positive definite or negative definite, there is no reference to extremes and how to check whether a stationary point $\mathrm{grad}\,(f) = \vec{0}$ is an extreme. Therefore, the next theorem about positive definite matrices is most relevant. We will restrict the discussion to the positive definite matrices because the argument is the same for negative definite matrices.

> ### Positive Definite and Extreme Value
>
> Let A be a real symmetric $(n \times n)$ matrix. Then it is equivalent:
>
> $$A \text{ is positive definite} \quad \Longleftrightarrow \quad \vec{x}^{\,t} A \vec{x} > 0 \ \text{ for all } \ \vec{x} \neq \vec{0} \in \mathbb{R}^n.$$

Proof:

$\Rightarrow$: Since A is a real symmetric $(n \times n)$ matrix, it can be diagonalized. This means that there are eigenvectors $\vec{x}_1, \ldots, \vec{x}_n$ and eigenvalues $\alpha_1, \ldots, \alpha_n$ such that the matrix A can be decomposed into

$$A = S \cdot D \cdot S^t$$

where the diagonal matrix D consists of the eigenvalues and the transfer matrix S consists of the corresponding eigenvectors. So

$$\begin{aligned}
\vec{x}^{\,t} \cdot A \cdot \vec{x} &= \vec{x}^{\,t} \cdot (S \cdot D \cdot S^t) \cdot \vec{x} \\
&= (S^t \vec{x})^t \cdot D \cdot (S^t \vec{x}) \\
&= (\vec{y})^t \cdot D \cdot (\vec{y}) \\
&= \alpha_1 y_1^2 + \alpha_2 y_2^2 + \cdots + \alpha_n y_n^2 > 0
\end{aligned}$$

when introducing the vector $\vec{y} = S^t \vec{x}$. Since the positive definite matrix A has only positive eigenvalues $\alpha_i > 0$, the right side is positive for all vectors $\vec{y} \neq \vec{0}$ and consequently the left side is positive for all vectors $\vec{x} \neq \vec{0}$.

$\Leftarrow$: If $\vec{x}^{\,t} A \vec{x} > 0$ for all $\vec{x} \neq \vec{0} \in \mathbb{R}^n$, then this condition is also true for the eigenvectors. So let $\vec{x}_i \neq 0$ be an eigenvector to the eigenvalue α_i. Then

$$\vec{x}_i^{\,t} A \vec{x}_i = \vec{x}_i^{\,t} (A \vec{x}_i) = \vec{x}_i^{\,t} \alpha_i \vec{x}_i = \lambda_i \, |\vec{x}_i|^2 > 0.$$

Hence, it must be $\lambda_i > 0$. This argument holds true for all eigenvalues, so all eigenvalues are positive. $\qquad\square$

To identify a matrix A as positive definite, we can either check that for all non-zero vectors it is $\vec{x}^{\,t} A \vec{x} > 0$, which is usually impossible. Or we have to find all the eigenvalues and check that they are positive, which is impractical. So we need a simple criterion to check. This leads to the Sylvester criterion:

Sylvester's Criterion

Let $A = \begin{pmatrix} a_{11} & \cdots & a_{1n} \\ \vdots & & \vdots \\ a_{n1} & \cdots & a_{nn} \end{pmatrix}$ *be a real symmetric* $(n \times n)$ *matrix.*

A is **positive definite**, *if and only if all sub-determinants* (**principal minors or leading minors**) $\triangle_k := \det \begin{pmatrix} a_{11} & \cdots & a_{1k} \\ \vdots & & \vdots \\ a_{k1} & \cdots & a_{kk} \end{pmatrix}$ are greater than zero: $\triangle_k > 0$ for all $k = 1, \ldots, n$.

Positive semi-definite, negative (semi-)definite

A is positive semi-definite, iff all principal minors are greater than or equal to zero: $\triangle_k \geq 0$ *for all* $k = 1, \ldots, n$.

A is **negative definite**, *iff the matrix* $(-A)$ *is positive definite.*

A is negative semi-definite, iff the matrix $(-A)$ *is positive semi-definite.*

Otherwise A is indefinite.

This equivalence is called the Sylvester or Hurwitz criterion.

Proof of the Sylvester's criterion: We will only prove the first part for a positive definite matrix. All the other parts are simple modifications. Since A is a real symmetric $(n \times n)$ matrix, it can be diagonalized: There are eigenvectors $\vec{s}_1, \ldots, \vec{s}_n$ and eigenvalues $\lambda_1, \ldots, \lambda_n$ such that the matrix A can be decomposed into

$$A = S \cdot D \cdot S^t$$

where the diagonal matrix D consists of the eigenvalues and the transfer matrix S consists of the corresponding eigenvectors.

$\Rightarrow$: Let A be positive definite. Since all the eigenvalues are positive, we will see that $\det(A) > 0$:

$$A = S \cdot D \cdot S^t \quad \Rightarrow \quad \det(A) = \det(S \cdot D \cdot S^t) = \det(D) \cdot \det(S)^2$$
$$= \lambda_1 \cdot \ldots \cdot \lambda_n \cdot \det(S)^2 > 0.$$

Let k be any number $1 \le k < n$. When computing the principal minor $\Delta_k = \det(A_k)$, we check that the matrix A_k is also positive definite. For each vector $\vec{x}_k \in \mathbb{R}^k$ we define the associated vector $\vec{x}_n \in \mathbb{R}^n$:

$$\vec{x}_n = \begin{pmatrix} \vec{x}_k \\ 0 \\ \vdots \\ 0 \end{pmatrix}.$$

So for all $\vec{x}_k \in \mathbb{R}^k$ it is

$$\vec{x}_k^t \, A_k \, \vec{x}_k = \vec{x}_n^t \, A \, \vec{x}_n > 0,$$

since A is positive definite and therefore $\vec{x}^t \, A_k \, \vec{x} > 0$ for all $\vec{x} \in \mathbb{R}^k$. So A_k is positive definite, and by the same argument as for A, we conclude that the determinant of A_k is positive, so all the principal minors are

$$\Delta_k = \det(A_k) > 0 \quad \text{for all } 1 \le k \le n.$$

$\Leftarrow$: We prove this direction by mathematical induction. Let's assume that all principal minors $\Delta_k = \det(A_k) > 0$ for all $k = 1, \ldots, n$. Then we have to show that A_n is positive definite.

(1) $n = 1$: For $n = 1$ we have $A_1 = (a_{11})$ with $\Delta_1 = \det(A_1) > 0$. So a_{11} is the eigenvalue of A_1. Since it is positive, the matrix A_1 is positive definite.

(2) $n \to n+1$: Let's assume that all principal minors of A_n are positive and that A_n is positive definite. Then we must conclude that all eigenvalues of A_{n+1} are positive. Therefore, we decompose the $(n+1) \times (n+1)$ matrix A_{n+1} into the $n \times n$ matrix A_n in the upper left, the $(n+1)^{th}$ column $\vec{a}_n$ in the upper right, $\vec{a}_n^t$ in the lower left and the number $a_{n+1,n+1}$:

$$A_{n+1} = \left(\begin{array}{c|c} A_n & \vec{a}_n \\ \hline \vec{a}_n^{\,t} & a_{n+1,n+1} \end{array} \right)$$

By the induction assumption, A_n is positive definite, so there exists a transformation matrix S with $S^{-1} = S^t$ and $D_n = S^t A_n S$. D_n is the diagonal matrix consisting of the positive eigenvalues $\lambda_1, ..., \lambda_n$. According to the previous decomposition, we get with the expanded transformation matrix

$$S_{n+1} = \left(\begin{array}{c|c} S & \begin{matrix} 0 \\ \vdots \\ 0 \end{matrix} \\ \hline 0,\ldots,0 & 1 \end{array} \right) :$$

$$\begin{aligned}
S_{n+1}^t A_{n+1} S_{n+1} &= \left(\begin{array}{c|c} S^t & \vec{0} \\ \hline \vec{0}^{\,t} & 1 \end{array} \right) \left(\begin{array}{c|c} A_n & \vec{a}_n \\ \hline \vec{a}_n^{\,t} & a_{n+1,n+1} \end{array} \right) \left(\begin{array}{c|c} S & \vec{0} \\ \hline \vec{0}^{\,t} & 1 \end{array} \right) \\
&= \left(\begin{array}{c|c} S^t A_n S & \vec{b}_n \\ \hline \vec{b}_n^{\,t} & b' \end{array} \right) = \left(\begin{array}{c|c} D_n & \vec{b}_n \\ \hline \vec{b}_n^{\,t} & b' \end{array} \right) = D_{n+1}
\end{aligned}$$

$$(*)$$

The matrix D_{n+1} on the right side is symmetric, so it can be diagonalized with the transfer matrix

$$T_{n+1} = \left(\begin{array}{c|c} I_n & \begin{matrix} \beta_1 \\ \vdots \\ \beta_n \end{matrix} \\ \hline 0,\ldots,0 & 1 \end{array} \right)$$

where $\beta_i = -\frac{b_i}{\lambda_i}$. We evaluate the matrix products directly and get the diagonal matrix D_n extended by a number d along the diagonal

$$T_{n+1}^t D_{n+1} T_{n+1} = \left(\begin{array}{c|c} D_n & \vec{0} \\ \hline \vec{0}^{\,t} & d \end{array} \right).$$

If we replace D_{n+1} with the identity given in $(*)$, we get

$$T_{n+1}^t S_{n+1}^t A_{n+1} S_{n+1} T_{n+1} = \left(\begin{array}{c|c} D_n & \vec{0} \\ \hline \vec{0}^{\,t} & d \end{array} \right)$$

$$\Rightarrow (S_{n+1}\, T_{n+1})^{t}\, A_{n+1}\, (S_{n+1}\, T_{n+1}) = \left(\begin{array}{c|c} D_n & \vec{0} \\ \hline \vec{0}^{\,t} & d \end{array}\right).$$

Calculating the determinants of the left and right sides

$$\det((S_{n+1}\, T_{n+1})^{t})\, \det(A_{n+1})\, \det(S_{n+1}\, T_{n+1}) = \det\left(\begin{array}{c|c} D_n & \vec{0} \\ \hline \vec{0}^{\,t} & d \end{array}\right),$$

we get

$$\det((S_{n+1}\, T_{n+1}))^{2}\, \det(A_{n+1}) = \lambda_1 \cdot \ldots \cdot \lambda_n \cdot d.$$

Since the principal minor $\det(A_{n+1}) > 0$ and all the eigenvalues λ_1, ..., $\lambda_n > 0$ it must also be $d > 0$. Therefore, all the eigenvalues of A_{n+1} are positive, so the matrix A_{n+1} is positive definite. $\qquad\square$

Example 11.5. Given is the (4×4) matrix $A = \begin{pmatrix} 2 & -1 & 0 & 0 \\ -1 & 2 & -1 & 0 \\ 0 & -1 & 2 & -1 \\ 0 & 0 & -1 & 2 \end{pmatrix}.$

We show that this matrix is positive definite using various methods of investigation.

(1) We determine the eigenvalues of A and check that they are all positive. So we compute the characteristic polynomial $p(\lambda) = \det(A - \lambda\, I_4)$, which must be zero.

$$p(\lambda) = \det(A - \lambda\, I_4) = \begin{vmatrix} 2 - \lambda & -1 & 0 & 0 \\ -1 & 2 - \lambda & -1 & 0 \\ 0 & -1 & 2 - \lambda & -1 \\ 0 & 0 & -1 & 2 - \lambda \end{vmatrix}$$

$$= \cdots = \cdots = \lambda^4 - 8\,\lambda^3 + 21\,\lambda^2 - 20\,\lambda + 5 = 0.$$

The computation of the characteristic polynomial is time-consuming but straightforward. Not so the determination of the roots which is possible for (2×2), for some (3×3), and for rare (4×4) matrices. In our case, we obtain

$$\lambda_{1/2} = \frac{3}{2} \pm \frac{1}{2}\sqrt{5} \qquad \lambda_{3/4} = \frac{5}{2} \pm \frac{1}{2}\sqrt{5}$$

All eigenvalues are positive, so the matrix A is positive definite.

(2) To show that A is positive definite, it is sufficient to check that for all $\vec{x} \neq \vec{0} \in \mathbb{R}^4$ it is $\vec{x}^{\,t} A \vec{x} > 0$.

$$\vec{x}^{\,t} A \vec{x} = \begin{pmatrix} x_1 & x_2 & x_3 & x_4 \end{pmatrix} \cdot \begin{pmatrix} 2 & -1 & 0 & 0 \\ -1 & 2 & -1 & 0 \\ 0 & -1 & 2 & -1 \\ 0 & 0 & -1 & 2 \end{pmatrix} \cdot \begin{pmatrix} x_1 \\ x_2 \\ x_3 \\ x_4 \end{pmatrix}$$

$$= \begin{pmatrix} x_1 & x_2 & x_3 & x_4 \end{pmatrix} \cdot \begin{pmatrix} 2\,x_1 - x_2 \\ -x_1 + 2\,x_2 - x_3 \\ -x_2 + 2\,x_3 - x_4 \\ -x_3 + 2\,x_4 \end{pmatrix}$$

$$= \begin{aligned} (2\,x_1^2 - x_1 x_2 \qquad\qquad\qquad\qquad \\ -x_1 x_2 + 2\,x_2^2 - x_2 x_3 \qquad\qquad \\ -x_2 x_3 + 2\,x_3^2 - x_3 x_4 \quad \\ -x_3 x_4 + 2\,x_4^2) \end{aligned}$$

$$= x_1^2 + x_4^2 + (x_1 - x_2)^2 + (x_2 - x_3)^2 + (x_3 - x_4)^2 > 0$$

The sum of squares is non-negative and since at least one component of the vector $\vec{x}$ is non-zero, the sum is greater than zero. This proves that the matrix is positive definite.

(3) According to Sylvester's criterion, it is also sufficient to compute the principal minors of the matrix A, which is a straightforward procedure for the calculation of all determinants:

$$\Delta_1 = \det(2) = 2 > 0$$

$$\Delta_2 = \begin{vmatrix} 2 & -1 \\ -1 & 2 \end{vmatrix} = 4 - 1 = 3 > 0$$

$$\Delta_3 = \begin{vmatrix} 2 & -1 & 0 \\ -1 & 2 & -1 \\ 0 & -1 & 2 \end{vmatrix} = 2 \begin{vmatrix} 2 & -1 \\ -1 & 2 \end{vmatrix} + \begin{vmatrix} -1 & 0 \\ -1 & 2 \end{vmatrix} = 6 - 2 = 4 > 0$$

$$\Delta_4 = \begin{vmatrix} 2 & -1 & 0 & 0 \\ -1 & 2 & -1 & 0 \\ 0 & -1 & 2 & -1 \\ 0 & 0 & -1 & 2 \end{vmatrix} = \cdots = 5 > 0$$

So all principal minors are greater than zero and the matrix is positive definite. □

Note: Using the Sylvester criterion we can show with mathematical induction for the corresponding $(n \times n)$ matrix of Example 11.5 that all principal minors are positive: $\Delta_k = k + 1$ for all $k = 1, \ldots, n$.

11.6 Problems on Positive Definite Matrices

11.1 Determine all eigenvalues and eigenvectors of the matrices

a) $A = \begin{pmatrix} 2 & 0 & -2 \\ 0 & 4 & 0 \\ -2 & 0 & 5 \end{pmatrix}$ b) $B = \begin{pmatrix} -2 & -9 & 5 \\ -5 & -10 & 7 \\ -9 & -21 & 14 \end{pmatrix}$

Check whether the eigenvectors form a basis of the $\mathbb{R}^3$.

11.2 Compute all eigenvectors of $A = \begin{pmatrix} 1 & -2 \\ -2 & 4 \end{pmatrix}$ and check whether the eigenvectors form a basis of the $\mathbb{R}^2$. Can the matrix be diagonalized?

11.3 Given is the matrix $A = \begin{pmatrix} -3 & 1 \\ 1 & -3 \end{pmatrix}$.

a) Determine all eigenvalues and eigenvectors for the matrix A.
b) Determine an orthonormal basis of eigenvectors.
c) Set up the matrix S and compute S^{-1} such that $A = S D S^{-1}$.

11.4 Compute a basis of eigenvectors in case of:

a) $A = \begin{pmatrix} 3 & 4 \\ -5 & -5 \end{pmatrix}$ b) $B = \begin{pmatrix} 3 & 1 & 1 \\ 1 & 5 & 1 \\ 1 & 1 & 3 \end{pmatrix}$ c) $C = \begin{pmatrix} 3 & -1 & 1 \\ -1 & 3 & -1 \\ 1 & -1 & 3 \end{pmatrix}$.

11.5 Given are the matrices from Problem 11.4. Which are positive definite?

11.6 Given is the matrix $A = \begin{pmatrix} 2 & 1 & -4 \\ 2 & 1 & 2 \\ -3 & 3 & 3 \end{pmatrix}$.

a) Determine all eigenvalues and eigenvectors for the matrix A.
b) Determine a basis of eigenvectors.
c) Set up the matrix S and compute S^{-1} such that $A = S D S^{-1}$.

11.7 Given is the matrix $A = \begin{pmatrix} 4 & 1 & 1 \\ 1 & 4 & 1 \\ 1 & 1 & 4 \end{pmatrix}$.

a) Can A be diagonalized?
b) Determine all eigenvalues and eigenvectors for the matrix A.
c) Determine an orthonormal basis of eigenvectors.
d) Set up the matrix S and compute S^{-1} such that $A = S D S^{-1}$.

Chapter 12

Integral Calculus for Multi-Variable Functions

12

In this chapter the concept of the definite integral is extended to include double and triple integrals. For each of these terms, the calculation of the integral value is reduced to an ordinary integral. First, double integrals are introduced in Section 12.1, e.g. to describe volumes, center of mass for planar surfaces, and surface moments. Then, in Section 12.2, the procedure is transferred to triple integrals to calculate the center of mass and moments of inertia of bodies.

12

12 Integral Calculus for Multi-Variable Functions

In this chapter the concept of the definite integral is extended to include double and triple integrals. For each of these terms, the calculation of the integral value is reduced to an ordinary integral. First, double integrals are introduced in Section 12.1, e.g. to describe volumes, center of mass for planar surfaces, and surface moments. Then, in Section 12.2, the procedure is transferred to triple integrals to calculate the center of mass and moments of inertia of bodies.

12.1 Double Integrals (Domain Integrals)

This section deals with two-dimensional integrals. The transition from the one-dimensional to the multi-dimensional case is not difficult in principle, but it is more complex in its construction, since it is not an interval but a two-dimensional domain that has to be described.

12.1.1 Definition

The definite integral of a positive function $f(x)$ in the interval $[a, b]$ is the area enclosing the curve $f(x)$ with the x-axis in $[a, b]$. The definite integral

$$\int_a^b f(x)\ dx$$

is constructed as the limit of the sum over all rectangles $(x_k - x_{k-1})\, f(\xi_k)$, where $a = x_0 < x_1 < \cdots < x_{n-1} < x_n = b$ is an interval subdivision with widths $\triangle x_k = (x_k - x_{k-1})$. For $n \to \infty$ the widths tend to zero (see Fig. 12.1) and the sum becomes the integral.

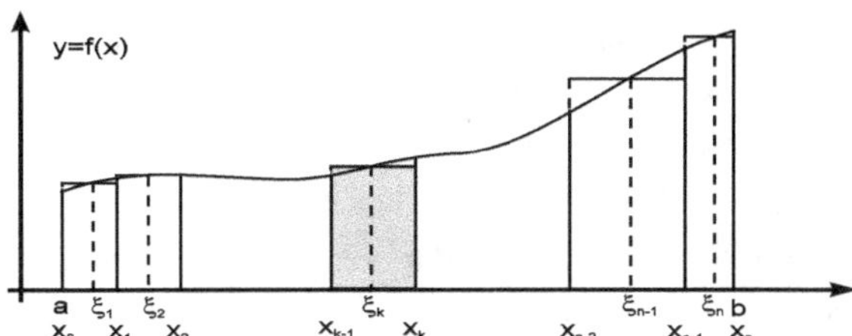

Figure 12.1. Sum of the rectangular areas

To apply this integral concept to two-dimensional domains, it is quite sufficient for our purposes to consider only a bounded, simply connected domain $G \subset \mathbb{R}^2$ in the (x, y)-plane, which has a smooth edge. In the following we will call G a *region or domain* and assume that $z = f(x, y)$ is a continuous, positive function in the region G. We want to find the volume between the functional surface $z = f(x, y)$ and G:

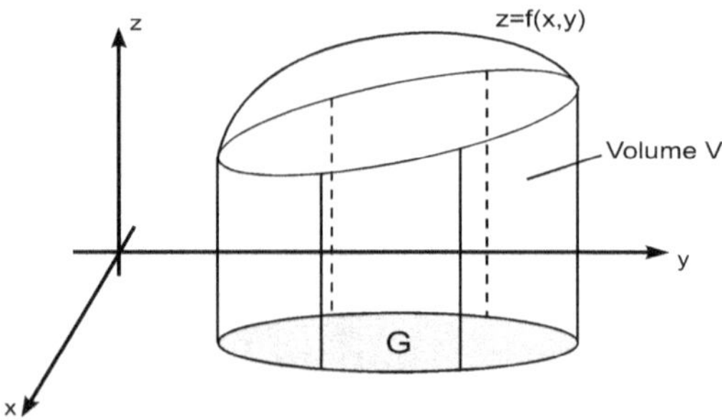

Figure 12.2. Function $f(x, y)$ over a two-dimensional region G

To find the volume V, the region G is divided into rectangular areas of lengths $\triangle x_i$ and widths $\triangle y_j$ (see Fig. 12.3). The number of subdivisions in the x-direction is n, the number of subdivisions in the y-direction is m. So i varies between 1 and n, j varies between 1 and m. For each rectangle with index (i, j), we select a point $P(\xi_i, \eta_j)$ and determine the corresponding function value $f(\xi_i, \eta_j)$. The volume of the cylinder above the rectangle is the base area $\Delta x_i \Delta y_j$ times the height $f(\xi_i, \eta_j)$:

$$V_{ij} = \Delta x_i \Delta y_j f(\xi_i, \eta_j).$$

Then the sum of all cylinder volumes over the region G is formed. (For rectangles outside G, the volume is set to zero.) The subtotal of all cylinder volumes is an approximation of the volume to be calculated

$$V \approx \sum_{j=1}^{m} \sum_{i=1}^{n} f(\xi_i, \eta_j) \, \Delta x_i \, \Delta y_j .$$

The finer the subdivision of the G domain, the more accurate the approximation. Therefore, we increase the number of subdivisions in the x- and y-directions. For $n, m \to \infty$, the subtotal tends to a limit, which is called the *Double Integral* or *Domain Integral.*

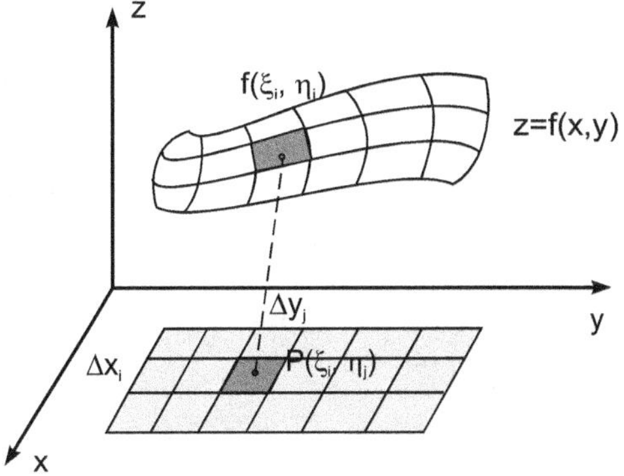

Figure 12.3. Construction of the double integral

Definition: (Double Integral, Domain Integral). *The limit value*

$$\iint\limits_{(G)} f\left(x,\,y\right)\,dG := \lim_{m\to\infty}\,\lim_{n\to\infty}\,\sum_{j=1}^{m}\sum_{i=1}^{n} f\left(\xi_i,\,\eta_j\right)\,\Delta x_i\,\Delta y_j$$

is called (if it exists) the Double Integral or Domain Integral of f over G. f is then called integratable over G.

Remarks:

(1) Also the notation $\displaystyle\int\limits_{(G)} f\left(x,y\right)\,dG$ is used with only one integral sign.

(2) We call

$(x,\,y)$	the integration variables
$f\left(x,\,y\right)$	the integrand
dG	the domain element
(G)	the integration domain.

(3) For continuous functions, the double integral is always defined.

(4) If the limit value exists, it is independent of the special domain decomposition.

(5) For this algebraic definition of the double integral $f\left(x,\,y\right) \geq 0$ is not needed, but for the geometric interpretation as a volume over the graph of f with G.

12.1.2 Calculating Double Integrals

To calculate double integrals $\iint\limits_{(G)} f\left(x,\,y\right)\,dG$, we split them into two successive ordinary integrals. To do this, the region G is covered by an axis-parallel rectangular grid with mesh sizes $\triangle x$ and $\triangle y$, and the sum is calculated over all cylinders whose base region is inside G:

$$V \approx \sum_{i=1}^{n} \left(\sum_{j=1}^{m} f\left(\xi_i,\,\eta_j\right) \triangle y \right) \triangle x \ .$$

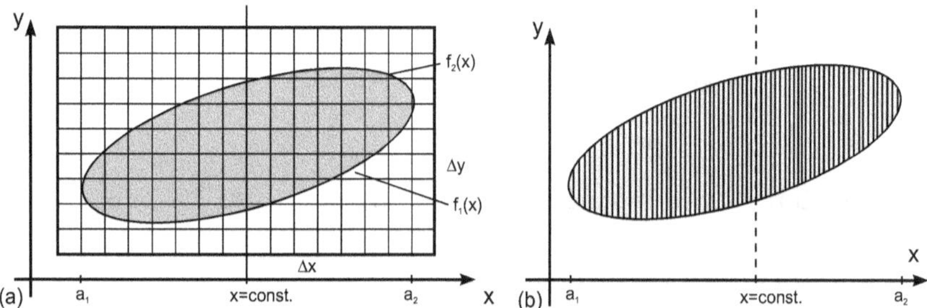

Figure 12.4. Calculation of the double integral using a decomposition in y-direction.

We decompose the domain G into strips parallel to the y-axis and represent the lower and upper limit of a stripe by functions $f_1\left(x\right)$ and $f_2\left(x\right)$. For fixed $\xi_i \in \left[x_{i-1},\, x_i\right]$, the inner sum converges for $\triangle y \to 0$ $(m \to \infty)$ to the integral

$$I_y\left(\xi_i\right) = \int_{f_1(\xi_i)}^{f_2(\xi_i)} f\left(\xi_i,\,y\right)\,dy \ .$$

This is a definite integral with the integration variable y. For fixed ξ_i, we integrate along the y-axis from $f_1\left(\xi_i\right)$ to $f_2\left(\xi_i\right)$

$$V \approx \lim_{m \to \infty} \sum_{i=1}^{n} \left(\sum_{j=1}^{m} f\left(\xi_i,\,\eta_j\right) \triangle y \right) \triangle x$$

$$= \sum_{i=1}^{n} I_y\left(\xi_i\right) \triangle x.$$

As $\triangle x \to 0$ for $n \to \infty$ the remaining sum converges to the integral

$$\lim_{n \to \infty} \sum_{i=1}^{n} I_y\left(\xi_i\right) \triangle x = \int_{a_1}^{a_2} I_y\left(x\right)\,dx \ .$$

This result is the double limit $n \to \infty$, $m \to \infty$, which is exactly the double integral of f over the domain G

$$\Rightarrow \iint\limits_{(G)} f\,(x,\,y)\ dx\,dy = \int_{x=a_1}^{a_2} \left(\int_{y=f_1(x)}^{f_2(x)} f\,(x,\,y)\ dy \right) dx.$$

Calculating Double Integrals

The calculation of the double integral $\iint\limits_{(G)} f\,(x,\,y)\ dG$ is done by two ordinary integrations, which are performed one after the other

$$\iint\limits_{(G)} f\,(x,\,y)\ dG = \int_{x=a_1}^{a_2} \underbrace{\left(\int_{y=f_1(x)}^{f_2(x)} f\,(x,\,y)\ dy \right)}_{x=const,\ \text{integration over } y}\ dx. \qquad \text{(D1)}$$

(1) The inner integral is integrated for fixed x with respect to y. The limits are $y_1 = f_1\,(x)$ and $y_2 = f_2\,(x)$. The result of the first integration is a function of the variable x only.

(2) The outer integral is obtained by integrating over x.

Remarks:

(1) The integration formula $(D1)$ corresponds to the decomposition of the area into strips parallel to the y-axis and and then summing all the strips in the x-direction (see Fig. 12.4 b). The individual strips parallel to the y-axis start at y_1 and end at y_2, which in turn depend on the position of the strip, i.e. on the x-coordinate: $y_1 = f_1(x)$ and $y_2 = f_2(x)$. The sum is made over all the strips between $x_{\min} = a_1$ and $x_{\max} = a_2$.

(2) If the roles of x and y are swapped, the left limit is defined by a function $g_1\,(y)$ and the right limit by $g_2\,(y)$. So the double integral is

$$\iint\limits_{(G)} f\,(x,\,y)\ dG = \int_{y=b_1}^{b_2} \underbrace{\left(\int_{x=g_1(y)}^{g_2(y)} f\,(x,\,y)\ dx \right)}_{y=const,\ \text{integration over } x}\ dy. \qquad \text{(D2)}$$

The inner integral is integrated for fixed y from $g_1(y)$ to $g_2(y)$ with respect to x. The result contains only the variable y. The outer integral is obtained by integrating over y (see Fig. 12.5, right).

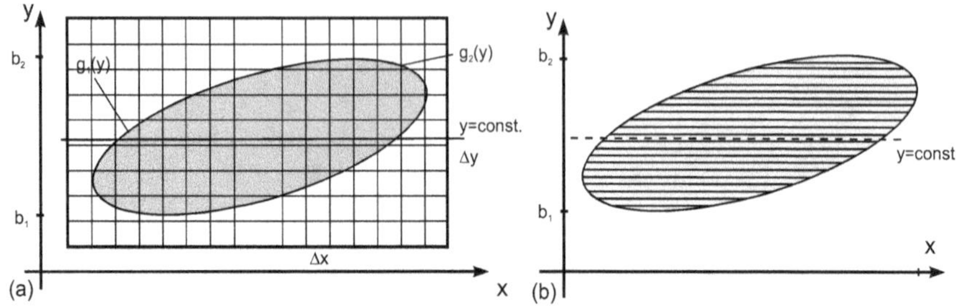

Figure 12.5. Calculation of the double integral using a decomposition in y-direction.

(3) Graphically, the integration formula $(D2)$ corresponds to a decomposition of the area into strips parallel to the x-axis and then summing all the strips in the y-direction. Here, the start and end points of the strips depend on y: $g_1(y)$ and $g_2(y)$. The summation takes into account all the strips from $y_{\min} = b_1$ to $y_{\max} = b_2$.

(4) The order of integration is determined by the order of the differentials dx and dy from inside to outside.

(5) For the representation of the double integral by two ordinary integrals it is assumed that the limits can be described by two functions $f_1(x)$ and $f_2(x)$ or $g_1(y)$ and $g_2(y)$. For complicated domains, G must be divided into appropriate subregions (see Fig. 12.6, left).

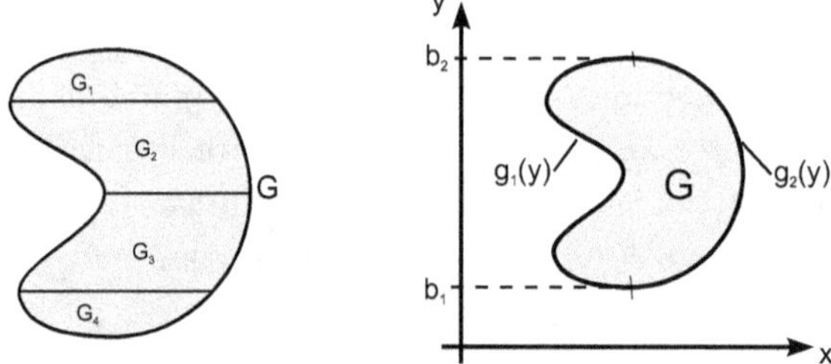

Figure 12.6. Subdivision into subregions

(6) The problem in calculating double integrals is to find the functions $f_1(x)$ and $f_2(x)$ or $g_1(y)$ and $g_2(y)$. Once these functions describing the limits have been found, the calculation is reduced to two ordinary integrals.

Examples 12.1 (Integrals with Specified Limits):

① The double integral

$$I = \int_{y=0}^{1} \int_{x=-2}^{y} x\,y \, dx \, dy$$

is determined by first specifying the inner integral (y fixed, x varying from $x = -2$ to y) and integrating along the variable x.

$$\int_{x=-2}^{y} x\,y \, dx = y \int_{x=-2}^{y} x \, dx = y \cdot \left[\frac{x^2}{2}\right]_{-2}^{y} = y\left[\frac{y^2}{2} - 2\right] = \frac{1}{2}y^3 - 2y$$

Then, the outer integration is performed on y

$$I = \int_{y=0}^{1} \left(\frac{1}{2} y^3 - 2\,y\right) dy = \left[\frac{1}{8} y^4 - y^2\right]_{0}^{1} = -\frac{7}{8} \,.$$

② To find the double integral

$$I = \int_{x=0}^{3} \int_{y=0}^{\pi} x^2 \sin{(y)} \, dy \, dx,$$

we start with the inner integral (x fixed, y varies from 0 to π)

$$\int_{y=0}^{\pi} x^2 \sin{(y)} \, dy = x^2 \int_{y=0}^{\pi} \sin{(y)} \, dy = x^2 \left[- \cos{(y)}\right]_0^{\pi} = 2\,x^2 \,,$$

then the outer integral is calculated

$$I = \int_{x=0}^{3} 2\,x^2 \, dx = \frac{2}{3} x^3 \Big|_{0}^{3} = 18. \qquad \square$$

12.1.3 Domain Decomposition

In the first examples on double integrals, the integration limits are already specified. In the next examples we start with an image of the domain G. Then, we have to decompose the domain into strips parallel to the x-axis or in strips parallel to the y-axis. This decomposition defines the inner and outer integration variables.

Example 12.2 (Domain Decomposition).

Given is a function $z = f(x, y)$, which is defined in the domain G (see Fig. 12.7). Find $\iint\limits_{(G)} f(x, y)\, dG$.

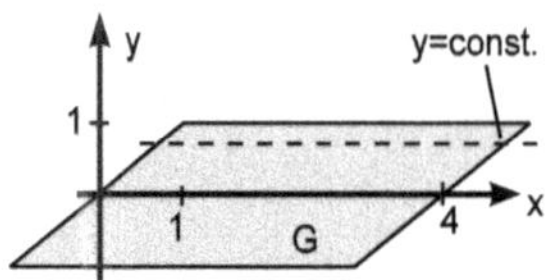

Figure 12.7. Domain G

To reduce the double integral to two simple integrals, we divide the domain G into strips parallel to the x-axis. The strips start at $x = y$ and end at $x = y + 4$. We have to consider all the strips from $y = -1$ to $y = 1$. This means that the integral formula $(D2)$ is chosen and x is set as the inner integration variable. So we keep $y = const$ and vary x between the values $g_1(y) = y$ and $g_2(y) = y + 4$ (see dotted line). The outer integral varies between $y = -1$ and $y = 1$.

$$\Rightarrow \iint\limits_{(G)} f(x, y)\, dG = \int_{y=-1}^{1} \underbrace{\left(\int_{x=y}^{y+4} f(x, y)\, dx \right)}_{y=const} dy \, . \qquad \square$$

Example 12.3 (With Domain Decomposition).

Find the double integral $\iint\limits_{(G)} f(x, y)\, dG$, if the domain G is a circle in the (x, y)-plane with radius R.

(i) We divide the domain into strips parallel to the x-axis: If the integration is performed with the formula $(D2)$, the inner integration for constant y is performed via the variable x.

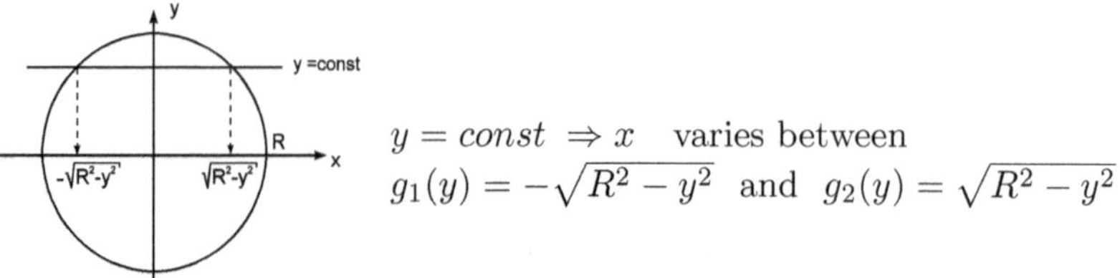

$y = const \Rightarrow x$ varies between
$$g_1(y) = -\sqrt{R^2 - y^2} \quad \text{and} \quad g_2(y) = \sqrt{R^2 - y^2}$$

Thus, for fixed y, the x-values vary between the lower limit $g_1(y) = -\sqrt{R^2 - y^2}$ and the upper limit $g_2(y) = \sqrt{R^2 - y^2}$. To specify the outer integral, y varies between $b_1 = -R$ and $b_2 = R$.

$$\Rightarrow \iint\limits_{(G)} f(x, y)\, dG = \int_{y=-R}^{R} \left(\int_{x=-\sqrt{R^2-y^2}}^{\sqrt{R^2-y^2}} f(x, y)\, dx \right) dy \, .$$

(ii) We divide the domain into strips parallel to the y-axis: If the integration is performed according to the formula $(D1)$, the inner integration for constant x is performed via the variable y.

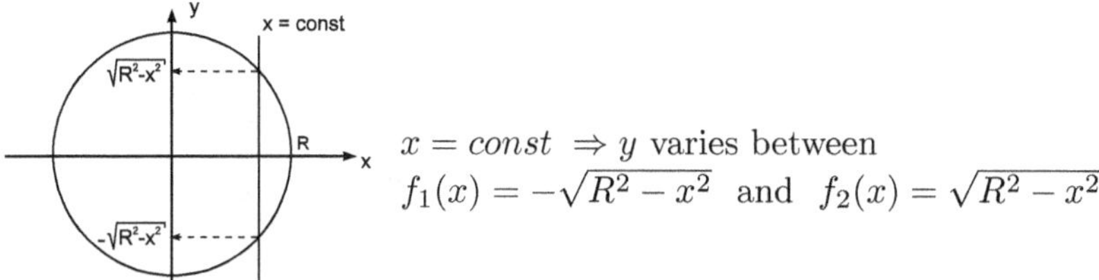

$x = const \Rightarrow y$ varies between
$$f_1(x) = -\sqrt{R^2 - x^2} \quad \text{and} \quad f_2(x) = \sqrt{R^2 - x^2}$$

For fixed x, the y values now vary between $f_1(x) = -\sqrt{R^2 - x^2}$ and $f_2(x) = \sqrt{R^2 - x^2}$. For the outer integral, x must vary between $-R$ and R

$$\Rightarrow \iint\limits_{(G)} f(x,\, y)\, dG = \int_{x=-R}^{R} \left(\int_{y=-\sqrt{R^2-x^2}}^{\sqrt{R^2-x^2}} f(x,\, y)\, dy \right) dx \; . \qquad \square$$

12.1.4 Applications

Double integrals are often used to calculate plane surfaces, to find the center of mass of plane surfaces, to find surface moments and to calculate volumes. Examples of each of these problems will be discussed below.

⊘ Calculating Areas

If we choose the function $f(x,\, y) = 1$, the double integral over G corresponds to the volume of the body with the base area G and the constant height 1. Numerically, this is exactly the area of G:

> **Area Calculation**
>
> ---
>
> The **Area** A of a flat region $G \subset \mathbb{R}^2$ is
>
> $$A = \iint\limits_{(G)} dG \, .$$

Example 12.4. Find the area between the straight line $f_1(x) = x + 2$ and the parabola $f_2(x) = 4 - x^2$.

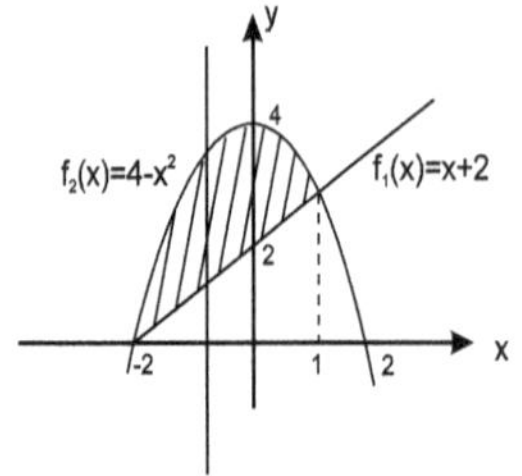

Figure 12.8. Area

To calculate the double integral, we divide the area into strips parallel to the y-axis; we therefore choose the formula $(D1)$: For fixed x, y varies from $f_1(x) = x + 2$ to $f_2(x) = 4 - x^2$; the remaining outer integral with respect to x is formed with the limits $a_1 = -2$ and $a_2 = +1$:

$$A = \iint\limits_{(G)} dG = \int_{x=-2}^{1} \underbrace{\left(\int_{y=x+2}^{4-x^2} dy \right)}_{x\ \text{const}} dx = \int_{x=-2}^{1} \Big[y \Big]_{x+2}^{4-x^2} dx$$

$$= \int_{x=-2}^{1} (-x^2 - x + 2)\, dx = \left[-\frac{1}{3} x^3 - \frac{1}{2} x^2 + 2x \right]_{-2}^{1} = 4.5. \qquad \square$$

Example 12.5. Search for the area of the region G shown in Fig. 12.9.

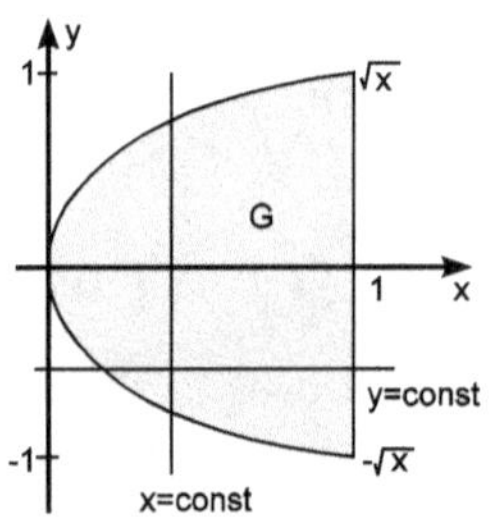

Figure 12.9.

(i) Dividing the domain into strips parallel to the y-axis: For $x = const$, y varies from $-\sqrt{x}$ to $\sqrt{x}$. With the formula $(D1)$ we get

$$A = \iint\limits_{(G)} dG = \int_{x=0}^{1} \underbrace{\left(\int_{y=-\sqrt{x}}^{\sqrt{x}} dy \right)}_{x=const} dx$$

$$= \int_{0}^{1} 2\sqrt{x}\, dx = \int_{0}^{1} 2 x^{\frac{1}{2}}\, dx = \frac{4}{3} \left[x^{\frac{3}{2}} \right]_{0}^{1} = \frac{4}{3}.$$

(ii) Dividing of the domain into strips parallel to the x-axis: With the formula $(D2)$ we get $(y = const \Rightarrow x$ varies from y^2 to $1)$

$$A = \iint\limits_{(G)} dG = \int_{y=-1}^{1} \underbrace{\left(\int_{x=y^2}^{1} dx \right)}_{y=const} dy$$

$$= \int_{y=-1}^{1} (1 - y^2)\, dy = \left[y - \frac{1}{3} y^3 \right]_{-1}^{1} = \frac{4}{3}. \qquad \square$$

⊙ Calculating the Center of Gravity for Plane Surfaces

In Section 12.2.2 we summarize formulas that determine the center of gravity coordinates of a surface under a graph f and the x-axis. For plane surfaces the following applies:

Center of Gravity for a Plane Surface

The coordinates of the center of gravity $S = (x_s, y_s)$ for a plane domain G are given by

$$x_s = \frac{1}{A} \iint\limits_{(G)} x\, dG, \qquad y_s = \frac{1}{A} \iint\limits_{(G)} y\, dG,$$

where A is the area of G.

Example 12.6. Given is the area from Example 12.5. We look for its center of gravity.

For the x-coordinate of S we calculate

$$x_s = \frac{1}{A} \iint\limits_{(G)} x\, dG = \frac{1}{A} \int_{x=0}^{1} \underbrace{\left(\int_{y=-\sqrt{x}}^{\sqrt{x}} x\, dy \right)}_{x=const} dx \,.$$

The inner integral over y is for fixed x

$$\int_{y=-\sqrt{x}}^{\sqrt{x}} x\, dy = x \int_{y=-\sqrt{x}}^{\sqrt{x}} dy = 2\,x\,x^{\frac{1}{2}} = 2\,x^{\frac{3}{2}}.$$

This gives for the outer integral

$$\int_{x=0}^{1} 2\,x^{\frac{3}{2}}\, dx = \frac{4}{5}\,x^{\frac{5}{2}}\Big|_{0}^{1} = \frac{4}{5} \curvearrowright x_s = \frac{1}{A}\cdot\frac{4}{5} = \frac{3}{5}\,.$$

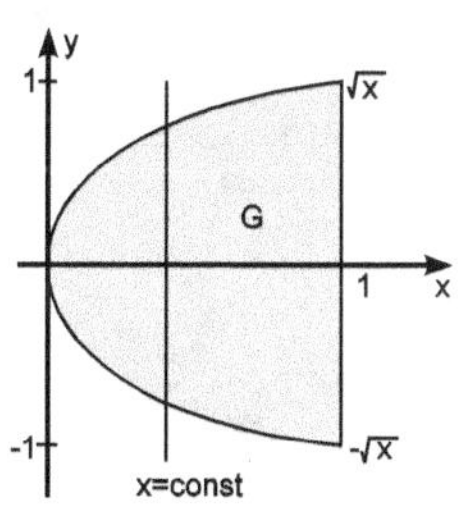

Due to the symmetry, $y_s = 0$.

$$\Rightarrow S = \left(\frac{3}{5}, 0 \right).$$

□

Example 12.7 (With Maple-Worksheet). Find the center of gravity coordinates for the domain from Example 12.4: With Example 12.4 the x-coordinate of the center of gravity is

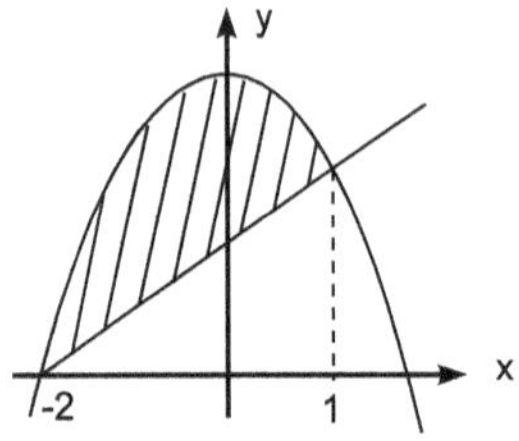

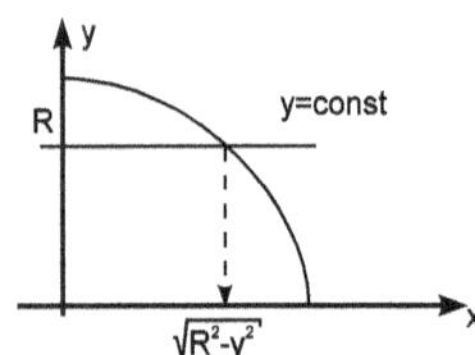

Figure 12.10.

$$x_s = \frac{1}{A} \iint\limits_{(G)} x \, dG$$

$$= \frac{1}{4.5} \int_{x=-2}^{1} \left(\int_{y=x+2}^{4-x^2} x \, dy \right) dx$$

$$x_s = \frac{2}{9} \int_{x=-2}^{1} x \left(4 - x^2 - x - 2\right) dx = \frac{2}{9} \int_{x=-2}^{1} \left(-x^3 - x^2 + 2x\right) dx$$

$$= \frac{2}{9} \left[-\frac{1}{4}x^4 - \frac{1}{3}x^3 + x^2 \right]_{-2}^{1} = -\frac{1}{2}.$$

Accordingly, we get for the y-component $y_s = 2.4$. $\square$

Example 12.8 (With Maple-Worksheet).

Figure 12.11.

The center of gravity coordinates of the quarter circle. To calculate the double integral we choose the formula $(D2)$: Here the inner integration is performed with constant y over the variable x from $g_1(y) = 0$ to $g_2(y) = \sqrt{R^2 - y^2}$. To find the outer integral over y, the variable y then varies from $0 \le y \le R$:

$$x_s = \frac{1}{A} \iint\limits_{(G)} x \, dG = \frac{1}{A} \int_{y=0}^{R} \left(\int_{x=0}^{\sqrt{R^2-y^2}} x \, dx \right) dy = \frac{1}{A} \int_{y=0}^{R} \frac{1}{2} \left(R^2 - y^2\right) dy;$$

$$y_s = \frac{1}{A} \iint\limits_{(G)} y \, dG = \frac{1}{A} \int_{y=0}^{R} \left(\int_{x=0}^{\sqrt{R^2-y^2}} y \, dx \right) dy = \frac{1}{A} \int_{y=0}^{R} y \sqrt{R^2 - y^2} \, dy.$$

We evaluate the integrals, taking into account that $A = \frac{\pi R^2}{4}$. The coordinates of the center of gravity are:

$$x_s = \frac{4}{3} \cdot \frac{R}{\pi} \quad \text{and} \quad y_s = \frac{4}{3} \cdot \frac{R}{\pi}$$ $\square$

⟩ Moments of Area

In mechanics, the description of bends requires the **moment of area** of plane surfaces. These moments are related to specific axes. A distinction is made between so-called *axial* moments, where the reference axis is in the plane of the surface and *polar* moments, where the axis is perpendicular to the plane of the surface. The following formulas apply

---

Moments of Area

Axial moment of area for the y-axis: $\displaystyle I_y = \iint\limits_{(G)} x^2 \, dG$

Axial moment of area for the x-axis: $\displaystyle I_x = \iint\limits_{(G)} y^2 \, dG$

Polar moment of area: $\displaystyle I_p = \iint\limits_{(G)} \left(x^2 + y^2\right) dG$

Example 12.9. Find the axial surface moments I_x and I_y of area G from Example 12.5.

The only difference in the calculation of the axial surface moments I_x and I_y is the function to be integrated. For I_x we need to use x^2 and for I_y we need to use y^2. Using the domain decomposition from Example 12.5, we calculate

$$
I_x = \iint\limits_{(G)} y^2 \, dG = \int_0^1 \left(\int_{-\sqrt{x}}^{\sqrt{x}} y^2 \, dy \right) dx
$$

$$
= \int_0^1 \left[\frac{1}{3} y^3 \right]_{-\sqrt{x}}^{\sqrt{x}} dx = \frac{2}{3} \int_0^1 x^{\frac{3}{2}} \, dx
$$

$$
= \frac{2}{3} \cdot \frac{2}{5} \left[x^{\frac{5}{2}} \right]_0^1 = \frac{4}{15}.
$$

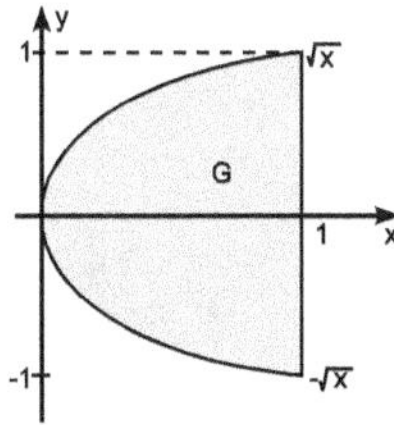

Figure 12.12.

$$
I_y = \iint\limits_{(G)} x^2 \, dG = \int_0^1 \left(\int_{-\sqrt{x}}^{\sqrt{x}} x^2 \, dy \right) dx = \int_0^1 x^2 \left(\int_{-\sqrt{x}}^{\sqrt{x}} dy \right) dx
$$

$$
= 2 \int_0^1 x^2 \, x^{\frac{1}{2}} \, dx = 2 \int_0^1 x^{\frac{5}{2}} \, dx = 2 \cdot \frac{2}{7} \left[x^{\frac{7}{2}} \right]_0^1 = \frac{4}{7}. \qquad \square
$$

⊙ Volume Calculation

By definition, the double integral is used to compute volumes, which covers a function graph of a positive function $z = f(x, y) > 0$ with the (x, y)-plane over a region G

$$V = \iint\limits_{(G)} f(x, y)\, dG.$$

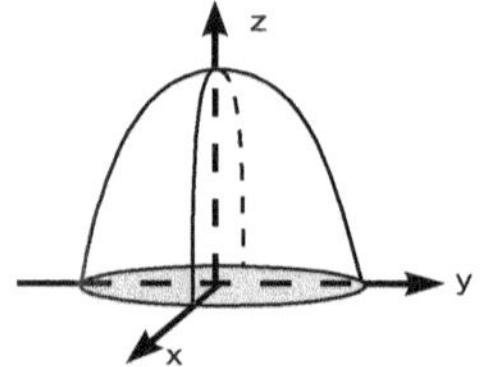

Figure 12.13.

Example 12.10 (With Maple-Worksheet).
Find the volume V covered by the graph

$$z = f(x, y) = 1 - x^2 - y^2$$

and the (x, y)-plane. The associated region G is the unit circle in the (x, y)-plane.

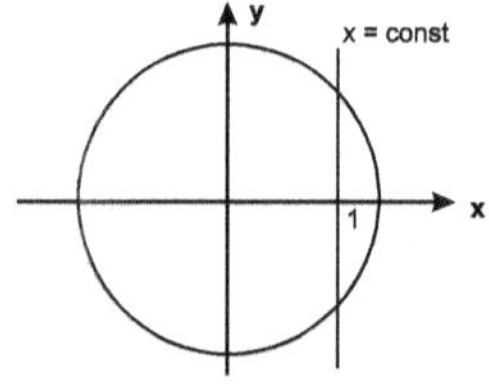

Figure 12.14.

We select strips parallel to the y-axis starting at $f_1(x) = -\sqrt{1 - x^2}$ and ending at $f_2(x) = \sqrt{1 - x^2}$. The limits for the x integration are $a_1 = -1$ and $a_2 = 1$. The integration formula $(D1)$ for double integrals is

$$V = \iint\limits_{(G)} f(x, y)\, dG$$

$$= \int_{x=-1}^{1} \left(\int_{y=-\sqrt{1-x^2}}^{\sqrt{1-x^2}} (1 - x^2 - y^2)\, dy \right) dx$$

$$= \int_{x=-1}^{1} \left[(1 - x^2)\, y - \frac{1}{3} y^3 \right]_{y=-\sqrt{1-x^2}}^{\sqrt{1-x^2}} dx$$

$$= \int_{x=-1}^{1} \frac{4}{3} (1 - x^2)^{3/2}\, dx$$

$$= \left[\frac{1}{3} (1 - x^2)^{3/2} + \frac{1}{2} (1 - x^2)^{1/2} + \frac{1}{2} \arcsin(x) \right]_{x=-1}^{1}$$

$$= \frac{1}{2} \pi. \qquad \qquad \square$$

12.2 Triple Integrals

The term *triple integral* for functions $f(x, y, z)$ over a three-dimensional domain $G \subset \mathbb{R}^3$ is introduced and calculated in a similar way to the double integral. Since a function of three variables cannot be represented graphically, the triple integral over a function $f(x, y, z)$ has no geometric meaning at first. Only for the case $f(x, y, z) = 1$, the triple integral corresponds to the volume of the domain G. Triple integrals are also reduced to 3 successive ordinary integrations, where $3! = 6$ different integration sequences are now possible!

12.2.1 Definition and Calculation of Triple Integrals

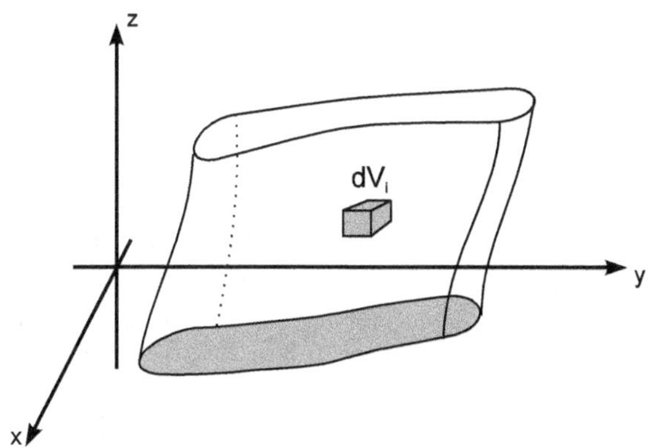

Figure 12.15. Volume element for defining triple integrals

To define the triple integral, the body is divided into small partial volumes dV_i $(i = 1, \ldots, n)$ and in each volume a point $P(x_i, y_i, z_i)$ is selected at which the function f is evaluated. Finally, the product of the function value and the volume element

$$f(x_i, y_i, z_i)\, dV_i$$

is calculated and summed over all volumes to

$$Z_n = \sum_{i=1}^{n} f(x_i, y_i, z_i)\, dV_i \ . \qquad \text{(Subtotal)}$$

The number of subtotals increases (at the same time as $dV_i \to 0$). When the subtotal Z_n for $n \to \infty$ reaches a limit value, then we call it a *Triple Integral* or a *Three-Dimensional Integral*.

Definition: (Triple Integral; Three-Dimensional Integral).
Let $G \subset \mathbb{R}^3$ and $f : G \to \mathbb{R}$ be continuous. The limit value

$$\iiint\limits_{(G)} f(x, y, z)\, dG = \lim_{n \to \infty} \sum_{i=1}^{n} f(x_i, y_i, z_i)\, dV_i$$

is called the **Triple Integral.**

The calculation of triple integrals is based on simple integrations, although the description of the integration limits is of course more complicated than for double integrals. The method of calculation is illustrated in the following example.

Example 12.11.

$$I = \iiint\limits_{(G)} f(x, y, z)\, dG = \int_0^1 \int_0^x \int_{-y^2}^{x^2} (1 + x)\, dz\, dy\, dx \ .$$

The order of integration is determined by the order of the differentials dz, dy, dx from the inside to the outside. The innermost integral is integrated over the variable z. The other variables x and y are kept constant.

$$I_1 = \int_{z=-y^2}^{z=x^2} (1 + x)\, dz = \Big[(1 + x)\, z\Big]_{z=-y^2}^{z=x^2}$$
$$= (1 + x)\, x^2 - (1 + x)\left(-y^2\right) \ .$$

I_1 no longer contains the variable z, but only x and y. The remaining double integral

$$I = \int_0^1 \int_0^x \left[(1 + x)\, x^2 + (1 + x)\, y^2\right] dy\, dx$$

is calculated by first taking the innermost integral over y

$$I_2 = \int_{y=0}^{y=x} \left[(1 + x)\, x^2 + (1 + x)\, y^2\right] dy$$
$$= \left[(1 + x)\, x^2\, y + \frac{1}{3}\, (1 + x)\, y^3\right]_{y=0}^{y=x} = \frac{4}{3}\, x^3 + \frac{4}{3}\, x^4.$$

I_2 contains only the variable x used in the last integration

$$I = \int_{x=0}^{x=1} \left(\frac{4}{3}\, x^3 + \frac{4}{3}\, x^4\right) dx = \left[\frac{1}{3}\, x^4 + \frac{4}{15}\, x^5\right]_0^1 = \frac{3}{5} \ . \qquad \square$$

12.2.2 Applications

The application examples are intended to illustrate the calculation of triple integrals. Usually the volumes, masses, center of mass coordinates and moments of inertia of bodies are to be calculated using triple integrals. In special cases, the calculation can be greatly simplified by introducing an adapted coordinate system; in some cases, this is only possible with special coordinate systems.

The most important systems for the applications are polar, cylindrical and spherical coordinates.

(1) **Polar Coordinates:** For polar coordinates, a point in the (x, y)-plane is uniquely specified by the angle φ, $0 \le \varphi < 2\pi$, and the radius $r \ge 0$. The transformation equations are

$$x = r \cos\varphi , \qquad y = r \sin\varphi .$$

A double integral in polar coordinates is

$$\iint\limits_{(x,y)} f(x,\, y)\, dx\, dy = \iint\limits_{(r,\varphi)} f(r\cos\varphi,\, r\sin\varphi)\, r\, dr\, d\varphi.$$

(2) **Cylindrical Coordinates:** A point in $(x,\, y,\, z)$-space is uniquely defined by the specification of its polar coordinates $(r,\, \varphi)$ in the $(x,\, y)$-plane and its z-component. A triple integral in cylindrical coordinates is

$$\iiint\limits_{(x,y,z)} f(x,\, y,\, z)\, dx\, dy\, dz = \iiint\limits_{(r,\varphi,z)} f(r\cos\varphi,\, r\sin\varphi,\, z)\, r\, dr\, d\varphi\, dz.$$

(3) **Spherical Coordinates:** By specifying two angles φ and ϑ, and the distance from the origin, any point in $\mathbb{R}^3$ is well-defined:

$$x = r \cos\varphi \cos\vartheta, \; y = r \sin\varphi \cos\vartheta, \; z = r \sin\vartheta.$$

A triple integral in spherical coordinates is

Triple Integral in Spherical Coordinates

$$\iiint\limits_{(x,y,z)} f\left(x,\, y,\, z\right)\, dx\, dy\, dz =$$

$$\iiint\limits_{(r,\varphi,\vartheta)} f\left(r\cos\varphi\cos\vartheta,\, r\sin\varphi\cos\vartheta,\, r\sin\vartheta\right)\cdot r^2\cos\vartheta\, dr\, d\varphi\, d\vartheta.$$

The following is a summary of the most important formulas from the physics of rigid bodies for volume, mass, center of gravity coordinates and moments of inertia.

Volume and Mass of a Body

If $K \subset \mathbb{R}^3$ is a rigid body with a location dependent density $\rho = \rho\left(x,\, y,\, z\right)$, then

$$V = \iiint\limits_{(K)} dx\, dy\, dz \qquad \text{Volume of a body and}$$

$$M = \iiint\limits_{(K)} \rho\left(x,\, y,\, z\right)\, dx\, dy\, dz \quad \text{Mass of a body.}$$

Center of Mass

The **coordinates of the center of gravity** $S\left(x_s,\, y_s,\, z_s\right)$ are

$$x_s = \frac{1}{M}\iiint\limits_{(K)} x\cdot\rho\left(x,\, y,\, z\right)\, dx\, dy\, dz$$

$$y_s = \frac{1}{M}\iiint\limits_{(K)} y\cdot\rho\left(x,\, y,\, z\right)\, dx\, dy\, dz$$

$$z_s = \frac{1}{M}\iiint\limits_{(K)} z\cdot\rho\left(x,\, y,\, z\right)\, dx\, dy\, dz.$$

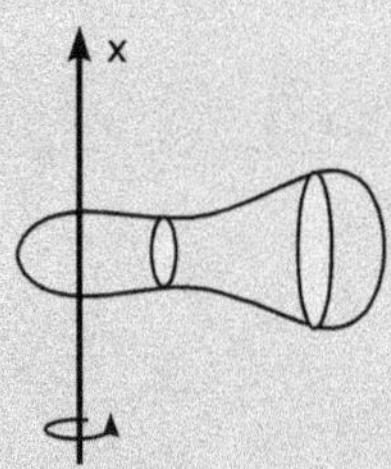

Figure 12.16.

When a rigid body rotates about the axis of rotation x, the integral

$$I_x = \iiint\limits_{(K)} \rho\,(x,\,y,\,z)\,\left(y^2 + z^2\right)\,dx\,dy\,dz$$

is called the **moment of inertia relative to the x-axis**. The moments of inertia with respect to the y and z axes are defined accordingly.

To find the moment of inertia of a body K with respect to any axis A, *Steiner's theorem* apply:

Steiner's Theorem/Parallel Axis Theorem

For an axis of rotation A parallel to the center of gravity axis S at a distance d, the following applies

$$I_A = I_S + M\,d^2\,,$$

where I_S is the moment of inertia with respect to the center of gravity axis and M is the mass of the body. According to this theorem, it is sufficient to consider only the axes through the center of gravity.

Application Example 12.12 (Volume Calculation).

We are looking for the volume of a body created by rotating x^2 around the z-axis. For the calculation we introduce cylindrical coordinates:

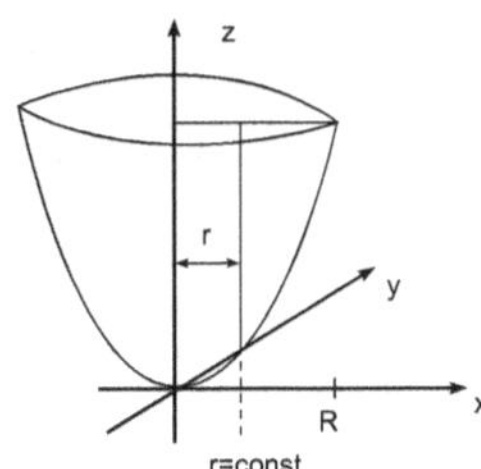

φ-Integration: $\varphi = 0$ to $\varphi = 2\pi$
                        independent on r and z.

z-Integration: for constant r
                        z from $z = r^2$ to $z = R^2$.

r-Integration: $r = 0$ to $r = R$.

$$V = \iiint\limits_{(K)} dx\,dy\,dz = \iiint\limits_{(K)} r\,d\varphi\,dr\,dz$$

$$= \int_{r=0}^{R} \int_{z=r^2}^{R^2} \int_{\varphi=0}^{2\pi} r\,d\varphi\,dz\,dr$$

$$= \int_{r=0}^{R} \int_{z=r^2}^{R^2} r\,2\pi\,dz\,dr = \int_{r=0}^{R} \left[r\,2\pi\,z\right]_{z=r^2}^{z=R^2} dr$$

$$= 2\pi \int_{r=0}^{R} \left(R^2\,r - r^3\right)\,dr = 2\pi \left[\frac{1}{2}\,R^2\,r^2 - \frac{1}{4}\,r^4\right]_{r=0}^{R} = \frac{\pi}{2}\,R^4.$$

Alternatively, the integration can also be performed slice by slice:

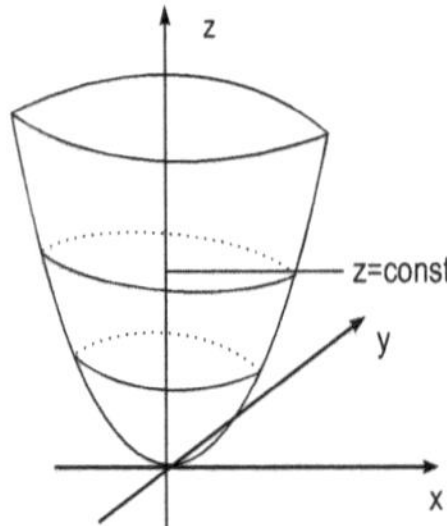

φ-Integration: $\varphi = 0$ to $\varphi = 2\pi$
independent of r and z.

r Integration: for constant z
r from $r = 0$ to $r = \sqrt{z}$.

z Integration: $z = 0$ to $z = R^2$.

$$\Rightarrow V = \int_{z=0}^{R^2} \int_{r=0}^{\sqrt{z}} \int_{\varphi=0}^{2\pi} r\,d\varphi\,dr\,dz = \frac{\pi}{2}\,R^4. \qquad \square$$

Application Example 12.13 (Center of Gravity Coordinates).

Find the coordinates of the center of mass of a sphere with density $\rho = 1$ and radius R.

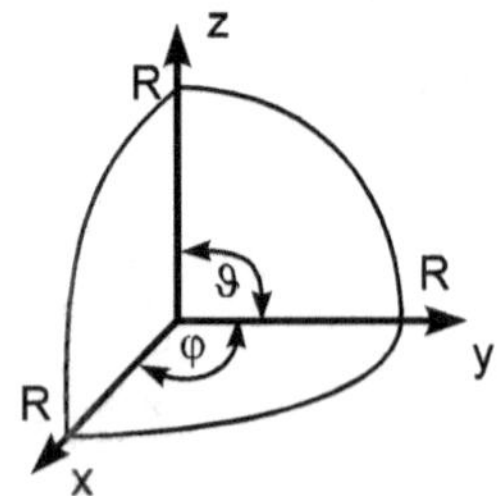

r-Integration: $r = 0$ to $r = R$
independent of ϑ and φ.

φ-Integration: $\varphi = 0$ to $\varphi = \frac{\pi}{2}$
independent of r and ϑ.

ϑ-Integration: $\vartheta = 0$ to $\vartheta = \frac{\pi}{2}$
independent of r and φ.

(i) Calculation of the volume in spherical coordinates:

$$V = \iiint\limits_{(K)} r^2 \cos\vartheta \, dr \, d\varphi \, d\vartheta$$

$$V = \int_{\vartheta=0}^{\pi/2} \int_{\varphi=0}^{\pi/2} \left(\int_{r=0}^{R} r^2 \cos\vartheta \, dr \right) d\varphi \, d\vartheta$$

$$= \int_{\vartheta=0}^{\pi/2} \cos\vartheta \int_{\varphi=0}^{\pi/2} \left[\frac{r^3}{3} \right]_{r=0}^{r=R} d\varphi \, d\vartheta = \frac{1}{3} R^3 \int_{\vartheta=0}^{\pi/2} \cos\vartheta \int_{\varphi=0}^{\pi/2} d\varphi \, d\vartheta$$

$$= \frac{1}{3} R^3 \int_{\vartheta=0}^{\pi/2} \cos\vartheta \cdot \frac{\pi}{2} \, d\vartheta = \frac{1}{6} R^3 \pi \int_{\vartheta=0}^{\pi/2} \cos\vartheta \, d\vartheta = \frac{1}{6} R^3 \pi.$$

(ii) Calculation of the center of gravity coordinate x_s:

$$x_s = \frac{1}{V} \iiint\limits_{(K)} x \, r^2 \cos\vartheta \, dr \, d\varphi \, d\vartheta$$

$$= \frac{1}{V} \int_{\vartheta=0}^{\pi/2} \int_{\varphi=0}^{\pi/2} \int_{r=0}^{R} r \cos\varphi \cos\vartheta \, r^2 \cos\vartheta \, dr \, d\varphi \, d\vartheta$$

$$= \frac{1}{V} \int_{\vartheta=0}^{\pi/2} \cos^2\vartheta \int_{\varphi=0}^{\pi/2} \cos\varphi \int_{r=0}^{R} r^3 \, dr \, d\varphi \, d\vartheta$$

$$= \frac{1}{V} \int_{\vartheta=0}^{\pi/2} \cos^2\vartheta \int_{\varphi=0}^{\pi/2} \cos\varphi \, \frac{1}{4} R^4 \, d\varphi \, d\vartheta = \frac{1}{V} \int_{\vartheta=0}^{\pi/2} \cos^2\vartheta \, \frac{1}{4} R^4 \, d\vartheta$$

$$= \frac{1}{V} \cdot \frac{1}{4} R^4 \left[\frac{1}{2}\vartheta + \frac{1}{2}\cos\vartheta \sin\vartheta \right]_0^{\pi/2} = \frac{1}{V} \cdot \frac{1}{16} R^4 \pi = \frac{3}{8} R.$$

(iii) Similarly, we calculate $y_s = z_s = \frac{3}{8} R$. $\square$

Application Example 12.14 (**Cube's Moment of Inertia**).

Find the moment of inertia of a homogeneous cube about the z-axis.

$$I_z = \iiint\limits_{(K)} (x^2 + y^2) \, \rho \, dx \, dy \, dz$$

$$= \rho \int_{x=0}^{x=l} \left(\int_{y=0}^{y=l} \left(\int_{z=0}^{z=l} (x^2 + y^2) \, dz \right) dy \right) dx$$

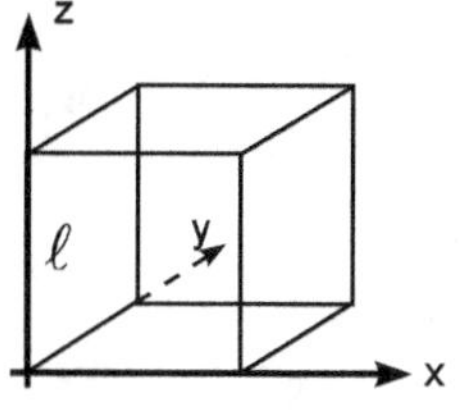

Figure 12.17. Cube

$$= \rho \int_{x=0}^{l} \int_{y=0}^{l} (x^2 + y^2) \left(\int_0^l dz \right) dy\, dx$$

$$= \rho\, l \int_{x=0}^{l} \int_{y=0}^{l} (x^2 + y^2)\, dy\, dx$$

$$= \rho\, l \int_{x=0}^{l} \left[x^2 y + \frac{y^3}{3} \right]_{y=0}^{l} dx = \rho\, l \int_{x=0}^{l} \left(x^2 l + \frac{1}{3} l^3 \right) dx$$

$$= \rho\, l \left[\frac{1}{3} x^3 l + \frac{1}{3} l^3 x \right]_0^l = \rho\, l \left(\frac{1}{3} l^4 + \frac{1}{3} l^4 \right) = \frac{2}{3} \rho\, l^5.$$

With the mass $M = \rho \cdot V = \rho \cdot l^3$ follows $\boxed{I_z = \tfrac{2}{3} M\, l^2}$.

Application Example 12.15 (Cylinder's Moment of Inertia).

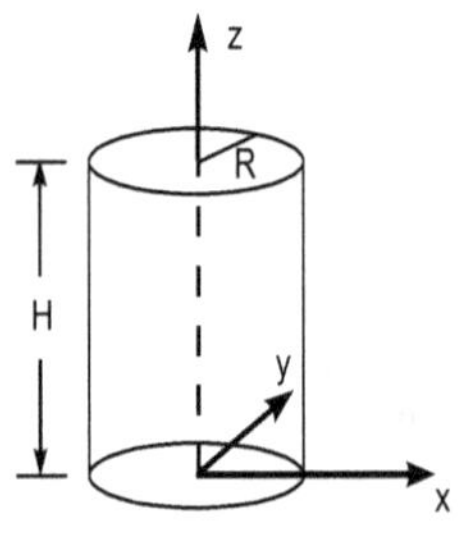

Figure 12.18. Cylinder

We need to find the moment of inertia of a cylinder of height H and base area πR^2 with respect to the z-axis. To calculate the moment of inertia, we use cylinder coordinates

$$I_z = \rho \iiint\limits_{(K)} (x^2 + y^2)\, dx\, dy\, dz$$

$$= \rho \iiint\limits_{(K)} r^2\, r\, dr\, d\varphi\, dz$$

$$= \rho \int_{z=0}^{H} \int_{\varphi=0}^{2\pi} \int_{r=0}^{R} r^3\, dr\, d\varphi\, dz = \rho\, \frac{R^4}{4} \cdot 2\pi \cdot H.$$

With $\rho = \frac{M}{V} = \frac{M}{\pi R^2 H}$ we get $\boxed{I_z = M\, \tfrac{1}{2} R^2}$. $\qquad\qquad\square$

12.3 Problems on Integral Calculus

12.1 Calculate the following double integrals

a) $\displaystyle \int_{x=0}^{1} \int_{y=1}^{l} \frac{x^2}{y}\, dy\, dx$ b) $\displaystyle \int_{x=0}^{3} \int_{y=0}^{1-x} \left(25 - x^2 - y^2\right) dy\, dx$

c) $\displaystyle \int_{y=0}^{\pi} \int_{x=\pi/2}^{y-1} \sin\left(x + y\right) dx\, dy$ d) $\displaystyle \int_{y=0}^{\pi} \int_{x=\pi}^{y} x \cdot \cos\left(x + y\right) dx\, dy$

12.2 Determine the double integral over the region G for the function $z = x - y$ by applying both integral formula $(D1)$ and $(D2)$:

$$I = \iint\limits_{G} (x - y)\, dG \quad = \quad \int_{x=0}^{1} \int_{y=0}^{x} (x - y)\, dy\, dx$$
$$= \quad \int_{y=0}^{1} \int_{x=y}^{1} (x - y)\, dx\, dy$$

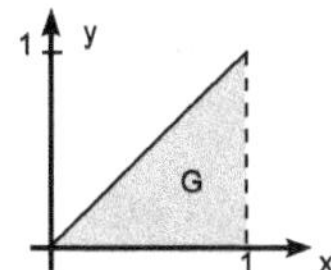

12.3 Show that the values of the integrals I_1 and I_2 are equal

$$I_1 = \int_{x=0}^{2} \int_{y=0}^{x^2} 2\, x\, y\, dy\, dx \qquad I_2 = \int_{y=0}^{4} \int_{x=\sqrt{y}}^{2} 2\, x\, y\, dx\, dy$$

12.4 Determine the area of the semicircle with radius $R = 2$ and center $(2,\, 0)$ in the upper half-plane by calculating the next integral in polar coordinates

$$\iint\limits_{(G)} 1\, dx\, dy = \int_{r=0}^{2} \int_{\varphi=0}^{\pi} r\, d\varphi\, dr \ .$$

12.5 Given are the graphs of $y = -x\,(x - 3)$ and $y = -2\,x$.
a) What area do they include?
b) What are the coordinates of the center of mass?

12.6 Determine the axial surface moments I_x and I_y and the polar surface moment I_p of a quarter circle with radius R.

12.7 Determine the center of mass of the triangular surface (Fig. a).

12.8 Determine the center of mass of the semicircle with radius R (Fig. b).

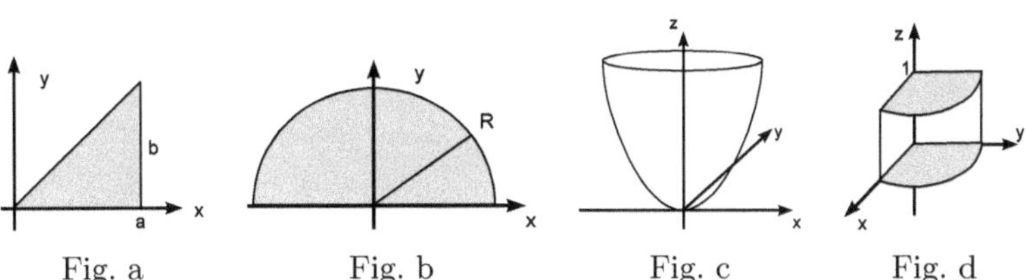

Fig. a   Fig. b   Fig. c   Fig. d

12.9 Calculate the triple integrals

a) $\displaystyle \int_{z=0}^{1} \int_{y=z-1}^{z} \int_{x=y}^{y+1} x^2 \, dx \, dy \, dz$

b) $\displaystyle \int_{z=-l}^{l} \int_{x=z}^{z^2} \int_{y=x-z}^{x+z} x \, y \, z \, dy \, dx \, dz$

c) $\displaystyle \int_{\varphi=0}^{\pi} \int_{\vartheta=-\pi/2}^{\pi/2} \int_{r=0}^{R} r^2 \, \cos\vartheta \, \sin\varphi \, dr \, d\vartheta \, d\varphi$

d) $\displaystyle \int_{x=-R}^{R} \int_{y=0}^{R} \int_{r=0}^{\sqrt{x^2+y^2}} r \, dr \, dy \, dx$

12.10 Determine the center of mass coordinate z_s as well as the moments of inertia of the body of rotation resulting from rotation of x^2 on the z axis. Enter cylindrical coordinates to describe the body (see Fig. c).

12.11 Determine the mass moments of inertia of a hemisphere ($z > 0$).

12.12 Compute the integral

$$I = \iiint\limits_{G} x^2 \, y \, dx \, dy \, dz$$

where
$$G = \left\{ (x, y, z) : x \geq 0, \, y \geq 0, \, x^2 + y^2 \leq 1, \, 0 \leq z \leq 1 \right\}.$$
(For the calculation introduce cylindrical coordinates; see Fig. d.)

12.13 Find the center of mass of the cylinder from problem 12.12.

Chapter 13
First-Order Differential Equations

13

In Chapter 13 we look at first-order linear differential equations. Important for the mathematical representation of the solution is that the general solution of the differential equation can be split into the general solution of the *homogeneous* problem plus a *particular* solution of the inhomogeneous differential equation. This special structure of the solution applies both to systems of linear differential equations and to differential equations of n-th order.

The methods we introduce for solving the linear differential equations are extended to non-linear differential equations and to the numerical solution of first-order differential equations.

13

13 First-Order Differential Equations

In Chapter 13 we look at first-order linear differential equations. Important for the mathematical representation of the solution is that the general solution of the differential equation can be split into the general solution of the *homogeneous* problem plus a *particular* solution of the inhomogeneous differential equation. This special structure of the solution applies both to systems of linear differential equations and to differential equations of n-th order.

The methods we introduce for solving the linear differential equations are extended to non-linear differential equations and to the numerical solution of first-order differential equations.

Differential equations are indispensable in science and engineering because they express many laws of nature. Differential equations are the result of a mathematical-physical modelling that describes the occurring phenomena. However, it is not only solving differential equations within mathematics that is important for engineers, but also setting up the model that ultimately leads to a differential equation. Therefore, in each section we will first describe the modelling of application-relevant examples and then go on to systematically solve the different types of differential equations.

> **Definition:** A **differential equation** is an equation in which, in addition to the function (or functions) sought, derivatives of that function(s) also occur. An **ordinary differential equation** is an equation in which only functions and their *ordinary* derivatives occur, in contrast to **partial differential equations** in which *partial* derivatives are also contained in the equation being studied.

In this and the next chapter we will only be dealing with **ordinary differential equations**, so we will omit the addition ordinary. Partial differential equations will be covered in Volume 3.

13.1 Introduction Problems

We start with problems from physics, electronics and other applications and their modelling. These examples show how to go from the physical problem to a mathematical model equation, which is in these cases a first-order differential equation.

Application Example 13.1 (RL-Circuit).

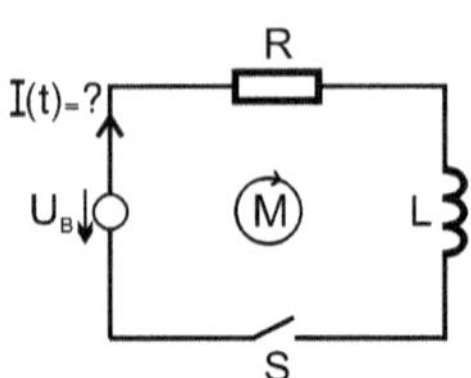

A resistor R, a coil with inductance L and a battery with voltage U_B are connected in series with a switch S. The switch is initially open and is closed at $t = 0$. Hence, the current is zero: $I(0) = 0$. What is the behavior of the current $I(t)$ as a function of time for $t > 0$?

Figure 13.1. RL-Circuit

To solve this problem, we first set up a model for the current. According to the mesh rule for the mesh M, the voltage drop along R plus the voltage drop along L equals the applied voltage U_B:

$$U_R + U_L = U_B.$$

Using Ohm's law ($U_R = R \cdot I(t)$) and the law of induction ($U_L = L \frac{dI(t)}{dt}$) we get

$$R I(t) + L \frac{dI(t)}{dt} = U_B$$

$$\Rightarrow \quad \frac{d}{dt} I(t) + \frac{R}{L} I(t) = \frac{1}{L} U_B \quad \text{with } I(0) = 0.$$

This is an *ordinary, first-order, linear differential equation* for the current $I(t)$. We are looking for a function $I(t)$ that satisfies the above differential equation with the initial condition.

We confirm by evaluation, that

$$I(t) = \frac{U_B}{R} \left(1 - e^{-\frac{R}{L} t}\right)$$

is the solution by inserting the derivative of $I(t)$

$$\dot{I}(t) = \frac{U_B}{R} \cdot \frac{R}{L} e^{-\frac{R}{L} t}$$

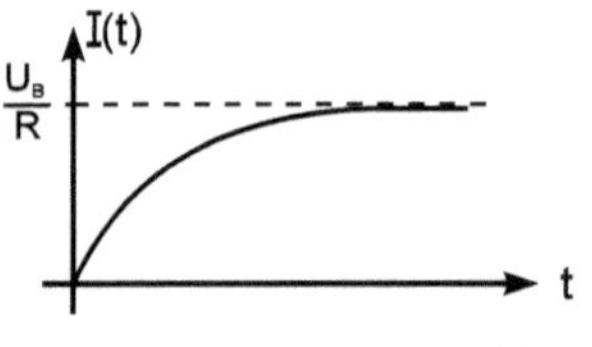

Figure 13.2. Current $I(t)$

and the function into the differential equation. We get

$$\Rightarrow \dot{I}(t) + \tfrac{R}{L} I(t) = \tfrac{U_B}{L} e^{-\frac{R}{L}t} + \tfrac{R}{L} \tfrac{U_B}{R} \left(1 - e^{-\frac{R}{L}t}\right) = \tfrac{U_B}{L}.$$

The function $I(t)$ thus satisfies the differential equation and also the required initial condition $I(0) = \tfrac{U_B}{R}\left(1 - e^{0}\right) = 0.$ $\qquad\square$

Application Example 13.2 (Barometric Formula).

The air pressure $p(h)$ at the height h above sea level is caused by the weight of the air column over the area A. The pressure difference $p(h) - p(h + dh)$ is equal to the weight G of the vertical column of air with cross-section A, which is located between h and $h + dh$. For small dh, the density $\rho(h)$ in this column of air is constant:

$$A\left(p(h) - p(h + dh)\right) = G = dm\,g = \rho(h)\,A \cdot dh\,g.$$

Dividing by dh and taking the limit $dh \to 0$ provides

$$p'(h) = \lim_{dh \to 0} \frac{p(h + dh) - p(h)}{dh} = -g\,\rho(h).$$

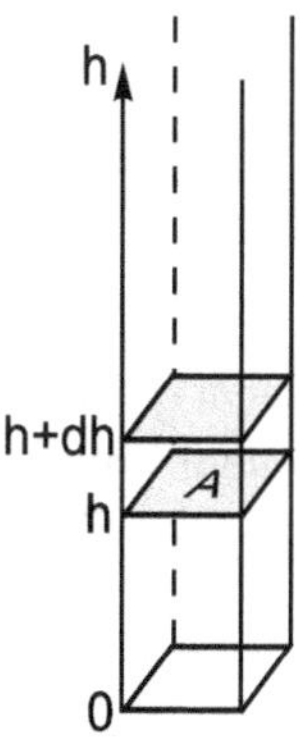

Figure 13.3.

If the air is considered to be an ideal gas, then the relationship between ρ, the pressure p and the temperature T is: $\quad \rho = \alpha\,\dfrac{p}{T}$ with a constant α.

(i) If we consider the temperature T to be constant and independent of the height, and introduce the constant $\beta = g\tfrac{\alpha}{T}$, we obtain the first-order linear differential equation with the initial condition $p(h_0) = p_0$

$$p'(h) = -\frac{\alpha\,g}{T}\,p(h) = -\beta\,p(h).$$

The solution of this differential equation is

$$p(h) = p_0\,e^{-\beta\,(h - h_0)} \qquad \textbf{(Barometric Formula)}.$$

(ii) The above differential equation is only valid in a small range because of the assumption $T = const$. In reality, the temperature decreases with increasing altitude. The simplest model assumption is that T has a linear decrease in temperature:

$$T(h) = T_0 - b\,(h - h_0).$$

This leads to the first-order linear differential equation

$$p'(h) = \frac{-\alpha g}{T_0 - b(h - h_0)}\, p(h) \qquad \text{with } p(h_0) = p_0.$$

It can be confirmed by differentiation that

$$p(h) = p_0 \left(1 - \frac{b}{T_0}(h - h_0)\right)^{\frac{\alpha g}{b}}$$

is the solution of the differential equation. $\qquad\qquad\square$

Application Example 13.3 (Radioactive Decay).

Let $n(t)$ be the number of atoms of a radioactive substance at time t. The amount of this substance that decays in a time period dt is proportional to the amount of substance and the time period dt:

$$n(t + dt) - n(t) \sim -dt \cdot n(t).$$

If the proportionality constant $\lambda > 0$ is introduced, then

$$n(t + dt) - n(t) = -\lambda\, dt\, n(t).$$

As the number of radioactive atoms decreases, a minus sign appears on the right-hand side of the equation. Division by dt and then by $dt \to 0$ gives

$$n'(t) = \lim_{dt \to 0} \frac{n(t + dt) - n(t)}{dt} = -\lambda\, n(t) \qquad \textbf{(Radioactive Decay).}$$

This is a *first-order linear differential equation* with the initial condition $n(0) = N$. The solution of the differential equation is

$$n(t) = N\, e^{-\lambda t},$$

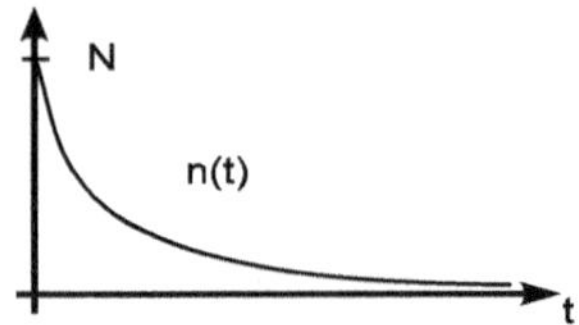

Figure 13.4. Radioactive decay   which can be checked again by inserting the function $n(t)$ directly into the differential equation. $\qquad\square$

Now we will clarify how to systematically find solutions to first-order linear differential equations. Mathematically, the general problem is as follows:

General Problem: We consider the **first-order linear differential equation**

$$y'(x) = h(x)\, y(x) + f(x), \qquad\qquad (D1)$$

where $h(x)$ and $f(x)$ are given continuous functions on an interval I.

For $f(x) \neq 0$ $(D1)$ is an **inhomogeneous** differential equation.
For $f(x) = 0$ $(D1)$ is a **homogeneous** differential equation.

In the case of an inhomogeneous differential equation, $f(x) \neq 0$ is called the **inhomogeneity** or the **interference function**.

13.2 Solving Homogeneous Differential Equations

First, the homogeneous problem

$$y'(x) = h(x)\, y(x)$$

is treated with the initial condition $y(x_0) = y_0$. The solution of this differential equation is calculated using the method of **separation of variables**. This involves replacing $y'(x)$ by $\frac{dy}{dx}$ and separating the variables by formally multiplying the equation by dx and dividing it by y:

$$\frac{dy}{dx} = h(x)\, y(x) \;\Rightarrow\; \frac{dy}{y} = h(x)\, dx.$$

Symbolic integration gives

$$\int_{y_0}^{y} \frac{d\tilde{y}}{\tilde{y}} = \int_{x_0}^{x} h(\tilde{x})\, d\tilde{x} \;\Rightarrow\; \ln \tilde{y}\big|_{y_0}^{y} = \ln \frac{y}{y_0} = \int_{x_0}^{x} h(\tilde{x})\, d\tilde{x}.$$

If we apply the exponential function on both sides, the solution is

$$y(x) = y_0\, e^{\int_{x_0}^{x} h(\tilde{x})\, d\tilde{x}}. \qquad\qquad (H)$$

With the formula (H) we can solve any first-order homogeneous differential equation by calculating the definite integral $\int_{x_0}^{x} h(\tilde{x})\, d\tilde{x}$. In applications, the method Separation of Variables is often used instead of applying this solution formula.

Application Example 13.4 (RL-Circuit).

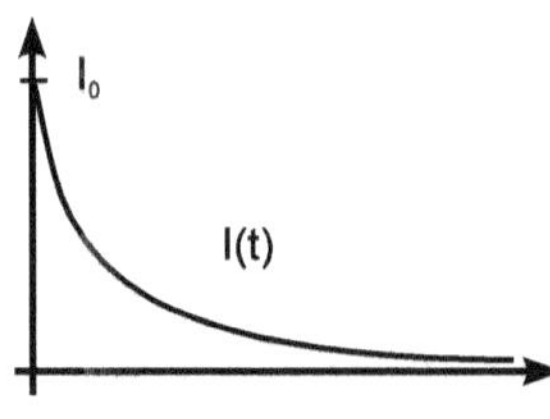

Figure 13.5. Current $I(t)$

The RL-circuit discussed in Example 13.1 will be investigated. At time $t_0 = 0$ the battery is bypassed. A modified model equation then applies

$$\frac{d}{dt} I(t) + \frac{R}{L} I(t) = 0 \quad \text{with} \quad I(0) = I_0.$$

$$\hookrightarrow \quad \dot{I}(t) = -\frac{R}{L} I(t).$$

The solution formula (H) for homogeneous differential equations gives

$$I(t) = I_0\, e^{\int_{t_0}^{t} (-\frac{R}{L})\, d\tau} = I_0\, e^{-\frac{R}{L}(t-t_0)} = I_0\, e^{-\frac{R}{L} t}$$

which is shown in Fig. 13.5. We can see that the current decays exponentially towards zero. The exponential decay is a typical behavior of a discharge problem.   □

Application Example 13.5 (Barometric Formula).

The Example 13.2 (i) and (ii) are also solved using the formula (H):

(i) $p'(h) = -\beta\, p(h) \quad \hookrightarrow \quad p(h) = p(h_0)\, e^{\int_{h_0}^{h} -\beta\, d\tilde{h}} = p(h_0)\, e^{-\beta(h-h_0)}.$

(ii) $p'(h) = -\frac{\alpha g}{T_0 - b(h-h_0)}\, p(h):$

$$\hookrightarrow p(h) = p(h_0)\, e^{\int_{h_0}^{h} \frac{-\alpha g}{T_0 - b(\tilde{h}-h_0)}\, d\tilde{h}} = p(h_0)\, e^{\frac{\alpha g}{b} \ln(T_0 - b(\tilde{h}-h_0))\big|_{h_0}^{h}}.$$

Inserting the upper and lower limits

$$\frac{\alpha g}{b} \ln\left(T_0 - b\left(\tilde{h} - h_0\right)\right)\bigg|_{h_0}^{h} = \frac{\alpha g}{b} \left[\ln\left(T_0 - b(h-h_0)\right) - \ln(T_0)\right]$$

$$= \frac{\alpha g}{b} \ln \frac{T_0 - b(h-h_0)}{T_0} = \ln\left(1 - \frac{b}{T_0}(h-h_0)\right)^{\frac{\alpha g}{b}}$$

gives us the solution $p(h)$

$$p(h) = p(h_0)\left(1 - \frac{b}{T_0}(h-h_0)\right)^{\frac{\alpha g}{b}}.$$ □

13.3 Solving Inhomogeneous Differential Equations

Now we are going to study inhomogeneous differential equations

$$y'(x) = h(x)\, y(x) + f(x)$$

with initial condition $y(x_0) = y_0$, and calculate the solution of the differential equation by the method of **variation of constants**. The solution of the corresponding *homogeneous* problem $y'(x) = h(x)\, y(x)$ is

$$y(x) = c\, e^{\int_{x_0}^{x} h(\tilde{x})\, d\tilde{x}}.$$

To obtain a solution to the *inhomogeneous* differential equation, we start from this solution and vary the constant c, allowing c to be a function $c(x)$. We therefore choose a **product approach**

$$y(x) = c(x) \cdot \varphi(x) \qquad \text{with} \quad \varphi(x) = e^{\int_{x_0}^{x} h(\tilde{x})\, d\tilde{x}}.$$

We differentiate $y(x)$ using the product rule

$$y'(x) = c'(x)\, \varphi(x) + c(x)\, \varphi'(x)$$
$$= c'(x)\, \varphi(x) + c(x)\, \varphi(x) \cdot h(x)$$

since $\varphi'(x) = e^{\int_{x_0}^{x} h(\tilde{x})\, d\tilde{x}} \cdot h(x) = \varphi(x)\, h(x)$. If we replace $c(x) \cdot \varphi(x) = y(x)$ and insert the approach into the differential equation, we obtain

$$y'(x) = c'(x)\, \varphi(x) + h(x) \cdot y(x) = f(x) + h(x) \cdot y(x).$$

So $y(x)$ is the solution of the inhomogeneous differential equation when

$$c'(x)\, \varphi(x) = f(x) \;\Rightarrow\; c'(x) = \frac{f(x)}{\varphi(x)}.$$

The following integration returns

$$c(x) = c_0 + \int_{x_0}^{x} \frac{f(\tilde{x})}{\varphi(\tilde{x})}\, d\tilde{x}.$$

So the solution $y(x) = c(x) \cdot \varphi(x)$ of the inhomogeneous problem is

$$y(x) = \left(c_0 + \int_{x_0}^{x} \frac{f(\tilde{x})}{\varphi(\tilde{x})}\, d\tilde{x} \right) \cdot \varphi(x).$$

The constant c_0 must be specified so that $y(x_0) = y_0 \quad \Rightarrow \quad c_0 = y_0$.

First-Order Linear Differential Equations

Let $h, f : I \to \mathbb{R}$ be continuous functions. The first-order linear differential equation

$$y'(x) = h(x)\, y(x) + f(x)$$

$$y(x_0) = y_0$$

has **exactly one** solution on the interval I. This unique solution is determined in two steps:

(1) First, a solution of the homogeneous DEq is calculated

$$\varphi(x) = e^{\displaystyle \int_{x_0}^{x} h(\tilde{x})\, d\tilde{x}} \qquad\qquad (H1)$$

(2) and then $\varphi(x)$ is inserted into the solution expression for $y(x)$

$$y(x) = \varphi(x)\left(y_0 + \int_{x_0}^{x} \frac{f(\tilde{x})}{\varphi(\tilde{x})}\, d\tilde{x}\right). \qquad\qquad (I)$$

Comment on the method: Whenever partial solutions of a differential equation are known, an attempt is made to construct further or other solutions of the differential equation by considering the partial information in a special approach. In the case of the inhomogeneous differential equation, the partial information is the knowledge of the homogeneous solution $\varphi(x)$. The idea of variation of constants is that the inhomogeneity of the differential equation causes the amplitude of the homogeneous solution to vary. Therefore, the homogeneous solution $\varphi(x)$ is multiplied by a location dependent amplitude $c(x)$. This unknown amplitude is determined by inserting the approach into the differential equations. In some applications the ready-made solution formulas are used or the solution methods are applied instead:

(1) Solving the homogeneous differential equation $y'(x) = h(x)\, y(x)$ by separating the variables or directly by $y(x) = c\, e^{\int_{x_0}^{x} h(\tilde{x})\, d\tilde{x}}$.

(2) Solving the inhomogeneous DEq by variation of the constants

$$y(x) = c(x) \cdot e^{\int_{x_0}^{x} h(\tilde{x})\, d\tilde{x}}. \qquad\qquad \square$$

Proof of the theorem on first-order linear differential equations:

To check that the given formula provides a solution to the initial value problem, i.e. that the differential equation and the initial condition $y\,(x_0) = y_0$ are satisfied, we just have to substitute the expression into the differential equation. This will confirm that the right-hand side is equal to the left-hand side.

To prove that $y\,(x)$ is the unique solution of the differential equation with initial condition, we have to argue as follows: Let $y_2\,(x)$ also be a solution. Then it applies to the difference

$$d\,(x) = y\,(x) - y_2\,(x)$$

by differentiation

$$d'\,(x) = y'\,(x) - y_2'\,(x) = h\,(x)\,y\,(x) + f\,(x) - [h\,(x)\,y_2\,(x) + f\,(x)]$$
$$= h\,(x)\,(y\,(x) - y_2\,(x)) = h\,(x) \cdot d\,(x)$$

with $d\,(x_0) = y\,(x_0) - y_2\,(x_0) = 0$. So $d\,(x)$ is the solution of the homogeneous differential equation

$$d'\,(x) = h\,(x)\,d\,(x) \quad \text{with} \ \ d\,(x_0) = 0. \tag{$*$}$$

The next step is to show that the difference is $d\,(x) = 0$ for all $x \in I$: For this purpose, we define the function

$$u\,(x) := d\,(x)\,e^{-\int_{x_0}^{x} h\,(\tilde{x})\,d\tilde{x}}.$$

With the product and chain rule, the derivative is

$$u'\,(x) = d'\,(x)\,e^{-\int_{x_0}^{x} h\,(\tilde{x})\,d\tilde{x}} + d\,(x)\,e^{-\int_{x_0}^{x} h\,(\tilde{x})\,d\tilde{x}}\,(-h\,(x))$$
$$= \underbrace{[d'\,(x) - h\,(x)\,d\,(x)]}_{=0} \cdot e^{-\int_{x_0}^{x} h\,(\tilde{x})\,d\tilde{x}} = 0,$$

because $d\,(x)$ is the solution of the homogeneous differential equation $(*)$.

So $u\,(x)$ is a constant function. The constant is determined by evaluating $u(x)$ at the position x_0: With $u\,(x_0) = const = d\,(x_0) = 0$ we conclude $u\,(x) = 0$. So d is the zero function: $d\,(x) = 0$, and consequently $y\,(x) = y_2\,(x)$ for all $x \in I$. $\qquad\qquad\square$

Interpretation of the Solution Formula

By expanding the solution formula (I), we obtain the following representation

$$y\left(x\right) = \underbrace{y_0\,\varphi\left(x\right)}_{\text{homogeneous}} + \underbrace{\varphi\left(x\right)\int_{x_0}^{x}\frac{f\left(\tilde{x}\right)}{\varphi\left(\tilde{x}\right)}\,d\tilde{x}}_{\text{particular}}\ .$$

The general solution of the inhomogeneous differential equation can be written as the sum of the **general solution of the homogeneous problem** and **a special solution of the inhomogeneous problem**. A special solution of the inhomogeneous differential equation is called a **particular solution**.

Example 13.6 (Sample). Given is the differential equation

$$y'\left(x\right) = 2\,x\,y\left(x\right) + x^3 \quad \text{with } y\left(0\right) = y_0.$$

Find the solution $y(x)$ that also satisfies the initial condition.

> **Procedure:** We solve this differential equation in three steps. First we find a solution to the homogeneous differential equation, then we compute a special solution to the inhomogeneous differential equation, and finally we put both parts together.

(1) Solving the **homogeneous** differential equation $y'\left(x\right) = 2\,x\,y\left(x\right)$. For this step we insert $h(x) = 2\,x$ into the formula $(H1)$:

$$\varphi\left(x\right) = e^{\int_0^x 2\,\tilde{x}\,d\tilde{x}} = e^{x^2}.$$

(2) Solving the **inhomogeneous** differential equation $y'\left(x\right) = 2\,x\,y\left(x\right) + x^3$ with formula (I):

$$y\left(x\right) = e^{x^2}\left(y_0 + \int_0^x \frac{t^3}{e^{t^2}}\,dt\right) = e^{x^2}\,y_0 + e^{x^2}\int_0^x \frac{t^3}{e^{t^2}}\,dt.$$

First, we calculate the indefinite integral

$$\int t^3\,e^{-t^2}\,dt$$

with the substitution ($\xi = t^2$, $d\xi = 2t\,dt$)

$$\int t^3\, e^{-t^2}\, dt = \int t^3\, e^{-\xi}\, \frac{d\xi}{2t} = \frac{1}{2} \int \xi\, e^{-\xi}\, d\xi$$

and integrate by parts

$$\frac{1}{2} \int \xi\, e^{-\xi}\, d\xi = \frac{1}{2}\left[-\xi\, e^{-\xi} + \int e^{-\xi}\, d\xi\right] = \frac{1}{2}[-\xi\, e^{-\xi} - e^{-\xi}] + C.$$

With back substitution ($\xi = t^2$) and insertion of the limits, we obtain

$$\int_0^x t^3\, e^{-t^2}\, dt = \frac{1}{2}\left[-t^2\, e^{-t^2} - e^{-t^2}\right]_0^x = \frac{1}{2} - \frac{1}{2}\, e^{-x^2}\, (x^2 + 1).$$

(3) The **general solution** of the differential equation is therefore

$$y(x) = y_0\, e^{x^2} + e^{x^2}\left[\frac{1}{2} - \frac{1}{2}\, e^{-x^2}\, (x^2+1)\right] = y_0\, e^{x^2} + \frac{1}{2}\, e^{x^2} - \frac{1}{2}\, (x^2+1) \ \square$$

Application Example 13.7 (RL-Circuit).

The solution formula (I) for inhomogeneous linear differential equations treats the problem of the RL-circuit from the Example 13.1:

$$\dot{I}(t) = -\frac{R}{L}\, I(t) + \frac{1}{L}\, U_B \quad \text{with} \quad I(0) = 0.$$

(1) The solution of the **homogeneous** differential equation $\dot{I}(t) = -\frac{R}{L}\, I(t)$ is according to formula (H)

$$I_h(t) = c\, e^{-\frac{R}{L}\, t}.$$

(2) And, with $I_0 = 0$, the solution of the **inhomogeneous** differential equation is obtained with the formula (I).

$$I(t) = e^{-\frac{R}{L}\, t}\left(I_0 + \int_{t_0}^t \frac{U_B}{L}\, \frac{1}{e^{-\frac{R}{L}\, \tau}}\, d\tau\right)$$

$$= e^{-\frac{R}{L}\, t}\, \frac{U_B}{L} \int_0^t e^{\frac{R}{L}\, \tau}\, d\tau = e^{-\frac{R}{L}\, t}\, \frac{U_B}{L}\, \frac{L}{R}\, e^{\frac{R}{L}\, \tau}\Big|_0^t$$

$$\Rightarrow I(t) = e^{-\frac{R}{L}t}\frac{U_B}{R}\left(e^{\frac{R}{L}t} - 1\right) = \frac{U_B}{R}\left(1 - e^{-\frac{R}{L}t}\right).$$

This is the time behavior of the current already discussed in the Example 13.1. $\square$

Application Example 13.8 (Modelling an RC-Circuit).

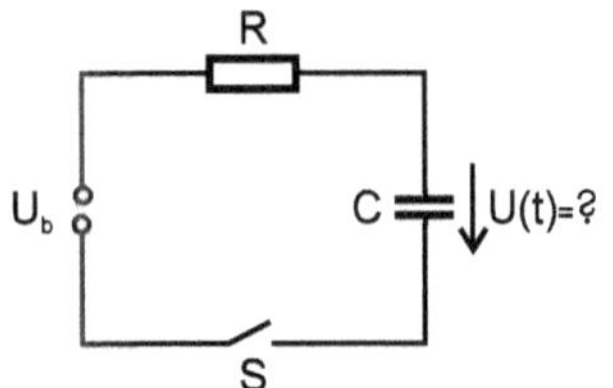

Figure 13.6. RC-circuit

A resistor R, a capacitor with capacity C and a voltage source $U_b(t)$ are connected in series with a switch. The switch is initially open and is closed at $t = 0$. Find the voltage $U(t)$ across the capacitor as a function of the time for $t > 0$?

According to the mesh rule, the voltages are

$$U_R(t) + U(t) = U_b(t). \tag{$*$}$$

The ohmic resistance is $U_R(t) = R \cdot I(t)$. For the capacitance we know the relationship between voltage and charge $U(t) = \frac{1}{C}Q(t)$, so

$$U(t) = \frac{1}{C}Q(t) = \frac{1}{C}\int I(t)\, dt \;\Rightarrow\; \dot{U}(t) = \frac{1}{C}I(t) \;\Rightarrow\; I(t) = C \cdot \dot{U}(t),$$

where

$$U_R(t) = R \cdot I(t) = RC \cdot \dot{U}(t).$$

Inserted into the equation $(*)$, we get

$$RC\,\dot{U}(t) + U(t) = U_b(t) \qquad \text{with}\;\; U(0) = 0$$

$$\hookrightarrow \dot{U}(t) = -\frac{1}{RC}U(t) + \frac{1}{RC}U_b(t) \;\; \text{with}\;\; U(0) = 0. \qquad \square$$

The next two examples solve this problem for different input voltages. The Example 13.9 discusses the case of a constant applied voltage $U_b(t) = \hat{U}_b$ whereas in the Example 13.10 we chose an alternating voltage $U_b(t) = \hat{U}_b \sin(\omega t)$.

Application Example 13.9 (RC-Circuit with DC Voltage).

For a **constant battery voltage** $U_b(t) = \hat{U}_b$, the solution is analogous to the procedure in Example 13.7

$$U(t) = \hat{U}_b \left(1 - e^{-\frac{1}{RC}t}\right).$$

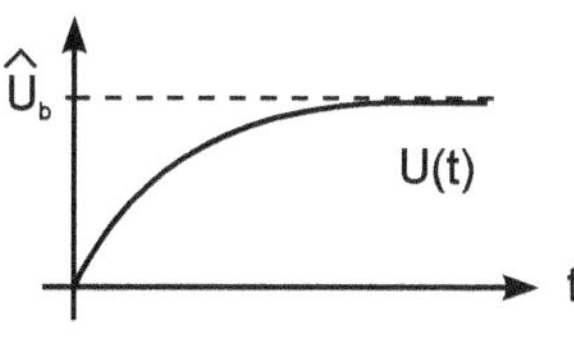

Figure 13.7. Charging curve

The voltage across the capacitor $U(t)$, and hence the charge $Q(t) = C \cdot U(t)$, grows asymptotically with time to reach the final voltage $\hat{U}_b$ and the final charge $Q_\infty = C \cdot \hat{U}_b$. $\qquad\square$

Application Example 13.10 (Sample/Application Example).

If the input voltage in Fig. 13.6 is a **AC voltage**

$$U_b(t) = \hat{U}_b \sin(\omega t)$$

with peak value $\hat{U}_b$ and frequency ω, the differential equation takes the form

$$\dot{U}(t) = -\frac{1}{RC} U(t) + \frac{1}{RC} \hat{U}_b \sin(\omega t) \quad \text{with} \quad U(0) = 0.$$

(1) According to the formula $(H1)$, we first solve the homogeneous differential equation

$$\dot{U}(t) = -\frac{1}{RC} U(t) \quad \text{with} \quad U_h(t) = e^{\int_0^t -\frac{1}{RC}\,dt} = e^{-\frac{1}{RC}t}.$$

(2) Then the solution formula (I) is used

$$U(t) = e^{-\frac{1}{RC}t} \left(U(0) + \frac{\hat{U}_b}{RC} \int_0^t \frac{\sin(\omega\tau)}{e^{-\frac{1}{RC}\tau}}\,d\tau\right).$$

To calculate the integral, we partially integrate twice,

$$\int_0^t \sin(\omega\tau)\, e^{\frac{1}{RC}\tau}\,d\tau$$

$$= \left[\sin(\omega\tau) \cdot RC \cdot e^{\frac{1}{RC}\tau}\right]_0^t - \int_0^t \omega \cos(\omega\tau)\, RC\, e^{\frac{1}{RC}\tau}\,d\tau$$

$$= RC \sin(\omega t)\, e^{\frac{1}{RC}\, t} - \omega\, RC \left\{ \left[\cos(\omega \tau)\, e^{\frac{1}{RC}\, \tau}\, RC \right]_0^t + \int_0^t \omega \sin(\omega \tau)\, e^{\frac{1}{RC}\, \tau}\, RC\, d\tau \right\}$$

$$= RC \sin(\omega t)\, e^{\frac{1}{RC}\, t} - \omega\, (RC)^2 \left(\cos(\omega t)\, e^{\frac{1}{RC}\, t} - 1 \right)$$
$$\qquad -\omega^2\, (RC)^2 \int_0^t \sin(\omega \tau)\, e^{\frac{1}{RC}\, \tau}\, d\tau.$$

Since the remaining integral on the right-hand side is equal to the integral to be calculated, we add the term to both sides of the equation $\omega^2\, (RC)^2 \int_0^t \sin(\omega \tau)\, e^{\frac{1}{RC}\, \tau}\, d\tau$ and divide by the factor $1 + (\omega\, RC)^2$:

$$\int_0^t \sin(\omega \tau)\, e^{\frac{1}{RC}\, \tau}\, d\tau = \frac{1}{1 + (\omega\, RC)^2} \left\{ RC \sin(\omega t)\, e^{\frac{1}{RC}\, t} \right.$$
$$\left. -\omega\, (RC)^2 \cos(\omega t)\, e^{\frac{1}{RC}\, t} + \omega\, (RC)^2 \right\}.$$

(3) We insert this result into the solution formula taking into account the initial condition $U(0) = 0$. We obtain the final result

$$U(t) = e^{-\frac{1}{RC}\, t}\, \frac{\hat{U}_b}{1 + (\omega\, RC)^2} \left\{ \sin(\omega t)\, e^{\frac{1}{RC}\, t} - \omega\, RC \cos(\omega t)\, e^{\frac{1}{RC}\, t} + \omega\, RC \right\}$$
$$= \frac{\hat{U}_b}{1 + (\omega\, RC)^2} \left\{ \sin(\omega t) - \omega\, RC \cos(\omega t) + \omega\, RC\, e^{-\frac{1}{RC}\, t} \right\}.$$

Discussion: The solution $U(t)$ consists of an exponential decay and a periodic term.

$$U(t) = \underbrace{\frac{\hat{U}_b\, \omega\, RC}{1 + (\omega\, RC)^2}\, e^{-\frac{1}{RC}\, t}}_{\text{Short Range}} + \underbrace{\frac{\hat{U}_b}{1 + (\omega\, RC)^2}\, (\sin(\omega t) - \omega\, RC \cos(\omega t))}_{\text{Long Time Behavior}}.$$

The exponential term reflects the **initial conditions**. Whereas, the **long term behavior** of the solution is determined by the periodic component. In Fig. 13.8 the solution is drawn for the parameters $RC = 10$, $\hat{U}_b = 1$ and $\omega = 1$. This shows the short range and the long range behavior.

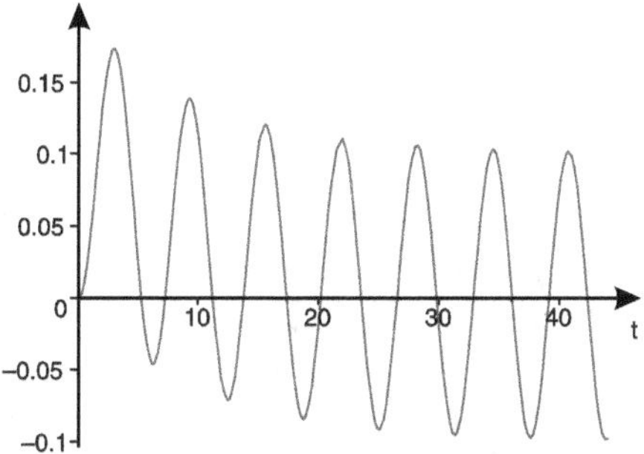

Figure 13.8. Solution of DEq

Physical Interpretation: The long term part of the solution is a pure harmonic oscillation with the amplitude A, the frequency ω and the phase φ:

$$\sin(\omega t) - \omega RC \cos(\omega t) = A \sin(\omega t + \varphi)$$

To determine A and φ we need two equations. For this, we expand the right-hand side using the addition theorem

$$A \sin(\alpha + \beta) = A \sin(\alpha)\cos(\beta) + A \cos(\alpha)\sin(\beta).$$

With $\alpha = \omega t$ and $\beta = \varphi$, the coefficient comparison yields

$$-\omega RC = A \sin\varphi. \tag{1}$$

$$1 = A \cos\varphi. \tag{2}$$

Dividing equation (1) by (2) gives

$$\tan\varphi = -RC\omega \;\Rightarrow\; \varphi.$$

Adding the squares of equations (1) and (2) gives

$$1 + (RC\omega)^2 = A^2 \cos^2\varphi + A^2 \sin^2\varphi = A^2(\cos^2\varphi + \sin^2\varphi) = A^2$$

$$\Rightarrow\; A = \sqrt{1 + (RC\omega)^2}.$$

So the solution $U(t)$ is finally

$$U(t) = \text{Short Range} + \frac{\hat{U}_b}{\sqrt{1 + (\omega RC)^2}} \sin(\omega t + \varphi). \qquad \square$$

13.4 Linear Differential Equations with Constant Coefficients

With the two formulas (H) and (I) any first-order linear differential equation can be solved. However, the evaluation of the integrals can be very time-consuming. In many cases it is not necessary to use this integral representation of the solution, because a *particular* solution can be obtained with a special approach. This is especially the case for the differential equations with **constant** coefficients. For the concrete problem, an approach for the particulate solution with free parameters is chosen according to the type of *inhomogeneity*. These parameters are then determined by inserting the approach into the differential equation.

13.4.1 Homogeneous Solution

For the solution of the *homogeneous* linear differential equations with constant coefficient α,

$$y'(x) = \alpha \cdot y(x),$$

the following approach is chosen

$$y_h(x) = c\,e^{\lambda x}.$$

We insert this function in the differential equation:

$$c\lambda e^{\lambda x} = \alpha \cdot c\,e^{\lambda x} \hookrightarrow \lambda = \alpha.$$

Therefore, the general solution of the homogeneous differential equation is

$$y_h(x) = c\,e^{\alpha x}.$$

13.4.2 Particular Solution

Here we consider the *inhomogeneous* linear differential equation

$$y'(x) = \alpha\,y(x) + f(x)$$

with $\alpha \neq 0$. Depending on the inhomogeneity (*interference function*) f, a *particular solution* $y_p(x)$ can be found by a simple approach. For common inhomogeneities, the approach functions are given in Tab. 13.1.

Table 13.1: Approach functions for particular solutions.

Inhomogeneity	Approach function	Parameter
$f(x) = \sum\limits_{i=0}^{n} a_i\, x^i$ (Polynomial of degree n)	$y_p(x) = \sum\limits_{i=0}^{n} A_i\, x^i$	$A_0, ..., A_n$
$f(x) = a \cdot \sin(\omega x)$ (Sine function)	$y_p(x) = A \sin(\omega x) + B \cos(\omega x)$	$A,\ B$
$f(x) = a \cdot \cos(\omega x)$ (Cosine function)	$y_p(x) = A \sin(\omega x) + B \cos(\omega x)$	$A,\ B$
$f(x) = a\, e^{\mu x}\ (\mu \neq \alpha)$ (Exponential function)	$y_p(x) = A\, e^{\mu x}$	A

Note: If the inhomogeneity $f(x)$ consists of a sum of several individual functions, the appropriate approach function is selected for each inhomogeneity individually and the unknown parameters are determined by inserting the approach function into the differential equation. Finally, all the particular solutions are summed and a special solution of the inhomogeneous problem is obtained.

Example 13.11 (Sample).

Find the general solution to the linear differential equation

$$y'(x) = 4 \cdot y(x) + x^3.$$

(1) To solve the **homogeneous** differential equation $y'(x) = 4 \cdot y(x)$, we use the approach

$$y_h(x) = c\, e^{\lambda x}.$$

Inserting this into the differential equation yields

$$c\lambda e^{\lambda x} = 4c e^{\lambda x} \ \hookrightarrow\ \lambda = 4 \ \Rightarrow\ y_h(x) = c\, e^{4x}.$$

(2) A particular solution of the **inhomogeneous** differential equation

$$y'(x) = 4 \cdot y(x) + x^3$$

is given according to Table 13.1 by the approach

$$y_p(x) = a\, x^3 + b\, x^2 + c\, x + d,$$

because the inhomogeneity $f(x) = x^3$ is a polynomial of degree 3. This function inserted into the inhomogeneous differential equation yields

$$3\,a\,x^2 + 2\,b\,x + c \;=\; 4\,a\,x^3 + 4\,b\,x^2 + 4\,c\,x + 4\,d + x^3$$

$$= (4\,a + 1)\,x^3 + 4\,b\,x^2 + 4\,c\,x + 4\,d.$$

We compare the coefficients of both sides for descending powers in x. This gives the following results

$$
\begin{aligned}
x^3: &\quad 4a + 1 = 0 &\hookrightarrow&\quad a = -\tfrac{1}{4}\\
x^2: &\quad 4b = 3a = -\tfrac{3}{4} &\hookrightarrow&\quad b = -\tfrac{3}{16}\\
x^1: &\quad 4c = 2b = -\tfrac{3}{8} &\hookrightarrow&\quad c = -\tfrac{3}{32}\\
x^0: &\quad 4d = c = -\tfrac{3}{32} &\hookrightarrow&\quad d = -\tfrac{3}{128}.
\end{aligned}
$$

A particular solution is therefore

$$y_p(x) = -\frac{1}{4}\,x^3 - \frac{3}{16}\,x^2 - \frac{3}{32}\,x - \frac{3}{128}.$$

(3) So the **general solution** of the inhomogeneous differential equation is

$$y(x) = y_h(x) + y_p(x) = c\,e^{4x} - \frac{1}{4}\,x^3 - \frac{3}{16}\,x^2 - \frac{3}{32}\,x - \frac{3}{128}. \qquad \square$$

Example 13.12 (With Maple-Worksheet). Given is the differential equation from Example 13.10 for an AC voltage $U_b(t) = \hat{U}_b \sin(\omega t)$ with the parameters $R \cdot C = 1$

$$\dot{U}(t) = -U(t) + \hat{U}_b \sin(\omega t); \quad U(0) = 0.$$

This is a first-order linear differential equation with a constant coefficient.

(1) The **homogeneous** differential equation $\dot{U}(t) = -U(t)$ is solved by the approach $U_h(t) = c\,e^{\lambda t}$. Inserting this in the differential equation results in $\lambda = -1$.

$$\Rightarrow U_h(t) = c\,e^{-t}.$$

(2) To solve the **inhomogeneous** differential equation, the approach for a particular solution is chosen according to Table 13.1:

$$U_p(t) = A \sin(\omega t) + B \cos(\omega t). \qquad (*)$$

We insert $U_p(t)$ into the inhomogeneous differential equation to find the unknown constants A and B:

$$A\,\omega\,\cos(\omega t) - B\,\omega\,\sin(\omega t) = -A\,\sin(\omega t) - B\,\cos(\omega t) + \hat{U}_b\,\sin(\omega t).$$

We rearrange the right-hand side of the equation by multiples of $\cos(\omega t)$ and $\sin(\omega t)$.

$$(-B\,\omega)\sin(\omega t) + (A\,\omega)\cos(\omega t) = \left(\hat{U}_b - A\right)\sin(\omega t) - B\,\cos(\omega t).$$

This identity is only true if the coefficients of the sine and cosine functions on both sides of the equation match. Comparing the coefficients of $\cos(\omega t)$ and $\sin(\omega t)$ leads to the system of linear equations

$$
\begin{aligned}
\cos(\omega t): \quad A\,\omega \;&=\; -B & (1)\\
\sin(\omega t): \quad -B\,\omega \;&=\; \hat{U}_b - A & (2)
\end{aligned}
$$

from which we can find A and B. We insert (1) into (2)

$$-B\,\omega = \hat{U}_b + \frac{1}{\omega}B \;\Rightarrow\; B = \frac{-\omega}{\omega^2 + 1}\,\hat{U}_b.$$

With (1) then follows

$$A = \frac{\hat{U}_b}{\omega^2 + 1}.$$

So a particular solution according to $(*)$ is

$$U_p(t) = \frac{\hat{U}_b}{\omega^2 + 1}\left(\sin(\omega t) - \omega\,\cos(\omega t)\right).$$

(3) The **general solution** is

$$U(t) = U_h(t) + U_p(t) = c\,e^{-t} + \frac{\hat{U}_b}{\omega^2 + 1}\left(\sin(\omega t) - \omega\,\cos(\omega t)\right).$$

(4) Finally, the constant c is determined by the initial condition $U(0) = 0$:

$$0 = c + \frac{\hat{U}_b}{\omega^2 + 1}\,(0 - \omega) \;\Rightarrow\; c = \frac{\omega\,\hat{U}_b}{\omega^2 + 1}.$$

The solution of the problem is therefore given by

$$\boxed{U(t) = \frac{\hat{U}_b}{\omega^2 + 1}\left(\omega\,e^{-t} + \sin(\omega t) - \omega\,\cos(\omega t)\right)}$$

which matches with the result from Example 13.10. $\square$

13.5 First-Order Non-linear Differential Equations

The method of separation of variables is not only used to solve homogeneous linear differential equations, but some *non-linear* differential equations can also be solved with this method. Substitution methods and power series approaches to solving non-linear differential equations are also introduced.

13.5.1 Differential Equation with Separable Variables

We consider the differential equation

$$y'(x) = f(x)\, g(y)$$

with given continuous functions f and g. If g is not a linear function, then the differential equation is non-linear! Nevertheless, it can be solved by the method of **separation of variables** already introduced for solving homogeneous first-order *linear* differential equations. This type of differential equation is therefore called a *separable* differential equation. The differential equation is rewritten as follows

$$\frac{dy}{dx} = f(x) \cdot g(y) \qquad |: g(y) \cdot dx$$

$$\hookrightarrow \frac{dy}{g(y)} = f(x) \cdot dx.$$

The left side of the equation now contains only the variable y and the right side only the variable x. The subsequent integration gives

$$G(y) = \int \frac{dy}{g(y)} = \int f(x)\, dx.$$

The root function of the left integral $G(y)$ is then resolved with respect to y, which is possible in many cases.

Example 13.13. The differential equation $y'(x) = e^{y(x)} \cos x$ with $y(0) = y_0$ is solved by separating the variables:

$$\frac{dy}{dx} = e^y \cos x \qquad |: e^y \cdot dx$$

$$\frac{dy}{e^y} = \cos x \, dx.$$

As there is an initial value problem, the definite integral is chosen on both sides of the equation. The integration over $\tilde{y}$ starts at y_0 and the integration over $\tilde{x}$ starts at $x_0 = 0$:

$$\int_{y_0}^{y} e^{-\tilde{y}}\, d\tilde{y} = \int_{0}^{x} \cos \tilde{x}\, d\tilde{x}.$$

The integration is performed on both sides and then solved for y

$$-e^{-\tilde{y}}\Big|_{y_0}^{y} = \sin \tilde{x}\Big|_{0}^{x} \qquad \hookrightarrow \qquad -e^{-y} + e^{-y_0} = \sin x$$

$$\hookrightarrow \quad e^{-y} = e^{-y_0} - \sin x \qquad \Rightarrow \quad y(x) = -\ln\left(e^{-y_0} - \sin x\right).$$

Remark: If the indefinite form $\int e^{-y}\, dy = \int \cos x\, dx$ is chosen instead of the definite integral, an integration constant C must be taken into account. This constant C will be finally determined by the initial condition $y(0) = y_0$. $\qquad\qquad\qquad\qquad\qquad\qquad\qquad\qquad\qquad$ □

Application Example 13.14 **(Free Fall with Air Resistance).**

We study the rate of descent $v(t)$ of a body with mass m under air resistance. The forces acting on the body are

(1) the gravity $m\,g$,

(2) the air resistance $-k\,v^2$, assuming a quadratic dependence of the frictional force on the velocity,

where g is the acceleration due to gravity and k is the coefficient of friction. According to Newton's law of motion, the accelerating force $m\frac{dv}{dt}$ is equal to the sum of all the forces acting on the mass

$$m\,\frac{dv(t)}{dt} = m\,g - k\,v^2(t).$$

Assuming a free fall from rest, the initial condition is $v(0) = 0$. This first-order *non-linear* differential equation is solved by separating the variables:

$$\frac{dv}{dt} = g - \frac{k}{m}\,v^2 \qquad \Big| : \left(g - \frac{k}{m}\,v^2\right)\ \cdot dt$$

$$\hookrightarrow \quad \frac{dv}{g - \frac{k}{m}\,v^2} = dt.$$

We integrate the left and right sides independently

$$\int_0^t d\tilde{t} = \int_0^v \frac{d\tilde{v}}{g - \frac{k}{m}\tilde{v}^2} = \frac{1}{g}\int_0^v \frac{d\tilde{v}}{1 - \frac{k}{mg}\tilde{v}^2}.$$

The left integral gives $\int_0^t d\tilde{t} = t$. To calculate the right integral we substitute $\xi = \sqrt{\frac{k}{mg}}\,v$. So $d\xi = \sqrt{\frac{k}{mg}}\,dv$ and we get

$$\int \frac{d\tilde{v}}{1 - \frac{k}{mg}\tilde{v}^2} = \sqrt{\frac{mg}{k}}\int \frac{d\xi}{1 - \xi^2} = \sqrt{\frac{mg}{k}}\,\operatorname{artanh}(\xi) + C.$$

After substituting the corresponding integral, we obtain the expression

$$\frac{1}{g}\int_0^v \frac{d\tilde{v}}{1 - \frac{k}{mg}\tilde{v}^2} = \frac{1}{g}\sqrt{\frac{mg}{k}}\,\operatorname{artanh}\left(\sqrt{\frac{k}{mg}}\,\tilde{v}\right)\Bigg|_0^v$$

$$= \sqrt{\frac{m}{kg}}\,\operatorname{artanh}\left(\sqrt{\frac{k}{mg}}\,v\right).$$

All in all it follows

$$t = \sqrt{\frac{m}{kg}}\,\operatorname{artanh}\left(\sqrt{\frac{k}{mg}}\,v\right).$$

This equation is solved by applying the function tanh:

$$\sqrt{\frac{kg}{m}}\,t = \operatorname{artanh}\left(\sqrt{\frac{k}{mg}}\,v\right) \hookrightarrow \tanh\left(\sqrt{\frac{kg}{m}}\,t\right) = \sqrt{\frac{k}{mg}}\,v$$

$$\Rightarrow v(t) = \sqrt{\frac{mg}{k}}\,\tanh\left(\sqrt{\frac{kg}{m}}\,t\right).$$

The free fall velocity $v(t)$ of a body with mass m under air resistance is shown in Fig. 13.9:

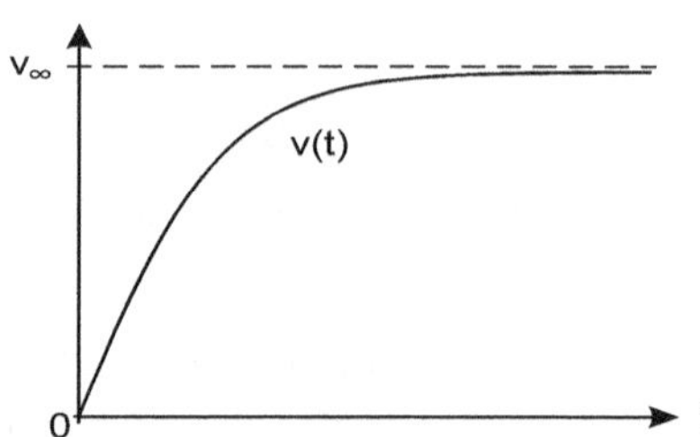

Figure 13.9. Free fall velocity under air resistance

Discussion: As $t \to \infty$ the final velocity $v_\infty = \lim_{t\to\infty} v(t) = \sqrt{\frac{mg}{k}}$ is reached, because $\lim_{x\to\infty} \tanh(x) = 1$. The body will then fall at a constant velocity because the frictional force and the gravitational force cancel each other out.   □

13.5.2 Solving Differential Equations by Substitution

(1) Differential equations of the type

$$y'\left(x\right) = f\left(\frac{y\left(x\right)}{x}\right)$$

can be transformed by the **substitution**

$$u\left(x\right) := \frac{y\left(x\right)}{x}$$

into a differential equation for $u\left(x\right)$, which can often be solved by separating the variables. To obtain the differential equation for $u(x)$, both $\frac{y(x)}{x}$ and $y'(x)$ must be replaced by terms in $u(x)$ and $u'(x)$. From $u\left(x\right) = \frac{y(x)}{x}$ follows $y\left(x\right) = x \cdot u\left(x\right)$ and the product rule gives

$$y'\left(x\right) = u\left(x\right) + x\,u'\left(x\right).$$

If we replace $y'\left(x\right)$ and $\frac{y}{x}$, we get the differential equation for $u\left(x\right)$:

$$u\left(x\right) + x \cdot u'\left(x\right) = f\left(u\right)$$

$$\Rightarrow x \cdot u' = f\left(u\right) - u.$$

Separating the variables gives

$$\frac{du}{f\left(u\right) - u} = \frac{dx}{x}$$

and the following integration

$$\int \frac{du}{f\left(u\right) - u} = \int \frac{dx}{x} = \ln|x| + C.$$

Integrating the left-hand side gives an expression for $u(x)$ which, if possible, is solved for $u(x)$. **Back-substitution** gives the solution $y\left(x\right)$.

Example 13.15. Given is the initial value problem

$$x^2\,y'\left(x\right) = y^2\left(x\right) + x \cdot y\left(x\right) \quad \text{with } y\left(1\right) = -1.$$

Division by x^2

$$y'(x) = \left(\frac{y(x)}{x}\right)^2 + \left(\frac{y(x)}{x}\right)$$

and subsequent substitution $\boxed{u(x) := \dfrac{y(x)}{x}}$ gives

$$y(x) = x \cdot u(x) \;\hookrightarrow\; y'(x) = u(x) + x \cdot u'(x) \;,$$

and

$$u(x) + x \cdot u'(x) = u^2(x) + u(x)$$

$$\Rightarrow \boxed{\; u'(x) = \frac{1}{x}\,u^2(x) \quad \text{with } u(1) = \frac{y(1)}{1} = -1. \;}$$

Separating the variables provides

$$\frac{du}{u^2} = \frac{dx}{x} \;\hookrightarrow\; \int_{-1}^{u}\frac{d\tilde{u}}{\tilde{u}^2} = \int_{1}^{x}\frac{d\tilde{x}}{\tilde{x}} \;\hookrightarrow\; -\tilde{u}^{-1}\Big|_{-1}^{u} = \ln|\tilde{x}|\Big|_{1}^{x}$$

$$\hookrightarrow -\frac{1}{u} - 1 = \ln x \;\hookrightarrow\; u(x) = \frac{1}{-1 - \ln x}.$$

Back-substitution gives finally the solution

$$y(x) = -\frac{x}{1 + \ln x}. \qquad\qquad \square$$

(2) Differential equations of the type

$$y'(x) = f(a\,x + b\,y + c)$$

are transformed by the **substitution**

$$u(x) := a\,x + b\,y + c$$

into a differential equation for $u(x)$, which can also be solved by separating the variables: For $u(x) = a\,x + b\,y(x) + c$ follows

$$u'(x) = a + b\,y'(x) \;\hookrightarrow\; y'(x) = \frac{1}{b}\,(u'(x) - a)$$

for $b \neq 0$. So the differential equation for $u(x)$ is

$$\boxed{u'(x) - a = b\,f(u).}$$

Subsequent separation of the variables leads to the solution. The special case $b = 0$ is integrated directly without substitution to get $y(x)$.   □

Example 13.16. The differential equation

$$y'(x) = \frac{1}{1 + x - y(x)}$$

is replaced by the substitution

$$\boxed{u(x) := 1 + x - y(x)}$$

$(\hookrightarrow u'(x) = 1 - y'(x))$ in the differential equation

$$1 - u'(x) = \frac{1}{u(x)} \quad \Rightarrow \quad u'(x) = 1 - \frac{1}{u(x)} = \frac{u(x) - 1}{u(x)}$$

Separation of the variables gives for $u \neq 1$

$$\frac{u}{u - 1}\,du = dx.$$

Integration with partial fraction decomposition $\dfrac{u}{u - 1} = 1 + \dfrac{1}{u - 1}$ gives

$$\int \left(1 + \frac{1}{u - 1}\right) du = \int dx \hookrightarrow u + \ln|u - 1| = x + C.$$

The back substitution $u = 1 + x - y$ results in

$$1 + x - y + \ln|x - y| = x + C$$

$$\Rightarrow \ln|x - y| = y + C - 1.$$

This is an implicit equation for $y(x)$ that can **not** be solved for $y(x)$

$$|x - y(x)| = e^{C-1}\,e^{y(x)}.$$

Note that in the special case of $u = 1$, which was excluded in the above considerations, $y(x) = x$ is the solution due to the definition of $u(x)$.   □

13.5.3 Power Series Approach

Another way of finding the solution to a differential equation is to take a **power series approach** to the solution at the expansion point x_0 (see Section 9.2):

$$y(x) = a_0 + a_1(x - x_0) + a_2(x - x_0)^2 + \cdots + a_n(x - x_0)^n + \cdots$$

$$= \sum_{n=0}^{\infty} a_n(x - x_0)^n.$$

This approach includes the unknown coefficients $a_0, a_1, \ldots, a_n, \ldots$, which are determined by inserting the function $y(x)$ into the differential equation. Since power series can be differentiated as often as desired within their convergence range, this method can also be used for higher order differential equations.

> The unknown function and all its derivatives appearing in the differential equation are replaced by the power series approach or its derivatives, and an attempt is made to calculate the coefficients a_i ($i \in \mathbb{N}_0$) by comparing the coefficients. The convergence range of the series is then determined.

Example 13.17. Given is the differential equation

$$y'(x) + 2xy(x) - 2x^2 - 1 = 0.$$

We choose a power series approach at the point $x_0 = 0$:

$$y(x) = a_0 + a_1 x + a_2 x^2 + \cdots + a_n x^n + \cdots = \sum_{n=0}^{\infty} a_n x^n$$

$$\hookrightarrow y'(x) = a_1 + 2a_2 x + 3a_3 x^2 + \cdots + n a_n x^{n-1} + \cdots = \sum_{n=1}^{\infty} n a_n x^{n-1}.$$

Inserting this into the differential equation, we get

$$\begin{aligned}
&\left(a_1 + 2a_2 x + 3a_3 x^2 + 4a_4 x^3 + \cdots + (n+1)a_{n+1} x^n + \cdots\right) \\
&+ 2x\left(a_0 + a_1 x + a_2 x^2 + \cdots + a_{n-1} x^{n-1} + a_n x^n + \cdots\right) \\
&- 2x^2 - 1 = 0.
\end{aligned}$$

By reordering with respect to the powers of x, we continue

$$(a_1 - 1)\, x^0 + (2\, a_2 + 2\, a_0)\, x^1 + (3\, a_3 + 2\, a_1 - 2)\, x^2 + (4\, a_4 + 2\, a_2)\, x^3$$
$$+ \cdots + [(n + 1)\, a_{n+1} + 2\, a_{n-1}]\, x^n = 0.$$

We compare the coefficients starting from x^0

$$
\begin{array}{llll}
x^0: & a_1 - 1 = 0 & \Rightarrow & a_1 = 1 \\
x^1: & 2\, a_2 + 2\, a_0 = 0 & \Rightarrow & a_2 = -a_0 \\
x^2: & 3\, a_3 + 2\, a_1 - 2 = 0 & \Rightarrow & a_3 = 0 \\
x^3: & 4\, a_4 + 2\, a_2 = 0 & \Rightarrow & a_4 = \frac{(-2)}{4}\,(-a_0) \\
x^4: & 5\, a_5 + 2\, a_3 = 0 & \Rightarrow & a_5 = 0 \\
x^5: & 6\, a_6 + 2\, a_4 = 0 & \Rightarrow & a_6 = -\frac{2}{6}\, a_4 = \frac{(-2)\,(-2)}{6}\,(-a_0) \\
\vdots & & & \vdots \\
x^n: & (n + 1)\, a_{n+1} + 2\, a_{n-1} = 0. & &
\end{array}
$$

So for odd indices, we obtain $a_3 = a_5 = a_7 = \cdots = a_{2\,k+1} = 0$, $(k = 1, 2, 3, \ldots)$, and for even indices

$$
\begin{aligned}
a_{2\,k} &= \frac{(-2)\,(-2) \cdot \ldots \cdot (-2)}{2\,k\,(2\,k - 2) \cdot \ldots \cdot 4}\,(-a_0) = \frac{(-2)^{k-1}}{2^{k-1}\,k!}\,(-a_0) \\
&= \frac{(-1)^{k-1}}{k!}\,(-a_0) = \frac{(-1)^k}{k!}\, a_0 \qquad k = 1, 2, 3, \ldots.
\end{aligned}
$$

The solution has its representation in form of a power series

$$y(x) = a_0 + 1\,x - a_0\, x^2 + a_0\, \frac{1}{2!}\, x^4 - a_0\, \frac{1}{3!}\, x^6 + a_0\, \frac{1}{4!}\, x^8 \pm \cdots$$

$$y(x) = x + a_0 \sum_{k=0}^{\infty} (-1)^k\, \frac{x^{2\,k}}{k!}.$$

The unknown constant a_0 corresponds to the constant of the homogeneous part of the solution expected in a linear differential equation. The convergence range of the series is $\mathbb{R}$. We rewrite the power series in the form

$$y(x) = x + a_0 \sum_{k=0}^{\infty} \frac{1}{k!}\,(-x^2)^k$$

to identify the exponential function $e^t = \sum_{k=0}^{\infty} \frac{1}{k!}\, t^k$. In this special example, there exists a closed representation of this solution:

$$y(x) = x + a_0\, e^{-x^2}. \qquad\qquad \square$$

13.6 Numerical Solution of 1st Order DEq

As discussed in the previous section, non-linear differential equations can be solved analytically, e.g., if they are separable. However, for other first-order differential equations or systems of differential equations, an approximate method is required to solve them.

13.6.1 Directional Fields

Characteristic of first-order differential equations is that we know the derivative of the unknown function

$$y'(x) = f(x, y(x))$$

at each point (x, y) because it is given by the right-hand side of the differential equation $f(x, y(x))$. The differential equation can therefore be represented as a *directional field*, in which the slope of the function $y(x)$ (i.e. $f(x, y(x))$) is plotted as a vector at each point in the plane. For example, for the differential equation

$$y'(x) = -y(x) + 1,$$

the directional field is given by

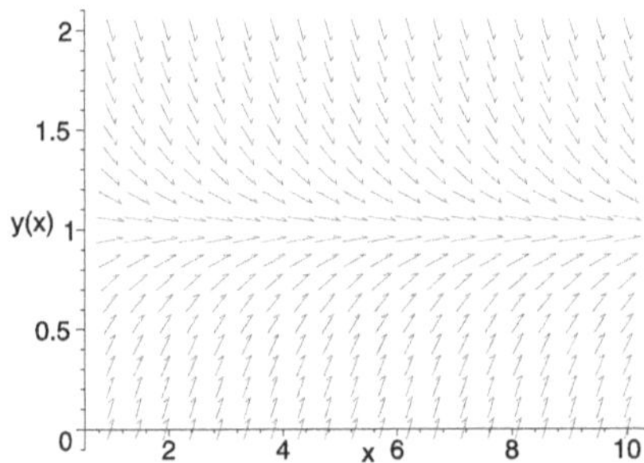

Figure 13.10. Directional field of the differential equation

If the initial value $y(x_0)$ of the differential equation is given, then we know the point in the plane $(x_0, y(x_0))$ where the solution begins. With this additional information we can construct the solution starting at the point $(x_0, y(x_0))$: We use the slope of the function at this point $f(x_0, y(x_0))$ to go to the next point at the position $x_0 + dx$ and thus get y at $x_0 + dx$. Then, both $y(x_0 + dx)$ and the slope $y'(x_0 + dx) = f(x_0 + dx, y(x_0 + dx))$ are known and $y(x_0 + 2dx)$ can be constructed next and so on.

In Fig. 13.11 shows the directional field of the differential equation along with the approximate solution for the initial value $y(4) = 0$.

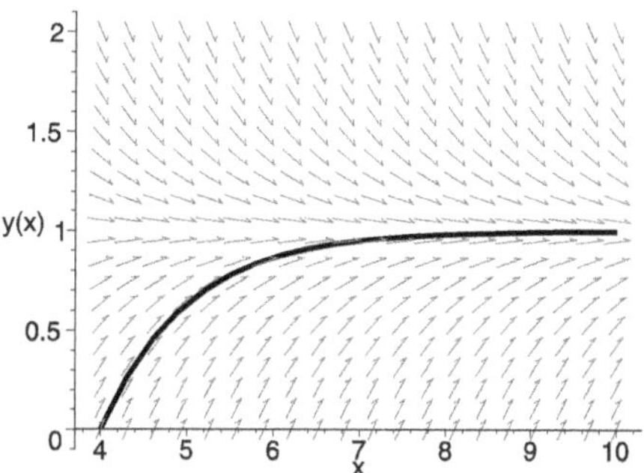

Figure 13.11. Construction of the solution using the directional field

This procedure leads to the Euler method for the numerical solution of first-order differential equations.

13.6.2 Euler's Line Traction Method

Starting from the initial value problem (IVP)

$$y'(t) = f(t, y(t)) \quad \text{with} \quad y(t_0) = y_0 . \tag{1}$$

We will solve this IVP numerically for times $t_0 < t \leq T$. To do this, the interval $[t_0, T]$ is divided into N sub-intervals of the length

$$h = dt = \frac{T - t_0}{N}.$$

The variables h or dt are called **step size** and **time step**, respectively. Intermediate times are defined at

$$t_j = t_0 + j \cdot dt \qquad j = 0, \ldots, N.$$

We will compute the solution only at these discrete times $t_0, t_1, t_2, \ldots, t_N$: Starting from the initial value y_0, we determine the approximations $y_1, y_2, \ldots, y_N$ for the function values $y(t_1)$, $y(t_2)$, $\ldots$, $y(t_N)$ of the solution of (1). This procedure is called the **discretization** of the IVP.

For the initial value (t_0, y_0) the exact slope $\tan \alpha$ of the solution function is known from equation (1)

$$y'(t_0) = \tan \alpha = f(t_0, y_0) .$$

For a small increment h, the function y is approximated in the interval $[t_0, t_0 + h]$ by its tangent (linearization) (see Fig. 13.12 a). For the functional value $y(t_1)$ the following approximation applies

$$y(t_1) = y(t_0 + h) \approx y(t_0) + y'(t_0) \cdot h.$$

We define

$$\boxed{y_1 := y_0 + f(t_0, y_0)\, h.}$$

So the functional value $y(t_1)$ at time t_1 is approximated by y_1. Continuing from this incorrect value y_1, we use (1) and evaluate the gradient $y_1' = f(t_1, y_1)$. With y_1 and y_1' we approximate at the next time $t_2 = t_1 + h$:

$$y(t_2) = y(t_1 + h) \approx y(t_1) + y'(t_1)\, h. \qquad \text{(Linearization)}$$

We set

$$\boxed{y_2 := y_1 + f(t_1, y_1)\, h.}$$

y_2 is an approximation to the exact value of $y(t_2)$ at the time t_2.

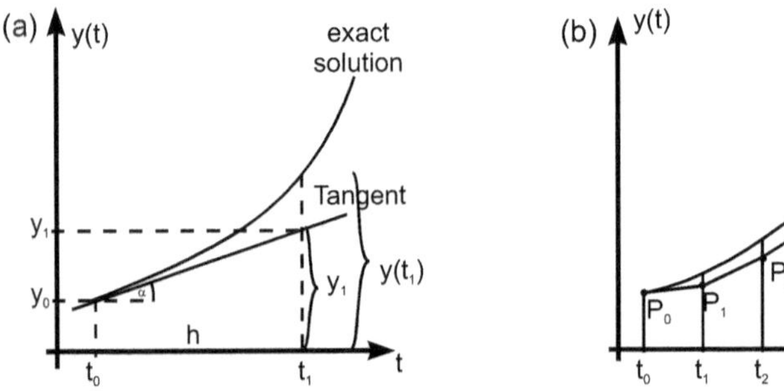

Figure 13.12. a) 1st step of approximation   b) Euler's polygon tracing method

The procedure described is repeated for the new point $P_2 = (t_2, y_2)$, which is usually not on the exact solution curve. In general, the approximation method can be described by

$$\text{new value} = \text{old value} + \text{change in solution}$$

or rather

$$y_{new} = y_{old} + f\left(t_{old},\, y_{old}\right) \cdot h.$$

Starting from the point $P_0 = (t_0,\, y_0)$, we update the values

$$y_{i+1} = y_i + h\,f\left(t_i,\, y_i\right) \qquad i = 0,\, 1,\, 2,\, \ldots,\, N - 1.$$

The solution curve is therefore composed of straight lines (see Fig. 13.12b). According to its geometrical meaning, this method is called the **Polygon Tracing Method** or, according to its inventor, **Euler's method.**

Note: For many examples, this simple procedure provides sufficiently accurate approximations $y_1,\, y_2,\, \ldots,\, y_N$ for the functional values $y\left(t_1\right),\, y\left(t_2\right),\, \ldots,\, y\left(t_N\right)$ that are being sought.

Example 13.18 (With MAPLE-Worksheet). The initial value problem

$$y'\left(t\right) = y\left(t\right) + e^t \quad \text{with} \;\; y\left(0\right) = 1$$

has the exact solution

$$y\left(t\right) = \left(t + 1\right) e^t.$$

Using the Euler method, approximate solutions of this differential equation are computed for increments of $h = 0.05$ and $h = 0.025$ in the interval $0 \le t \le 0.2$ and compared with the exact solution. The iteration rule is

$$y_{i+1} = y_i + h\left(y_i + e^{t_i}\right).$$

Especially for the step size $h = 0.05$ we compute

$$
\begin{aligned}
y_0 &= 1 \\
y_1 &= y_0 + h\left(y_0 + e^{0.00}\right) = 1.1 \\
y_2 &= y_1 + h\left(y_1 + e^{0.05}\right) = 1.207564 \\
y_3 &= y_2 + h\left(y_2 + e^{0.1}\right) = 1.323201 \\
y_4 &= y_3 + h\left(y_3 + e^{0.15}\right) = 1.447453
\end{aligned}
$$

The table below shows these approximate values are given for $h = 0.05$ and $h = 0.025$ along with the exact values.

Results:

t	$y\ (h = 0.05)$	$y\ (h = 0.025)$	y exact
0.00	1.000 000	1.000 000	1.000 000
0.05	1.100 000	1.101 883	1.103 835
0.10	1.207 564	1.211 552	1.215 688
0.15	1.323 201	1.329 535	1.336 109
0.20	1.447 453	1.456 396	1.465 683

Discussion: The comparison shows that the approximate values (second and third columns) improve with smaller increments. To confirm this behavior, select $h = 0.0125$ and $h = 0.00625$ and compare these values with the exact solution. $\square$

13.7 Problems on First-Order Differential Equations

13.1 What are the general solutions of the following linear first-order differential equations with constant coefficients?

a) $y' + 4y = 0$ b) $2y' + 4y = 0$ c) $-3y' = 8y$

d) $ay' - by = 0$ $(a \neq 0)$ e) $-3y' + 18y = 0$ f) $L\frac{dI}{dt} + RI = 0$

13.2 Solve the following initial value problems:

a) $2v + \dot{v} = 0$, $v(0) = 10\,\frac{m}{s}$.   When is $v(t) < 10\,\frac{cm}{s}$?

b) $y' + \lambda y = 0$, $y(0) = 1$.   What is the result for λ, if $y(1) = \frac{1}{2}$?

c) $-\frac{dN}{dt} = \frac{1}{\tau}N$, $N(0) = N_0$.

How big is τ when $N(1{,}819 \cdot 10^{11}) = \frac{1}{2}N_0$?

13.3 Determine the general solution of the following linear first-order differential equations:

a) $y' + xy = 4x$ b) $y' + \frac{y}{1+x} = e^{2x}$

c) $xy' + y = x \cdot \sin x$ d) $y' \cos x - y \sin x = 1$

e) $y' - 2\cos x\, y = \cos x$ f) $xy' - y = x^2 + 4$

13.4 Solve the following inhomogeneous differential equation:

a) $y'(t) + \frac{1}{RC}y(t) = U_0 \sin(\omega t)$ (RC alternating current circuit)

b) $y'(t) + \frac{R}{L}y(t) = U_0 e^{-2t}$ (RL alternating current circuit)

13.5 Determine a solution for the following first-order differential equation by separating the variables:

a) $x^2 y' = y^2$ b) $y'(1 + x^2) = xy$ c) $y' = 1 - y^2$

d) $y' = (1 - y)^2$ e) $y' \sin y = -x$ f) $y' = e^y \cos x$

13.6 Solve the following initial value problems:

a) $y' + \cos x \cdot y = 0$; $y\left(\frac{\pi}{2}\right) = 2\pi$

b) $x(x+1)y' = y$; $y(1) = \frac{1}{2}$

c) $y^2 y' + x^2 = 1$; $y(2) = 1$

d) $x^2 y' = y^2 + xy$; $y(1) = -1$ (Substitute: $u = \frac{y}{x}$)

e) $yy' = 2e^{2x}$; $y(0) = 2$

13.7 Solve by substitution $\left(u = \frac{y}{x}\right)$

a) $xy' = y + 4x$ b) $x^2 y' = \frac{1}{4}x^2 + y^2$

13.8 Solve the following differential equations by separating the variables:

a) $\frac{1}{x^5}y'(x) = \frac{1}{(y(x))^2}$ b) $(x+1)(x-1)y'(x) = 2y(x)$

c) $xy' + \frac{1}{x}y' = y$ d) $y^2 - 2yy' + 1 = 0$

e) $y' = \cos^2 y$ f) $2y' = y^4 \cdot \sqrt{x}$

13.9 Which solutions have the following initial value problems:
a) $y'(x) = \sin x \cdot y^2(x)$, $y(0) = 1$ b) $y + y' = e^x$, $y(0) = 1$
c) $\sin x \cdot y' = \cos x \cdot y$, $y\left(\frac{\pi}{2}\right) = \frac{\pi}{2}$ d) $y' \cdot (1 + x^2) = 2xy$, $y(1) = 4$

Applications

13.11 A chemical reaction $A + B \rightarrow X$ can be explained by

$$\frac{dx}{dt} = k\,(a - x)\,(b - x)$$

if the number of molecules of type A or B at the beginning of the reaction are a or b and $x(t)$ is the number of reaction molecules X at time t (k: reaction constant).
a) Solve the DG for $a \neq b$ and $x(0) = 0$.
b) When does the reaction stop assuming $a > b$?

13.12 The sinking velocity $v(t)$ of a particle of mass m in a liquid (k: friction factor, g: acceleration due to gravity) is described by

$$m\,\frac{dv}{dt} + k\,v = m\,g.$$

a) Determine the speed and position at times $t > 0$ for the initial values $v(0) = v_0$ and $s(0) = 0$.
b) What is the maximum speed $v_{\max}$ the particle can reach?

13.13 Solve task 13.12 if the medium has a resistance equal to $k\,v^2$ and $v(0) = 0$.

13.14 A body has at $t = 0$ the temperature T_0 and is subsequently cooled by passing air of constant temperature T_L according to

$$\frac{dT}{dt} = -a\,(T - T_L) \qquad (a > 0)$$

Determine $T(t)$ of the body temperature for $t > 0$. Which temperature is finally achieved?

13.15 The radial velocity distribution of stationary, laminar flows of a viscous incompressible fluid (viscosity η) along a pipe in which a pressure drop $\frac{\delta p}{\delta z}$ acts can be described by

$$-\frac{\Delta p}{\Delta z} + \eta\,\frac{1}{r}\,\frac{d}{dr}\left(r\,\frac{d}{dr}\,v_z(r)\right) = 0$$

How big is $v_z(r)$ if $v_z(R) = 0$ applies at the edge?
Note: After suitable transformations, integrate first from 0 to r and consider the integration constant at $r = 0$. Then integrate from R to r.

13.16 A body rolls an inclined plane (angle φ) and experiences frictional forces proportional to its speed and a pressure resistance proportional to the square of its speed. It applies:

$$m \cdot \dot{v} + R \cdot v + D \cdot v^2 = m \cdot g \cdot \sin \varphi.$$

The initial speed is $v(0) = 0$; the physical constants are $m = 1$, $g = 10$, $\varphi = \frac{\pi}{3}$, $D = \frac{4}{5}$, $R = 3$.
a) What is $v(t)$ with $D = 0$?
b) What is $v(t)$ with $R = 0$?
c) What is $v(t)$ with $D \neq 0$ and $R \neq 0$?

Chapter 14
Laplace Transform

An elegant method for solving differential equations makes use of the *Laplace transform*. The Laplace integral is particularly suitable for treating differential equations and systems of differential equations with initial conditions. The mathematical formulation of the Laplace transform of a time function $f(t)$ is

$$\mathcal{L}(f(t)) = F(s) = \int_0^\infty f(t)\ e^{-st}\ dt.$$

The *time function* $f(t)$ is associated with an *image function* $F(s)$, so that the Laplace transform is also called a *functional transformation*.

14

14 Laplace Transform

An elegant method for solving differential equations makes use of the *Laplace transform*. The Laplace integral is particularly suitable for treating differential equations and systems of differential equations with initial conditions. The mathematical formulation of the Laplace transform of a time function $f(t)$ is

$$\mathcal{L}(f(t)) = F(s) = \int_0^\infty f(t)\ e^{-st}\ dt.$$

The *time function* $f(t)$ is associated with an *image function* $F(s)$, so that the Laplace transform is also called a *functional transformation*.

When solving differential equations, it can be seen that the Laplace transform of the function $y(t)$ into the image domain $Y(s)$ transforms the differential equation into an algebraic equation. This algebraic equation for $Y(s)$ is generally easier to solve than the differential equation for $y(t)$. By the inverse transformation we finally obtain the desired function $y(t)$. This general procedure is shown schematically in the following diagram (see Fig. 14.1).

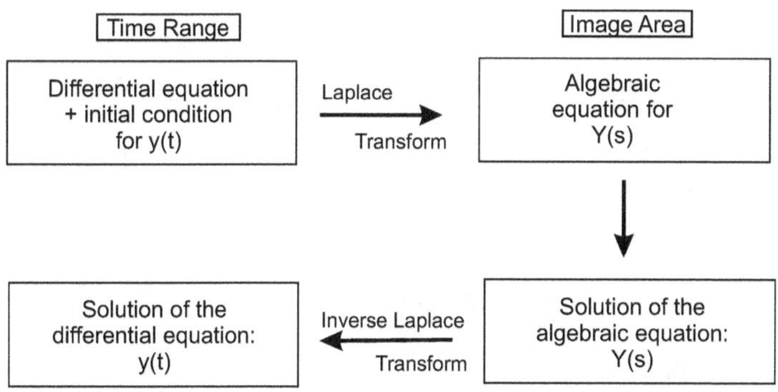

Figure 14.1. Laplace transform from the time domain to the image domain

A major advantage of solving linear differential equations with the Laplace transform is that the inhomogeneities of the differential equation do not have to be continuous **and** initial conditions are automatically satisfied.

Application Example 14.1 (Electrical Network).

An electrical network is shown in Fig. 14.2. We want to find the individual currents $I_1(t)$ and $I_2(t)$ in the two branches for the initial values $I_1(0) = I_2(0) = 0$.

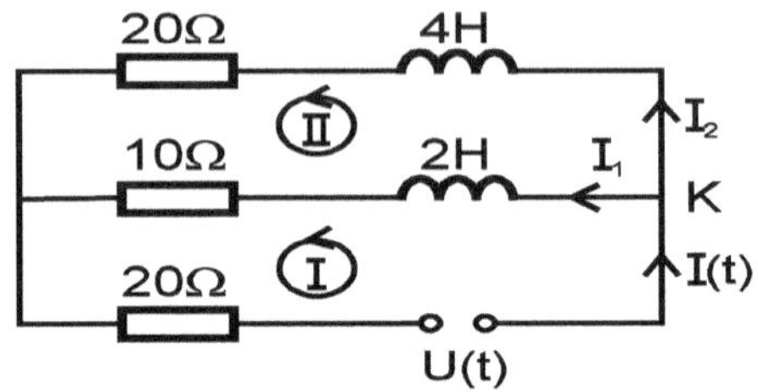

Figure 14.2. Electrical network

Setting up the model equations: We apply Kirchhoff's laws to the network, then according to the mesh rule

$$(I) \quad 20\,I(t) + 2\tfrac{d}{dt}I_1(t) + \;\;10\,I_1(t) = \;\;\;U(t)$$

$$(II)\; -10\,I_1(t) - 2\tfrac{d}{dt}I_1(t) + 4\tfrac{d}{dt}I_2(t) + 20\,I_2(t) = 0$$

and with the node rule

$$(K)\;\; I(t) = I_1(t) + I_2(t).$$

A system of first-order linear differential equation is obtained

$$20\,(I_1(t) + I_2(t)) + 2\,I_1'(t) + 10\,I_1(t) \;\;=\;\; U(t)$$

$$-10\,I_1(t) + 20\,I_2(t) - 2\,I_1'(t) + 4\,I_2'(t) \;\;=\;\; 0$$

with the initial conditions $I_1(0) = I_2(0) = 0$.

We will solve this system of first-order differential equations by using the Laplace transform, also taking into account the initial conditions. $\qquad\square$

14.1 Laplace Transform

The Laplace transform is an integral transformation that maps an arbitrary *time function* $f(t)$, $t \geq 0$, to an *image function* $F(s)$ according to

$$F(s) = \int_0^\infty f(t)\ e^{-st}\ dt.$$

Since the time integration starts at $t = 0$, we generally assume that $f(t) = 0$ for $t < 0$. In this textbook it is sufficient to work with $s \in \mathbb{R}$ so that the result is real-valued and $F(s)$ is a real function.

To ensure that the improper integral and thus the image function $F(s)$ is defined, the integral must be defined for the given s. A sufficient condition for this is that the function $f(t)$ has the following two properties:

Condition 1: $f : [0, \infty) \to \mathbb{R}$ is a **piecewise continuous** function: The domain of the function can be divided into a finite number of subintervals in which the function is continuous and bounded.

Condition 2: $f : [0, \infty) \to \mathbb{R}$ does not grow faster than an exponential function $e^{\alpha t}$ with appropriate $\alpha \in \mathbb{R}_{\geq 0}$: There exist a $T > 0$ and constants $\alpha \geq 0$, $M > 0$ such that

$$|f(t)| \leq M\ e^{\alpha t} \qquad \text{for}\ \ t \geq T\ .$$

f then has **at most of exponential growth** of the order α.

Fig. 14.3 shows a piecewise continuous function with at most exponential growth of order 1. In each subinterval $f(t)$ is continuous and for times $t > T$ it is $f(t) < e^t$.

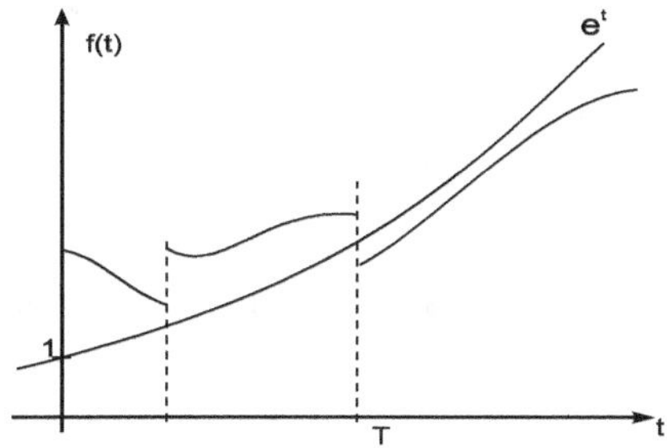

Figure 14.3. Function of at most exponential growth

Examples 14.2:

① Functions with at most exponential growth: $f(t) = const$, t^n, $\cos(t)$, $\sin(t)$, $e^{\alpha t}$, all bounded functions.

② Functions with growth greater than the exponential: e^{t^2}, $e^{\sin(t)\cdot t^3}$.

Theorem 14.1:

If $f : [0, \infty) \to \mathbb{R}$ has at most exponential growth of order α, then

$$\lim_{t \to \infty} e^{-st}\, f(t) = 0 \quad \text{for} \quad s > \alpha.$$

Proof: If $|f(t)| \le M\, e^{\alpha t}$, then for $s > \alpha$

$$\left| e^{-st}\, f(t) \right| \le e^{-st}\, M\, e^{\alpha t} = M\, e^{(\alpha - s)t} \to 0 \qquad \text{as } t \to \infty. \qquad \square$$

Laplace's Theorem:

Let $f : [0, \infty] \to \mathbb{R}$ be a piecewise continuous function with at most exponential growth of order α (i.e. $|f(t)| \le M\, e^{\alpha t}$ for $t > T$). Then the integral

$$\mathcal{L}(f(t)) := F(s) := \int_0^\infty f(t)\, e^{-st}\, dt \quad \text{for} \quad s > \alpha \qquad (*)$$

exists. $\mathcal{L}(f(t))$ is the **Laplace transform (image function)** of the time function $f(t)$.

Remarks:

(1) In general, $s = \delta + i\omega$ is a complex variable and $F(s)$ is a complex function. However, here s is considered to be a real variable and therefore $F(s)$ is a real-valued function, except for the specification of the inverse Laplace transform.

(2) The Laplace transform is used to handle time-varying events that start at time $t = 0$ and thus can be described by a function f with $f(t) = 0$ for $t < 0$.

(3) A pair of functions formed according to the formula $(*)$, $f(t)$ and $F(s)$, is called a **correspondence**. Therefore, the following symbolic notation is also used

$$f(t) \ \circ\!\!-\!\!\bullet \ F(s).$$

(4) We can also form the Laplace transform of functions which, instead of satisfying condition 1, satisfy the following more general condition: In every finite subinterval of $[0, \infty)$, f is piecewise continuous. This property is important for the Laplace transform of periodic functions.

Proof of the Laplace Theorem: We show that the integral

$$\int_0^\infty f(t)\, e^{-st}\, dt$$

has a finite value for every $s > \alpha$: Since f is piecewise continuous, we divide the interval $I = [0, \infty)$ into finitely many subintervals $I_1, \ldots, I_n$, so that f is continuous and bounded at each of these intervals $I_k = [t_{k-1}, t_k]$ $(k = 1, \ldots, n - 1)$ and $I_n = [t_n, \infty)$. Furthermore, $f(t)$ has at most exponential growth of the order α, i.e. there exist a T and constants α, M, so that

$$|f(t)| \le M\, e^{\alpha t} \qquad \text{for} \quad t > T \,.$$

We assume that $T > t_n$ and decompose the domain of f into

$$[0, \infty) = I_1 \cup I_2 \cup \cdots \cup I_{n-1} \cup [t_n, T] \cup [T, \infty) \,.$$

Then

$$\int_0^\infty f(t)\, e^{-st}\, dt = \int_0^{t_1} f(t)\, e^{-st}\, dt + \cdots + \int_{t_{n-1}}^{t_n} f(t)\, e^{-st}\, dt$$
$$+ \int_{t_n}^{T} f(t)\, e^{-st}\, dt + \int_T^\infty f(t)\, e^{-st}\, dt \,.$$

The first $n + 1$ integrals are finite, since f is continuous and bounded. The last integral is finite, since f has at most exponential growth:

$$\left| \int_T^\infty f(t)\, e^{-st}\, dt \right| \le \int_T^\infty e^{-st}\, |f(t)|\, dt \le M \int_T^\infty e^{-st}\, e^{\alpha t}\, dt$$

$$= M \int_T^\infty e^{-(s-\alpha)t}\, dt = M\, \frac{1}{-(s-\alpha)}\, e^{-(s-\alpha)t} \Big|_T^\infty$$

$$= \frac{M}{s-\alpha}\, e^{-(s-\alpha)T} \qquad \text{for} \quad s > \alpha.$$

Thus, all partial integrals are finite and $F(s)$ is defined for $s > \alpha$. $\qquad \square$

Examples 14.3:

① The Laplace transform of the **step function:** Given is the step function (Heaviside function)

$$S(t) := \begin{cases} 0 & \text{for } t < 0 \\ 1 & \text{for } t \geq 0 \end{cases}$$

For $s > 0$ we calculate:

$$\mathcal{L}(S(t)) = \int_0^\infty 1 \cdot e^{-st}\, dt = \left[-\frac{1}{s} e^{-st} \right]_0^\infty = \frac{1}{s} \Rightarrow$$

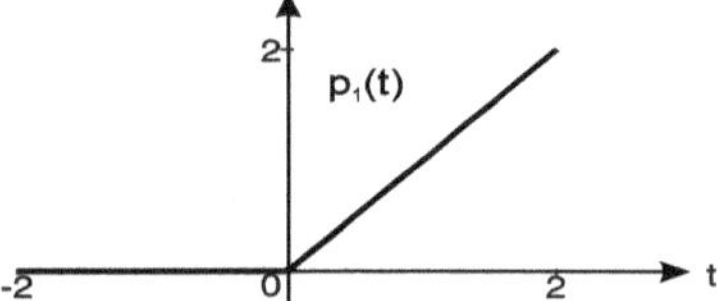

② Laplace transform of a **ramp function:**
(i) The Laplace transform of the function shown below

$$p_1(t) := \begin{cases} 0 & \text{for } t < 0 \\ t & \text{for } t \geq 0 \end{cases}$$

is obtained by integration by parts:

$$\mathcal{L}(p_1(t)) = \int_0^\infty t\, e^{-st}\, dt = t\, \frac{e^{-st}}{-s} \Big|_0^\infty + \frac{1}{s} \int_0^\infty e^{-st}\, dt$$

$$= -\frac{1}{s^2} e^{-st} \Big|_0^\infty = \frac{1}{s^2}\, .$$

The correspondence for this function $p_1(t) = t \cdot S(t)$ is (for $s > 0$)

$$t\, S(t) \; \circ\!\!-\!\!\bullet \; \frac{1}{s^2}\, .$$

(ii) The Laplace transform of the power function

$$p_n(t) := \begin{cases} 0 & \text{for } t < 0 \\ t^n & \text{for } t \geq 0 \end{cases} = t^n S(t) \qquad (n \in \mathbb{N})$$

is $\mathcal{L}(p_n(t)) = \frac{n!}{s^{n+1}}$. This result is obtained inductively by partial integration of

$$\mathcal{L}\left(p_{n+1}(t)\right) = \int_0^\infty t^{n+1}\, e^{-st}\, dt = t^{n+1}\, \frac{e^{-st}}{-s}\Big|_0^\infty + \frac{n+1}{s} \int_0^\infty t^n\, e^{-st}\, dt$$

$$= \frac{n+1}{s}\, \mathcal{L}\left(p_n(t)\right) = \frac{n+1}{s}\, \frac{n!}{s^{n+1}} = \frac{(n+1)!}{s^{n+2}}.$$

The start of the induction is given by ① or rather ②(i).

$$\Rightarrow \quad t^n\, S(t) \; \circ\!\!-\!\!\bullet \; \frac{n!}{s^{n+1}} \qquad\qquad s > 0 \qquad n \in \mathbb{N}_0\ .$$

③ The Laplace transform of the **exponential function:**

$$f(t) = \begin{cases} 0 & \text{for } t < 0 \\ e^{\alpha t} & \text{for } t \geq 0 \end{cases}$$

is $F(s) = \frac{1}{s-\alpha}$ for $s > \alpha$. Because

$$\mathcal{L}\left(f(t)\right) = \int_0^\infty e^{\alpha t}\, e^{-st}\, dt = \int_0^\infty e^{-(s-\alpha)t}\, dt$$

$$= \frac{1}{-(s-\alpha)}\, e^{-(s-\alpha)t}\Big|_0^\infty = \frac{1}{s-\alpha}\ .$$

$$\Rightarrow \quad e^{\alpha t}\, S(t) \; \circ\!\!-\!\!\bullet \; \frac{1}{s-\alpha} \qquad\qquad \text{for } s > \alpha\ .$$

④ The Laplace transform of the **shifted step function** $(\alpha > 0)$

$$S(t-\alpha) = \begin{cases} 0 & \text{for } t < \alpha \\ 1 & \text{for } t \geq \alpha \end{cases}$$

$$\mathcal{L}\left(S(t-\alpha)\right) = \int_0^\infty S(t-\alpha)\, e^{-st}\, dt = \int_\alpha^\infty 1 \cdot e^{-st}\, dt$$

$$= \frac{e^{-st}}{-s}\Big|_\alpha^\infty = \frac{e^{-\alpha s}}{s}$$

$$\Rightarrow \quad S(t-\alpha) \; \circ\!\!-\!\!\bullet \; \frac{e^{-\alpha s}}{s} \qquad\qquad \text{for } s > 0\ . \qquad\qquad \square$$

14.2 Inverse Laplace Transform

The Laplace transform calculates the image function $F(s)$ for a given time function $f(t)$. The inverse transformation from the image domain to the time domain, i.e. the integral of a given image function into the corresponding time function, is called the **Inverse Laplace Transform.**

The following symbols are used to indicate the correspondence between the image domain and the time domain

$$\mathcal{L}^{-1}\left(F\left(s\right)\right) = f\left(t\right) \quad \text{or} \quad F\left(s\right) \; \bullet\!\!-\!\!\circ \; f\left(t\right).$$

We use the correspondence of Fourier integrals from the chapter on Fourier Transform (see Volume 3) to construct an expression for the inverse transform

$$F\left(\omega\right) = \int_{-\infty}^{\infty} f\left(t\right) e^{i\omega t}\, dt \qquad (FT\,1)$$

$$f\left(t\right) = \frac{1}{2\pi} \int_{-\infty}^{\infty} F\left(\omega\right) e^{-i\omega t}\, d\omega \qquad (FT\,2).$$

The above correspondence means that $f\left(t\right)$ can be reconstructed from $F\left(\omega\right)$ using $(FT\,2)$. Since we have a time function $f\left(t\right)$ with $f\left(t\right) = 0$ for $t < 0$, we compute its Laplace transform by setting $s = \delta + i\omega$.

$$F\left(s\right) = F\left(\delta + i\omega\right) = \int_{0}^{\infty} f\left(t\right) e^{-st}\, dt$$

$$= \int_{0}^{\infty} f\left(t\right) e^{-(\delta+i\omega)t}\, dt = \int_{0}^{\infty} \left(f\left(t\right) e^{-\delta t}\right) e^{-i\omega t}\, dt \ .$$

From this representation, we obtain $(FT\,2)$

$$f\left(t\right) \cdot e^{-\delta t} = \frac{1}{2\pi} \int_{-\infty}^{\infty} F\left(\delta + i\omega\right) e^{i\omega t}\, d\omega$$

$$\Rightarrow \qquad f\left(t\right) = \frac{1}{2\pi} \int_{-\infty}^{\infty} F\left(\delta + i\omega\right) e^{\delta t} e^{i\omega t}\, d\omega$$

$$= \frac{1}{2\pi} \int_{\omega=-\infty}^{\omega=\infty} F\left(\delta + i\omega\right) e^{(\delta+i\omega)t}\, d\omega. \qquad (*)$$

We substitute the integration variable ω by $s = \delta + i\omega$, the differential by $d\omega = \frac{1}{i}\,ds$ and the integration limits by

ω	s
$-\infty$	$\delta - i\infty$
$+\infty$	$\delta + i\infty$

So we get

$$\frac{1}{2\pi i} \int_{s=\delta-i\infty}^{s=\delta+i\infty} F\left(s\right) e^{st}\, ds = \begin{cases} f\left(t\right) & \text{for } t \geq 0 \\ 0 & \text{for } t < 0 \end{cases} .$$

The expression $(*)$ represents the **inverse Laplace integral**. The integration takes place in the complex numbers domain and can be evaluated using methods of function theory. However, we will not enter into such a discussion in this context, but will introduce alternative methods in Section 14.4.□

Inverse Laplace Transform

If $F\left(s\right)$ is the Laplace transform of a function, then there exists exactly one continuous function $f\left(t\right)$ with at most exponential growth, such that $\mathcal{L}\left(f\left(t\right)\right) = F\left(s\right)$. $f\left(t\right)$ is given by

$$f\left(t\right) = \mathcal{L}^{-1}\left(F\left(s\right)\right) = \frac{1}{2\pi i} \int_{\delta-i\infty}^{\delta+i\infty} F\left(s\right) e^{st}\, ds .$$

The integration is performed in the complex. If necessary, it can be evaluated with the mathematical methods of function theory, which are not described in detail here. In practice, the back transformation is not done by evaluating the integral, but almost exclusively with the help of tables. This should be noted:

Note:

(1) If two image functions $F\left(s\right) = G\left(s\right)$ are identical, then the associated time functions $f\left(t\right) = \mathcal{L}^{-1}\left(F\left(s\right)\right)$ and $g\left(t\right) = \mathcal{L}^{-1}\left(G\left(s\right)\right)$ differ at most at discontinuities of f or g.

(2) Two **continuous** time functions f and g match if their image functions $\mathcal{L}(f(t))$ and $\mathcal{L}(g(t))$ match.

(3) Since the back transformation $\mathcal{L}^{-1}$ is essentially unambiguous, there is an unambiguous assignment between image and time functions, so that any correspondence

$$f(t) \; \circ\!\!-\!\!\bullet \; F(s)$$

can be read from left to right, but also from right to left

$$F(s) \; \bullet\!\!-\!\!\circ \; f(t) \; .$$

Examples 14.4:

① From the correspondence

$$t^2 \, S(t) \; \circ\!\!-\!\!\bullet \; \frac{2}{s^3} \qquad \text{for} \quad s > 0$$

follows the inverse Laplace transform

$$\frac{2}{s^3} \; \bullet\!\!-\!\!\circ \; t^2 \, S(t) \quad \text{or} \quad \mathcal{L}^{-1}\left(\frac{2}{s^3}\right) = t^2 \, S(t) \; .$$

② From the correspondence

$$S(t - \alpha) \; \circ\!\!-\!\!\bullet \; \frac{1}{s} e^{-\alpha s} \qquad \text{for} \quad s > 0$$

follows by back transformation

$$\mathcal{L}^{-1}\left(\frac{1}{s} e^{-\alpha s}\right) = S(t - \alpha) \quad \text{or} \quad \frac{1}{s} e^{-\alpha s} \; \bullet\!\!-\!\!\circ \; S(t - \alpha) \; . \qquad \square$$

14.3 Two Basic Properties

First we show that the Laplace transform is linear: The transform of a superposition of functions is equal to the superposition of the transformed functions. The second important property is that the transform of the derivative of a function is equal to the transform of the function multiplied by s minus $f(0)$. In the following section it is always assumed that the functions under consideration satisfy the conditions of Laplace's theorem, so that the Laplace transform of the functions is defined.

14.3.1 Linearity

We consider the superposition of two time functions f_1 and f_2 and calculate their Laplace transform:

$$\mathcal{L}\left(c_1 f_1(t) + c_2 f_2(t)\right) = \int_0^\infty \left(c_1 f_1(t) + c_2 f_2(t)\right) e^{-st}\, dt$$

$$= \int_0^\infty \left(c_1 f_1(t)\, e^{-st} + c_2 f_2(t)\, e^{-st}\right) dt$$

$$= c_1 \int_0^\infty f_1(t)\, e^{-st}\, dt + c_2 \int_0^\infty f_2(t)\, e^{-st}\, dt$$

$$= c_1\, \mathcal{L}\left(f_1(t)\right) + c_2\, \mathcal{L}\left(f_2(t)\right).$$

The above calculation steps are based on the linearity of the integral: The integral over a sum of functions is equal to the sum of the integrals, and that constant factors can be prepended to the integral. The Laplace transform assigns the same superposition of image functions to the superposition ($=$ linear combination) of original functions:

> **Linearity of the Laplace Transform**
>
> ---
>
> It holds: $\mathcal{L}\left(c_1 f_1(t) + c_2 f_2(t)\right) = c_1\, \mathcal{L}\left(f_1(t)\right) + c_2\, \mathcal{L}\left(f_2(t)\right).$
>
> Correspondence: $c_1 f_1(t) + c_2 f_2(t)\ \circ\!\!-\!\!\bullet\ c_1 F_1(s) + c_2 F_2(s).$

Examples 14.5:

① Find the Laplace transform of $f(t) = 2\,t^3 - 5\,t^2 + 3$. Using the addition theorem

$$F(s) = \mathcal{L}(2\,t^3 - 5\,t^2 + 3)$$
$$= 2\,\mathcal{L}(t^3) - 5\,\mathcal{L}(t^2) + 3\,\mathcal{L}(1)$$
$$= 2 \cdot \frac{3!}{s^4} - 5 \cdot \frac{2!}{s^3} + 3 \cdot \frac{1}{s} = \frac{12 - 10\,s + 3\,s^3}{s^4}.$$

② Find the Laplace transform of $f(t) = 4\,\sin(\omega t) + 5\,\cos(\omega t)$. With the addition theorem

$$F(s) = 4\,\mathcal{L}(\sin(\omega t)) + 5\,\mathcal{L}(\cos(\omega t))$$
$$= 4 \cdot \frac{\omega}{s^2 + \omega^2} + 5 \cdot \frac{s}{s^2 + \omega^2} = \frac{5\,s + 4\,\omega}{s^2 + \omega^2}. \qquad \square$$

The linearity of the Laplace transform also applies to the inverse transformation by the correspondence:

Addition Theorem of the Inverse Laplace Transform

$$\mathcal{L}^{-1}\left(c_1\,F_1\left(s\right) + c_2\,F_2\left(s\right)\right) = c_1\,\mathcal{L}^{-1}\left(F_1\left(s\right)\right) + c_2\,\mathcal{L}^{-1}\left(F_2\left(s\right)\right).$$

Examples 14.6:

① Find the time function to $F\left(s\right) = \dfrac{3\,s + 8}{s^2 + 16}$:

By decomposing the image function into the partial fractions

$$F\left(s\right) = \frac{3\,s + 8}{s^2 + 16} = 3 \cdot \frac{s}{s^2 + 4^2} + 2 \cdot \frac{4}{s^2 + 4^2}$$

and using the Example 14.5 ②, we obtain its time function

$$f\left(t\right) = 3\,\cos\left(4\,t\right) + 2\,\sin\left(4\,t\right) \ .$$

② Find the time function to $F\left(s\right) = \dfrac{5\,s^2 + 3\,s + 8}{s^3}$:

By decomposing the image function into partial fractions, we get

$$F\left(s\right) = 5 \cdot \frac{1}{s} + 3 \cdot \frac{1}{s^2} + 4 \cdot \frac{2}{s^3} \quad \Rightarrow \quad f\left(t\right) = 5 + 3\,t + 4\,t^2 \ . \qquad \square$$

14.3.2 Laplace Transform of the Derivative

To apply the Laplace transform to differential equations, it is necessary to be able to calculate the Laplace transform of the derivative of a function:

Laplace Transform of the Derivative

Let $f, f' : [0, \infty) \to \mathbb{R}$ be continuous and of at most exponential growth. Then

$$\mathcal{L}\left(f'\left(t\right)\right) = s\,\mathcal{L}\left(f\left(t\right)\right) - f\left(0\right).$$

Proof: Partial integration of the Laplace transform of f' gives:

$$\mathcal{L}\left(f'\left(t\right)\right) = \int_0^\infty f'\left(t\right) e^{-st}\, dt = \left[f\left(t\right) e^{-st}\right]_0^\infty + s \int_0^\infty f\left(t\right) e^{-st}\, dt$$

$$= \lim_{T \to \infty} f\left(T\right) e^{-sT} - f\left(0\right) + s\,\mathcal{L}\left(f\left(t\right)\right).$$

Since $f\left(t\right)$ has at most exponential growth, according to the Theorem 14.1, we have

$$\lim_{T \to \infty} f\left(T\right) e^{-sT} = 0$$

and therefore

$$\mathcal{L}\left(f'\left(t\right)\right) = s\,\mathcal{L}\left(f\left(t\right)\right) - f\left(0\right) . \qquad \square$$

Differentiating a function in the time domain is equivalent to multiplying the transform by s in the image domain and subtracting $f(0)$. So instead of a complicated arithmetic operation in the time domain, a simple multiplication takes place in the image domain.

Repeated application of the derivative rule leads inductively to the Laplace transform of the n-th derivative:

Laplace Transform of the n-th Derivative

Let $f, f', \ldots, f^{(n)} : [0, \infty) \to \mathbb{R}$ be continuous and of at most exponential growth, then

$$\mathcal{L}\left(f^{(n)}\left(t\right)\right) = s^n\, \mathcal{L}\left(f\left(t\right)\right) - s^{n-1} f\left(0\right) - s^{n-2} f'\left(0\right) - \cdots - f^{(n-1)}\left(0\right).$$

The image function of the n-th derivative of $f\left(t\right)$ is equal to the Laplace transform of $f\left(t\right)$ multiplied by s^n minus a polynomial in s of degree $n - 1$ with coefficients defined by f and its derivatives evaluated at $t = 0$.

Note: In the context of the Laplace transform, the function f always means that $f\left(t\right) = 0$ for $t < 0$. This results in the left limit being zero at the point $t = 0$, $f\left(-0\right) = 0$, and $f'\left(-0\right) = \cdots = f^{(n-1)}\left(-0\right) = 0$. If the function $f\left(t\right)$ or its derivatives have a jump at the point $t = 0$, then we have to use the right limits $f\left(+0\right)$, $f'\left(+0\right)$, $\ldots$, $f^{(n-1)}\left(+0\right)$ for the initial values $f\left(0\right), f'\left(0\right), \ldots, f^{(n-1)}\left(0\right)$.

⚠ **Caution:** This distinction between a value at a point and the limit value when approaching that point is important. These limits refer to initial values from which the function starts for $t > 0$ and which then guarantee a continuous connection of the function at $t = 0$.

Examples 14.7 (Calculating the LT via the Derivative):

① For the function $f(t) = e^{at}$ the Laplace transform is to be determined. Using the derivative theorem for

$$f'(t) = a\, e^{at}, \quad f(0) = 1$$

resulting in

$$\mathcal{L}\left(a\, e^{at}\right) = s\,\mathcal{L}\left(e^{at}\right) - 1 \;\Rightarrow\; a\,\mathcal{L}\left(e^{at}\right) = s\,\mathcal{L}\left(e^{at}\right) - 1.$$

$$\Rightarrow \mathcal{L}\left(e^{at}\right) = \frac{1}{s-a} \quad \text{or} \quad e^{at} \;\circ\!\!-\!\!\bullet\; \frac{1}{s-a} \quad \text{for } s > a\,.$$

② The trigonometric functions $\sin(\omega t)$ and $\cos(\omega t)$ satisfy the differential equations

$$f''(t) + \omega^2 f(t) = 0 \quad \text{with} \quad \begin{cases} f(0) = 0, \quad f'(0) = \omega \quad \text{for } \sin(\omega t) \\[2mm] f(0) = 1, \quad f'(0) = 0 \quad \text{for } \cos(\omega t). \end{cases}$$

Their Laplace transforms are determined by the derivative theorem. The differential equation is

$$\mathcal{L}\left(f''(t)\right) + \omega^2\,\mathcal{L}\left(f(t)\right) = \mathcal{L}(0) = 0\,.$$

For the sine function, taking into account the initial conditions, we get

$$s^2\,\mathcal{L}\left(f(t)\right) - \omega + \omega^2\,\mathcal{L}\left(f(t)\right) = 0$$

$$\Rightarrow \quad \mathcal{L}\left(\sin(\omega t)\right) = \frac{\omega}{s^2 + \omega^2} \quad \text{or} \quad \sin(\omega t) \;\circ\!\!-\!\!\bullet\; \frac{\omega}{s^2 + \omega^2}\,.$$

③ Analogous to the sine function is the cosine function

$$s^2\,\mathcal{L}\left(f(t)\right) - s\cdot 1 + \omega^2\,\mathcal{L}\left(f(t)\right) = 0$$

$$\Rightarrow \quad \mathcal{L}\left(\cos\left(\omega t\right)\right) = \frac{s}{s^2 + \omega^2} \quad \text{or} \quad \cos\left(\omega t\right) \; \circ\!\!-\!\!\bullet \; \frac{s}{s^2 + \omega^2}.$$

④ A generalization of the previous example is the Laplace transform of the function $f\left(t\right) = \sin\left(\omega t + \varphi\right)$. This function also satisfies the differential equation

$$f''\left(t\right) + \omega^2 f\left(t\right) = 0 \, ,$$

but with the initial conditions $f\left(0\right) = \sin\varphi$ and $f'\left(0\right) = \omega\cos\varphi$. Using the Laplace transform, the differential equation becomes

$$\mathcal{L}\left(f''\left(t\right)\right) + \omega^2 \mathcal{L}\left(f\left(t\right)\right) = 0 \, ,$$

$$\hookrightarrow \quad s^2 \mathcal{L}\left(f\left(t\right)\right) - s\sin\varphi - \omega\cos\varphi + \omega^2 \mathcal{L}\left(f\left(t\right)\right) = 0.$$

$$\Rightarrow \quad \begin{cases} \boxed{\; \mathcal{L}\left(\sin\left(\omega t + \varphi\right)\right) = \dfrac{s\sin\varphi + \omega\cos\varphi}{s^2 + \omega^2} \;} \\[2em] \text{or} \quad \boxed{\; \sin(\omega t + \varphi) \; \circ\!\!-\!\!\bullet \; \dfrac{s\sin\varphi + \omega\cos\varphi}{s^2 + \omega^2}. \;} \end{cases}$$

With $\varphi = \psi + \frac{\pi}{2}$ we get

$$\begin{cases} \boxed{\; \mathcal{L}\left(\cos\left(\omega t + \psi\right)\right) = \dfrac{s\cos\psi - \omega\sin\psi}{s^2 + \omega^2} \;} \\[2em] \text{or} \quad \boxed{\; \cos\left(\omega t + \psi\right) \; \circ\!\!-\!\!\bullet \; \dfrac{s\cos\psi - \omega\sin\psi}{s^2 + \omega^2}. \;} \end{cases}$$

⑤ The hyperbolic functions $\sinh\left(\omega t\right)$ and $\cosh\left(\omega t\right)$ are solutions of the differential equation

$$f''\left(t\right) - \omega^2 f\left(t\right) = 0 \quad \text{with} \quad \begin{cases} f\left(0\right) = 0, \quad f'\left(0\right) = \omega & \text{for } \sinh\left(\omega t\right) \\ f\left(0\right) = 1, \quad f'\left(0\right) = 0 & \text{for } \cosh\left(\omega t\right). \end{cases}$$

$$\Rightarrow \mathcal{L}\left(f''\left(t\right)\right) - \omega^2 \mathcal{L}\left(f\left(t\right)\right) = 0 \, .$$

With the initial conditions we get for $\sinh\left(\omega t\right)$

$$s^2 \mathcal{L}\left(f\left(t\right)\right) - \omega - \omega^2 \mathcal{L}\left(f\left(t\right)\right) = 0$$

$$\Rightarrow \quad \mathcal{L}\left(\sinh\left(\omega t\right)\right) = \frac{\omega}{s^2 - \omega^2} \quad \text{or} \quad \sinh\left(\omega t\right) \circ\!\!-\!\!\bullet \frac{\omega}{s^2 - \omega^2}.$$

⑥ Analogously, we obtain

$$\mathcal{L}\left(\cosh\left(\omega t\right)\right) = \frac{s}{s^2 - \omega^2} \quad \text{or} \quad \cosh\left(\omega t\right) \circ\!\!-\!\!\bullet \frac{s}{s^2 - \omega^2}.$$

14.4 Methods of the Inverse Transform

The inverse Laplace transform, i.e. the determination of the time function $f(t)$ for a given image function $F(s)$, is the most difficult and complicated step. The complex inverse formula is rarely used because it requires detailed knowledge of function theory. However, in many cases, the following procedures will lead to success.

⊘ The Use of Tables

The most common method of obtaining the original function for a given image function is to use tables. The advantage of tables is that in many cases the calculation has already been done and is reflected in the corresponding function pairs. The following table shows the Laplace transform of many elementary functions.

Note: The Laplace transform of the function t^p for $p > -1$ contains the gamma function $\Gamma(p+1)$ which is defined for $p > -1$ by

$$\Gamma\left(p+1\right) := \int_0^{\infty} x^p\, e^{-x}\, dx.$$

The gamma function has the properties

$$(1)\ \Gamma\left(p+1\right) = p\Gamma\left(p\right), \qquad (2)\ \Gamma\left(1\right) = 1, \qquad (3)\ \Gamma\left(\tfrac{1}{2}\right) = \sqrt{\pi}.$$

From the properties (1) and (2) it holds $\Gamma(n+1) = n!$ for $n \in \mathbb{N}$. This means that the gamma function is a generalization of the factorial to positive real numbers.

Table 14.1: Laplace transformed elementary functions

Function	Transformed $F(s)$		Function	Transformed $F(s)$
$S(t)$	$\dfrac{1}{s}$ $\quad s > 0$		$e^{at} t^n$	$\dfrac{n!}{(s-a)^{n+1}}$
t	$\dfrac{1}{s^2}$ $\quad s > 0$		$\sin(\omega t + b)$	$\dfrac{(\sin b)\, s + \omega\,(\cos b)}{s^2 + \omega^2}$
t^n	$\dfrac{n!}{s^{n+1}}$ $\quad s > 0$		$\cos(\omega t + b)$	$\dfrac{(\cos b)\, s - \omega\,(\sin b)}{s^2 + \omega^2}$
e^{at}	$\dfrac{1}{s-a}$ $\quad s > a$		$e^{bt} \sinh(at)$	$\dfrac{a}{(s-b)^2 - a^2}$
$\sin(\omega t)$	$\dfrac{\omega}{s^2 + \omega^2}$		$e^{bt} \cosh(at)$	$\dfrac{s-b}{(s-b)^2 - a^2}$
$\cos(\omega t)$	$\dfrac{s}{s^2 + \omega^2}$		$t \sin(\omega t)$	$\dfrac{2\omega s}{(s^2 + \omega^2)^2}$
$t^p,\ p > -1$	$\dfrac{\Gamma(p+1)}{s^{p+1}},\ s > 0$		$t \cos(\omega t)$	$\dfrac{s^2 - \omega^2}{(s^2 + \omega^2)^2}$
$e^{at} \sin(\omega t)$	$\dfrac{\omega}{(s-a)^2 + \omega^2}$			
$e^{at} \cos(\omega t)$	$\dfrac{s-a}{(s-a)^2 + \omega^2}$			
$\sinh(at)$	$\dfrac{a}{s^2 - a^2}$			
$\cosh(at)$	$\dfrac{s}{s^2 - a^2}$			

⊙ The Method of Partial Fraction Decomposition

In many applications of the Laplace transform, the resulting image functions are rational functions

$$F(s) = \frac{Z(s)}{N(s)}$$

of the variable s. The numerator $Z(s)$ and the denominator $N(s)$ are polynomials of degree($Z(s)$) < degree($N(s)$), and the polynomials $Z(s)$ and $N(s)$ have real coefficients. Depending on the type of poles of $F(s)$, we have the following cases for a partial fraction decomposition: (1) image function with only simple real poles, (2) image function with multiple real poles, (3) image function with simple complex poles, (4) image function with multiple complex poles.

(1) **Image function with only simple real poles:** In this case, the image function $F(s)$ is

$$F(s) = \frac{a_1}{s - s_1} + \frac{a_2}{s - s_2} + \cdots + \frac{a_n}{s - s_n} \ .$$

Each term is transformed back with the correspondence

$$\frac{a_i}{s - s_i} \quad \bullet\!\!-\!\!\circ \quad a_i\, e^{s_i t} \ .$$

(2) **Image function with multiple real poles:** If s_0 is a real pole of multiplicity k, then the decomposition is

$$\frac{1}{(s - s_0)^k} = \frac{b_1}{s - s_0} + \frac{b_2}{(s - s_0)^2} + \cdots + \frac{b_k}{(s - s_0)^k} \ .$$

Using the damping theorem (see Section 14.6.2) and the correspondence $t^i \;\circ\!\!-\!\!\bullet\; \dfrac{i!}{s^{i+1}}$ we get

$$\frac{b_i}{(s - s_0)^i} \quad \bullet\!\!-\!\!\circ \quad b_i\, e^{s_0 t}\, \frac{t^{i-1}}{(i-1)!} \ .$$

The time functions are determined from these correspondences.

(3) **Image function with simple complex poles:** If $s_0 = a + i\,b$ is a complex zero of $N(s)$, then the complex conjugate $s_0^* = a - i\,b$ is also a zero (*Fundamental Theorem of Algebra*, see Volume 1).

$$(s - s_0)(s - s_0^*) = s^2 - s(s_0 + s_0^*) + s_0\, s_0^* = s^2 - 2\,a\,s + a^2 + b^2$$

$$\Rightarrow F(s) = \frac{c_1\, s + c_2}{s^2 - 2\,a\,s + a^2 + b^2} + P(s) = F_0(s) + P(s) \ ,$$

where $P(s)$ is the sum of the remaining partial fractions. Using the shift rule (see Section 14.6.2) we obtain the correspondence belonging to $F_0(s)$

$$f_0(t) = e^{at} \left[c_1 \cos(b\,t) + \tfrac{1}{b}(c_2 + a\, c_1) \sin(b\,t) \right] \ .$$

(4) **Image function with complex poles of order k:** If $s_0 = a + i\,b$ is a k-fold complex zero of $N(s)$, then s_0^* is also a k-fold zero and the approach in (3) must be modified as follows

$$\frac{1}{(s-s_0)^k} = \frac{c_1\, s + d_1}{(s^2 - 2\,a\,s + a^2 + b^2)^1} + \frac{c_2\, s + d_2}{(s^2 - 2\,a\,s + a^2 + b^2)^2} + \cdots + \frac{c_k\, s + d_k}{(s^2 - 2\,a\,s + a^2 + b^2)^k} \ .$$

The shift rule is applied to return each expression analogously to (3).

14.5 Applications of the Laplace Transform

Linear differential equations and systems of linear differential equations with initial conditions are elegantly solved using the Laplace transform. In this section, the RL-circuit and an electrical network are discussed as examples.

Application Example 14.8 (RL-Circuit, with MAPLE-Worksheet).

An RL-circuit is shown in Fig. 14.4. Initially at $t = 0$ a constant voltage U_0 is applied. We look for the time behavior of the current $I(t)$ for the initial condition $I(0) = I_0$.

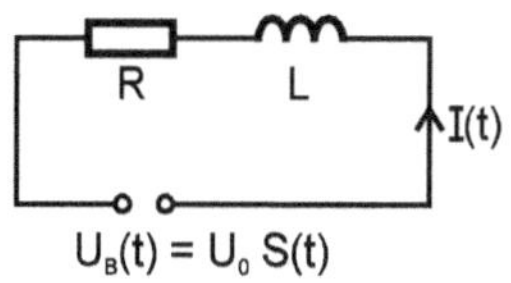

Figure 14.4. RL-Circuit

Applying the mesh rule, we get the model equation

$$LI'(t) + RI(t) = U_0\, S(t).$$

Step 1: We apply the Laplace transform to the differential equation. Because of the linearity, we get:

$$\mathcal{L}\left(LI'(t) + RI(t)\right) = \mathcal{L}\left(U_0\, S(t)\right)$$

$$\mathcal{L}\left(I'(t)\right) + \frac{R}{L}\mathcal{L}\left(I(t)\right) = \frac{1}{L}U_0\,\mathcal{L}\left(S(t)\right).$$

Step 2: We replace the Laplace transform of the derivative considering the initial condition and calculate the Laplace transform of the inhomogeneity.

$$s\,\mathcal{L}\left(I(t)\right) - I_0 + \frac{R}{L}\mathcal{L}\left(I(t)\right) = \frac{U_0}{L}\frac{1}{s}\,.$$

Step 3: We isolate the Laplace transform $\mathcal{L}\left(I(t)\right)$.

$$\left(s + \tfrac{R}{L}\right)\mathcal{L}\left(I(t)\right) = \frac{U_0}{L}\frac{1}{s} + I_0$$

$$\mathcal{L}\left(I(t)\right) = \frac{U_0}{L}\frac{1}{s + \frac{R}{L}}\frac{1}{s} + \frac{I_0}{s + \frac{R}{L}}\,.$$

Step 4: To the found Laplace transform we have to compute the corresponding time function. For a partial fraction decomposition

$$\mathcal{L}\left(I\left(t\right)\right) = \frac{U_0}{R}\left[\frac{1}{s} - \frac{1}{s + \frac{R}{L}}\right] + I_0\,\frac{1}{s + \frac{R}{L}}$$

we take into account the correspondences

$$\mathcal{L}(1) = \frac{1}{s}$$

$$\mathcal{L}(e^{-\frac{R}{L}t}) = \frac{1}{s + \frac{R}{L}}$$

and get

$$I\left(t\right) = \frac{U_0}{R}\left(1 - e^{-\frac{R}{L}t}\right) + I_0\,e^{-\frac{R}{L}t}.$$

Fig. 14.5 shows the slope of the current for different initial values of $I(0)$.

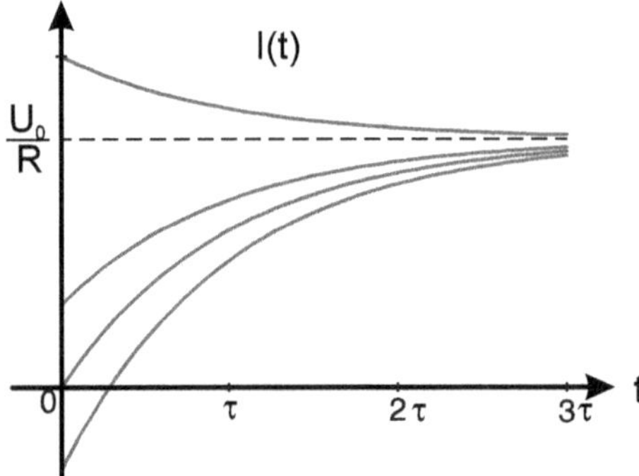

Figure 14.5. Slope of the current for different initial values I_0 with $\tau = R/L$

Discussion: Regardless of the initial value, the current $I(t)$ approaches the value $\frac{U_0}{R}$. The time constant is $\tau = \frac{R}{L}$. It determines how fast the final state is reached. $\qquad\square$

Application Example 14.9 (Electrical Network).

Given is the electrical network with constant voltage $U(t) = 120\,V$ discussed in the Example 14.1:

$$20\,(I_1(t) + I_2(t)) + 2\,I_1'(t) + 10\,I_1(t) = 120$$
$$-10\,I_1(t) + 20\,I_2(t) - 2\,I_1'(t) + 4\,I_2'(t) = 0$$

$$I_1(0) = I_2(0) = 0.$$

Using the same steps as for linear differential equations, the Laplace transform is also used for systems to find the solution directly:

Step 1: Apply the Laplace transform to the system.

$$
\begin{array}{rcccccccl}
20\,\mathcal{L}(I_1) & + & 20\,\mathcal{L}(I_2) & + & 2\mathcal{L}(I_1') & + & 10\,\mathcal{L}(I_1) & = & 120\,\mathcal{L}(1) \\
-10\,\mathcal{L}(I_1) & + & 20\,\mathcal{L}(I_2) & - & 2\mathcal{L}(I_1') & + & 4\mathcal{L}(I_2') & = & 0.
\end{array}
$$

Step 2: Replace the Laplace transform of the derivatives

$$
\begin{array}{rcl}
30\,\mathcal{L}(I_1) + 20\,\mathcal{L}(I_2) + 2\,\mathcal{L}(I_1)\cdot s & = & 120\cdot\dfrac{1}{s} \\
-10\,\mathcal{L}(I_1) + 20\,\mathcal{L}(I_2) - 2\,\mathcal{L}(I_1)\cdot s + 4\,\mathcal{L}(I_2)\cdot s & = & 0,
\end{array}
$$

because the initial conditions $I_1(0) = I_2(0) = 0$ disappear. The following linear system of equations is obtained for $\mathcal{L}(I_1)$ and $\mathcal{L}(I_2)$:

$$
\begin{array}{rcccl}
(30 + 2\,s)\,\mathcal{L}(I_1) & + & 20\,\mathcal{L}(I_2) & = & 120\cdot\dfrac{1}{s} \\
(-10 - 2\,s)\,\mathcal{L}(I_1) & + & (20 + 4\,s)\,\mathcal{L}(I_2) & = & 0.
\end{array}
$$

Step 3: Solve the system of linear equations:

$$\mathcal{L}(I_1) = \frac{60}{s\,(s+20)}; \qquad \mathcal{L}(I_2) = \frac{30}{s\,(s+20)}.$$

Step 4: Find the corresponding time function. A partial fraction decomposition for the two Laplace transforms gives

$$\mathcal{L}(I_1) = 3\left(\frac{1}{s} - \frac{1}{s+20}\right) \quad \text{and} \quad \mathcal{L}(I_2) = \frac{3}{2}\left(\frac{1}{s} - \frac{1}{s+20}\right)$$

$$\Rightarrow \quad I_1(t) = 3\left(1 - e^{-20\,t}\right) \quad \text{and} \quad I_2(t) = \frac{3}{2}\left(1 - e^{-20\,t}\right)$$

$$\Rightarrow I(t) = I_1(t) + I_2(t) = \frac{9}{2}\left(1 - e^{-20\,t}\right) \qquad \qquad \square$$

14.6 Properties of the Laplace Transform

This section introduces other properties of the Laplace transform that are used in many technical descriptions. These properties are often needed to apply the inverse Laplace transform to the correspondences given in the Table 14.1. In some cases, for example, the limit theorem can be used to make conclusions about the time function without calculating $f(t)$. The theorems are illustrated with examples.

14.6.1 Shift Theorem

The shift theorem makes a statement about the Laplace transform of a shifted time function $f(t - t_0)$: The function $f(t - t_0)$ is the function $f(t)$ shifted to the right by t_0 along the time axis (see Fig. 14.6).

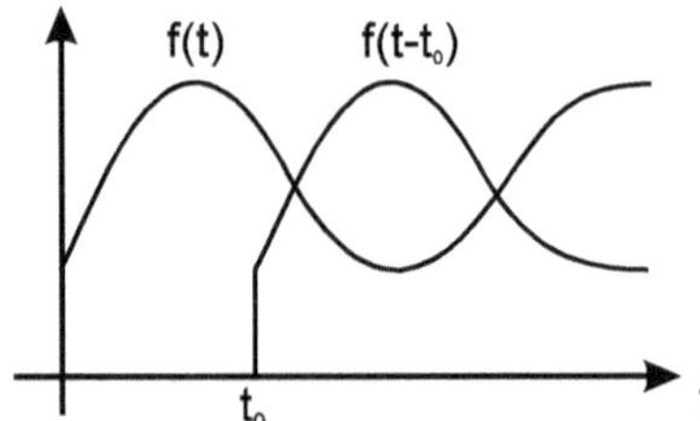

Figure 14.6. Original function $f(t)$ and shifted function $f(t - t_0)$

$$\mathcal{L}\left(f\left(t - t_0\right)\right) = \int_0^\infty f\left(t - t_0\right) e^{-st}\, dt = \int_{t_0}^\infty f\left(t - t_0\right) e^{-st}\, dt \ .$$

Substituting the integration variable $\tau = t - t_0$ gives

$$\mathcal{L}\left(f\left(t - t_0\right)\right) = \int_0^\infty f\left(\tau\right) e^{-s(\tau + t_0)}\, d\tau = e^{-s t_0} \int_0^\infty f\left(\tau\right) e^{-s\tau}\, d\tau$$

$$= e^{-s t_0}\, \mathcal{L}\left(f\left(t\right)\right) \ .$$

A shift of the time function $f(t)$ by t_0 leads to a multiplication of the image function $F(s)$ by the factor $e^{-s t_0}$:

Shift Theorem

If $F(s)$ is the Laplace transform of $f(t)$, then for $t_0 > 0$ we have

$$\mathcal{L}\left(f\left(t - t_0\right)\right) = e^{-s t_0} F(s) \ .$$

Correspondence: $f(t) \circ\!\!-\!\!\bullet F(s) \Rightarrow f(t - t_0) \circ\!\!-\!\!\bullet e^{-s t_0} F(s)$.

Examples 14.10:

① The step function shifted to the right by t_0: $S(t - t_0)$

$$S(t - t_0) = \begin{cases} 0 & \text{for } t < t_0 \\ 1 & \text{for } t > t_0 \end{cases}$$

According to the shift theorem of the Laplace transform, the image of the shifted step function is

$$F(s) = \mathcal{L}(S(t - t_0)) = e^{-st_0}\,\mathcal{L}(S(t)) = e^{-st_0}\,\frac{1}{s}\,,$$

which fits exactly with the Example 14.3 ④.

② Find the Laplace transform of a rectangular pulse with a pulse duration τ and pulse height A.

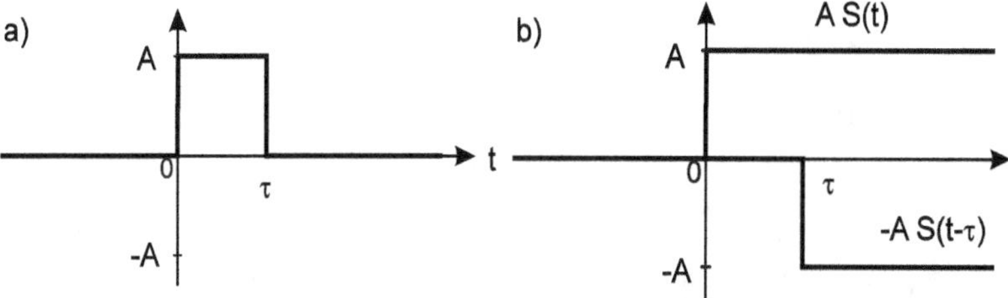

Figure 14.7. a) rectangular pulse and b) split into two step functions

The calculation of the Laplace transform of the rectangular function is based on the calculation of the transform of the step function from ①. Using Fig. 14.7 b), the rectangular signal is the difference between two step functions shifted with respect to each other:

$$f(t) = A\,S(t) - A\,S(t - \tau).$$

So we get

$$\mathcal{L}(f(t)) = A\mathcal{L}(S(t)) - A\mathcal{L}(S(t - \tau)) = A\,\frac{1}{s} - A\,e^{-s\tau}\,\frac{1}{s}$$

$$\Rightarrow \quad \mathcal{L}(f(t)) = \frac{A}{s}\left(1 - e^{-s\tau}\right). \qquad \square$$

⊘ Laplace transform of periodically continued functions

A time function $f(t)$ is created by *periodically continuing* the function

$$f_0(t) = \begin{cases} \text{defined} & \text{for } 0 \le t \le T \\ 0 & \text{else} \end{cases}$$

for $t \ge 0$. We want to find the Laplace transform $F(s)$ of the function $f(t)$ using the Laplace transform of $f_0(t)$.

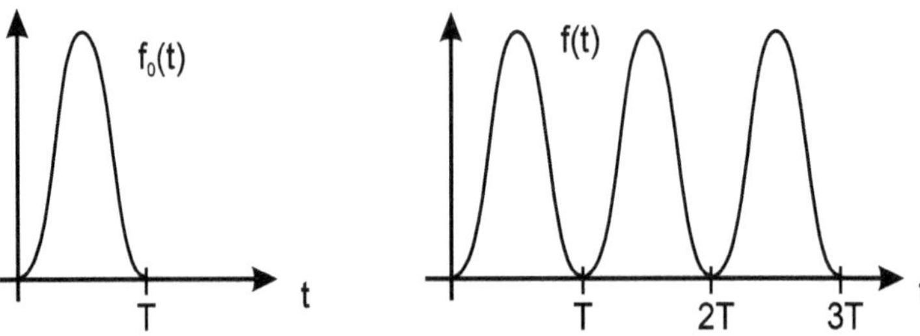

Figure 14.8. Periodically continued function

For the periodically continued time function $f(t)$ we have

$$f(t) = f_0(t)\, S(t) + f_0(t - T)\, S(t - T) + f_0(t - 2T)\, S(t - 2T)$$

$$+ f_0(t - 3T)\, S(t - 3T) + \cdots$$

With the given correspondence $f_0(t)\ \circ\!\!-\!\!\bullet\ F_0(s)$ and the displacement theorem, we get

$$F(s) = F_0(s)\left[1 + e^{-sT} + e^{-s\,2T} + e^{-s\,3T} + \cdots\right]$$

$$= F_0(s)\left[1 + e^{-sT} + \left(e^{-sT}\right)^2 + \left(e^{-sT}\right)^3 + \cdots\right].$$

By setting $q = e^{-sT}$ we can identify the expression in the brackets as a geometric series. Since $q = e^{-sT} < 1$, the series converges to $\frac{1}{1-q}$ and, therefore, the Laplace transform of the periodically continued time function $f(t)$ is

$$F(s) = F_0(s)\,\frac{1}{1 - e^{-sT}}.$$

Example 14.11. Calculate the Laplace transform of the periodic rectangular signal shown below.

$$f_0(t) = \begin{cases} 1 & \text{for } 0 \le t \le \frac{T}{2} \\[2mm] -1 & \text{for } \frac{T}{2} < t < T. \end{cases}$$

The function $f_0(t)$ can be described in the domain $0 \le t \le T$ as a superposition of shifted step functions

$$f_0(t) = \left[S(t) - S\left(t - \frac{T}{2}\right) \right] - \left[S\left(t - \frac{T}{2}\right) - S(t - T) \right]$$

$$= S(t) - 2S\left(t - \frac{T}{2}\right) + S(t - T).$$

Using the correspondence $S(t) \;\circ\!\!-\!\!\bullet\; \frac{1}{s}$ and the shift property, we get

$$\mathcal{L}(f_0(t)) = \frac{1}{s} - 2e^{-\frac{T}{2}s}\frac{1}{s} + e^{-Ts}\frac{1}{s} = \frac{1}{s}\left[1 - 2e^{-\frac{T}{2}s} + e^{-Ts} \right]$$

$$= \frac{1}{s}\left(1 - e^{-\frac{T}{2}s} \right)^2.$$

So

$$\mathcal{L}(f(t)) = \frac{1}{1 - e^{-sT}}\,\mathcal{L}(f_0(t)) = \frac{1}{s}\cdot\frac{\left(1 - e^{-\frac{T}{2}s}\right)^2}{1 - e^{-sT}}$$

$$= \frac{1}{s}\cdot\frac{\left(1 - e^{-\frac{T}{2}s}\right)^2}{\left(1 - e^{-\frac{T}{2}s}\right)\left(1 + e^{-\frac{T}{2}s}\right)} = \frac{1}{s}\cdot\frac{1 - e^{-\frac{T}{2}s}}{1 + e^{-\frac{T}{2}s}}. \qquad \square$$

14.6.2 Damping Theorem

The damping theorem gives a statement about the Laplace transform of a damped time function $f(t)\,e^{-at}$:

$$\mathcal{L}\left(e^{-at}f(t)\right) = \int_0^\infty e^{-st}e^{-at}f(t)\,dt = \int_0^\infty e^{-(s+a)t}f(t)\,dt = F(s+a):$$

Damping Property

If $F(s)$ is the Laplace transform of $f(t)$, then

$$\mathcal{L}\left(e^{-at}f(t)\right) = F(s+a).$$

Correspondence: $f(t) \;\circ\!\!-\!\!\bullet\; F(s) \Rightarrow e^{-at}f(t) \;\circ\!\!-\!\!\bullet\; F(s+a).$

Note: The constant a can be real or complex. However, a real damping of the time function $f(t)$ in the physical sense is only obtained for $a > 0$ or $\mathrm{Re}(a) > 0$. For $a < 0$ or $\mathrm{Re}(a) < 0$ the factor e^{-at} goes to infinity.

The Laplace transform of the time function $e^{-at} f(t)$ differs from the Laplace transform of $f(t)$ only in that s is replaced by $s + a$. **A shift of t_0 in the time domain causes a damping e^{-st_0} in the image domain (shift theorem) and vice verse a factor e^{-at} in the time domain causes a shift in the image domain (damping theorem).**

Examples 14.12:

① Find the Laplace transform of the time function $f(t) = e^{-3t} \sin(2t)$. With the correspondence

$$\sin(2t) \;\circ\!\!\!-\!\!\!\bullet\; \frac{2}{s^2 + 4}$$

and the damping property for $f(t)$, we obtain a shift of the argument by replacing s with $s + 3$

$$e^{-3t} \sin(2t) \;\circ\!\!\!-\!\!\!\bullet\; \frac{2}{(s+3)^2 + 4} = \frac{2}{s^2 + 6s + 13}\,.$$

② Given is the image function

$$F(s) = \frac{s+5}{s^2 + 2s + 10}\,.$$

To find the time function $f(t)$, we transform the image function into

$$F(s) = \frac{s+5}{(s+1)^2 + 9} = \frac{s+1}{(s+1)^2 + 3^2} + \frac{4}{3} \cdot \frac{3}{(s+1)^2 + 3^2}\,.$$

With the correspondences

$$\sin(\omega t) \;\circ\!\!\!-\!\!\!\bullet\; \frac{\omega}{s^2 + \omega^2} \qquad \text{and} \qquad \cos(\omega t) \;\circ\!\!\!-\!\!\!\bullet\; \frac{s}{s^2 + \omega^2}$$

and using the damping theorem, we get

$$f(t) = e^{-t} \cos(3t) + \frac{4}{3} \cdot e^{-t} \sin(3t)\,.$$

③ Find the time function for $F(s) = \dfrac{1}{(s+a)^2}$. From $\dfrac{1}{s^2} \;\bullet\!\!\!-\!\!\!\circ\; t$ we get with the damping theorem

$$f(t) = t\,e^{-at}\,. \qquad\qquad \square$$

14.6.3 Similarity Theorem

The similarity theorem makes a statement about the Laplace transform of stretched or compressed functions $f(at)$. The function $f(at)$ is obtained from the function $f(t)$ by stretching or compressing it along the time axis (see Fig. 14.9). If $0 < a < 1$, then $f(at)$ is stretched relative to f. If $a > 1$, however, then $f(at)$ is compressed relative to f.

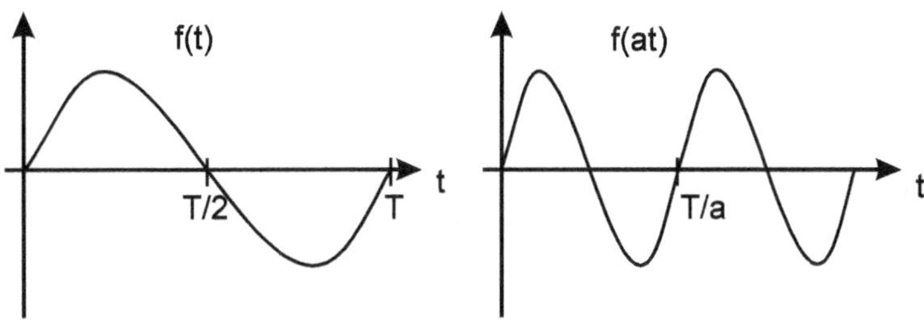

Figure 14.9. Function $f(t)$ and compressed function $f(at)$

Similarity Theorem

If $F(s)$ is the Laplace transform of $f(t)$, then

$$\mathcal{L}(f(at)) = \frac{1}{a} F\left(\frac{s}{a}\right) \qquad (a > 0).$$

Correspondence: $f(t) \; \circ\!\!\!-\!\!\!\bullet \; F(s) \Rightarrow f(at) \; \circ\!\!\!-\!\!\!\bullet \; \frac{1}{a} F\left(\frac{s}{a}\right).$

Proof:

$$\mathcal{L}(f(at)) = \int_0^\infty f(at)\, e^{-st}\, dt = \frac{1}{a} \int_0^\infty f(\tau)\, e^{-\frac{s}{a}\tau}\, d\tau = \frac{1}{a} F\left(\frac{s}{a}\right),$$

if the integral is calculated with the substitution $\tau = a \cdot t \;\left(dt = \frac{1}{a}\, d\tau\right).$ $\square$

14.6.4 Convolution Theorem

In applications often the problem arises: How to find the inverse transformation of an image function $F(s)$ which is a product of two image functions $F_1(s)$ and $F_2(s)$

$$F(s) = F_1(s) \cdot F_2(s)$$

with known correspondences

$$F_1(s) \; \bullet\!\!\!-\!\!\!\circ \; f_1(t) \quad \text{and} \quad F_2(s) \; \bullet\!\!\!-\!\!\!\circ \; f_2(t).$$

Then, the original function $f(t)$ is an integral over the time functions $f_1(t)$ and $f_2(t)$ of type

$$f(t) = \int_0^t f_1(\tau)\, f_2(t - \tau)\, d\tau\ ,$$

the so-called **convolution integral**. We introduce the notation $f(t) = (f_1 * f_2)(t)$ for this integral and call it the **convolution product**.

Convolution Theorem

If $F_1(s)$ and $F_2(s)$ are the Laplace transforms of $f_1(t)$ and $f_2(t)$, then the Laplace transform of the *convolution product*

$$(f_1 * f_2)(t) = \int_0^t f_1(\tau)\, f_2(t - \tau)\, d\tau$$

is given by $F_1(s) \cdot F_2(s)$:

$$\mathcal{L}(f_1 * f_2) = F_1(s) \cdot F_2(s)\,.$$

Correspondence:

$$\left.\begin{array}{l} f_1(t) \ \circ\!\!-\!\!\bullet\ F_1(s) \\ f_2(t) \ \circ\!\!-\!\!\bullet\ F_2(s) \end{array}\right\} \quad \Rightarrow \quad (f_1 * f_2)(t) \ \circ\!\!-\!\!\bullet\ F_1(s) \cdot F_2(s).$$

Note: In practical applications, we decompose $F(s)$ into a product $F(s) = F_1(s) \cdot F_2(s)$ of the image functions $F_1(s)$ and $F_2(s)$, where we know the associated time functions $f_1(t)$ and $f_2(t)$. Then $f(t) = (f_1 * f_2)(t)$.

Proof of the Convolution Theorem: We compute the Laplace transform of the convolution integral

$$(f_1 * f_2)(t) = \int_{\tau=0}^{\tau=t} f_1(\tau)\, f_2(t - \tau)\, d\tau\ :$$

$$\mathcal{L}(f_1 * f_2) = \int_{t=0}^{t=\infty} e^{-st} \left[\int_{\tau=0}^{\tau=t} f_1(\tau)\, f_2(t - \tau)\, d\tau \right] dt$$

$$= \int_{t=0}^{t=\infty} \left(\int_{\tau=0}^{\tau=t} e^{-st} f_1(\tau)\, f_2(t - \tau)\, d\tau \right) dt$$

$$= \int_{t=0}^{t=\infty} \left(\int_{\tau=0}^{\tau=\infty} e^{-st} f_1(\tau) \, f_2(t-\tau) \, S(t-\tau) \, d\tau \right) dt.$$

By adding the step function $S(t-\tau)$, the inner integral can be formally integrated from $\tau = 0$ to $\tau = \infty$, since $S(t-\tau) = 0$ for $\tau > t$. If the order of the integration is reversed, we get

$$\mathcal{L}(f_1 * f_2) = \int_{\tau=0}^{\tau=\infty} f_1(\tau) \left(\int_{t=0}^{t=\infty} e^{-st} f_2(t-\tau) \, S(t-\tau) \, dt \right) d\tau .$$

Applying the shift theorem to the inner integral

$$\int_{t=0}^{t=\infty} e^{-st} f_2(t-\tau) \, S(t-\tau) \, dt = e^{s\tau} F_2(s) ,$$

if $F_2(s)$ is the Laplace transform of $f_2(t) \cdot S(t) = f_2(t)$. This gives

$$\mathcal{L}(f_1 * f_2) = \int_{\tau=0}^{\tau=\infty} f_1(\tau) \, e^{-s\tau} F_2(s) \, d\tau$$

$$= F_2(s) \int_{\tau=0}^{\tau=\infty} f_1(\tau) \, e^{-s\tau} \, d\tau = F_2(s) \cdot F_1(s),$$

since the second integral is the Laplace transform of f_1. Using the integration variable τ instead of t is irrelevant for the definite integral. $\qquad \square$

Examples 14.13 (With MAPLE-Worksheet):

① Find the time function of

$$F(s) = \frac{1}{s(s-a)} = \frac{1}{s} \cdot \frac{1}{s-a} .$$

It is $\frac{1}{s} \circ\!\!-\!\!\bullet 1$ and $\frac{1}{s-a} \circ\!\!-\!\!\bullet e^{at}$. According to the convolution theorem, we obtain the correspondence

$$\frac{1}{s} \frac{1}{s-a} \bullet\!\!-\!\!\circ (1 * e^{at})(t) = \int_0^t 1 \cdot e^{a(t-\tau)} \, d\tau$$

$$= e^{at} \int_0^t e^{-a\tau} \, d\tau = \frac{1}{a} \left(e^{at} - 1 \right) .$$

② Find the time function $f(t)$ associated with

$$F(s) = \frac{s^2}{(s^2+\omega^2)^2} = \frac{s}{s^2+\omega^2} \cdot \frac{s}{s^2+\omega^2} .$$

It is

$$\frac{s}{s^2 + \omega^2} \quad \bullet\!\!-\!\!\circ \quad \cos(\omega t).$$

Using the convolution theorem, we obtain

$$\frac{s}{s^2 + \omega^2} \frac{s}{s^2 + \omega^2} \quad \bullet\!\!-\!\!\circ \quad f(t) = \cos(\omega t) * \cos(\omega t)$$

$$= \int_0^t \cos(\omega(t - \tau)) \cos(\omega\tau)\, d\tau \ .$$

Applying the addition theorem to $\cos(\omega t - \omega\tau)$ and using the formula $\sin x \cos x = \frac{1}{2}\sin 2x$ we obtain

$$f(t) = \frac{\sin(\omega t) + \omega t \cos(\omega t)}{2\,\omega} \ . \qquad\qquad \square$$

14.6.5 Limit Theorem

In some applications, only the behavior of the time function f at the beginning (i.e. for $t = 0$) and for large times t (i.e. for $t \to \infty$) is of interest and the exact timing in between is not needed. The start and end values of the time function can be obtained directly from the image function:

> **Limit Theorem**
>
> ---
>
> Let $F(s)$ be the image function of $f(t)$. Then:
>
> (1) **Initial value:** $f(0) = \lim_{t \to 0} f(t) = \lim_{s \to \infty}(s\,F(s))$.
>
> (2) **Final value:** $f(\infty) = \lim_{t \to \infty} f(t) = \lim_{s \to 0}(s\,F(s))$.

Proof: We start the proof of both equations with the derivative theorem

$$\mathcal{L}(f'(t)) = s\,F(s) - f(0) = \int_0^\infty f'(t)\,e^{-st}\,dt. \qquad (*)$$

(1)   We use the fundamental theorem of calculus

$$\lim_{s \to 0} \mathcal{L}(f'(t)) = \lim_{s \to 0} \int_0^\infty f'(t)\,e^{-st}\,dt = \int_0^\infty f'(t)\,\lim_{s \to 0} e^{-st}\,dt$$

$$= \int_0^\infty f'(t)\,dt = f(\infty) - f(0)\ .$$

If the limit value $s \to 0$ is applied to both sides of the equation $(*)$, then

$$\lim_{s \to 0} \left(s\, F\left(s\right) \right) - f\left(0\right) = f\left(\infty\right) - f\left(0\right) \Rightarrow f\left(\infty\right) = \lim_{s \to 0} \left(s\, F\left(s\right) \right).$$

(2) Because of

$$\lim_{s \to \infty} \mathcal{L}\left(f'\left(t\right)\right) = \lim_{s \to \infty} \int_0^\infty f'\left(t\right) e^{-st}\, dt = \int_0^\infty f'\left(t\right) \lim_{s \to \infty} e^{-st}\, dt = 0$$

we get with $(*)$

$$\lim_{s \to \infty} \left(s\, F\left(s\right) \right) - f\left(0\right) = 0 \ . \qquad\qquad \square$$

Examples 14.14:

① $F\left(s\right) = \dfrac{s}{s^2 + \omega^2}$. The corresponding time function has the initial value

$$f\left(0\right) = \lim_{s \to \infty} \left(s\, F\left(s\right) \right) = \lim_{s \to \infty} \frac{s^2}{s^2 + \omega^2} = 1$$

(see example 14.7 ③: $\cos\left(\omega t\right) \ \circ\!\!-\!\!\bullet \ \frac{s}{s^2+\omega^2}$).

② $F\left(s\right) = \dfrac{1}{s\left(s - 4\right)\left(s - 5\right)}$. The final value of the corresponding time function is

$$f\left(\infty\right) = \lim_{t \to \infty} f\left(t\right) = \lim_{s \to 0} \left(s\, F\left(s\right) \right) = \lim_{s \to 0} \frac{s}{s\left(s - 4\right)\left(s - 5\right)} = \frac{1}{20} \ . \qquad \square$$

14.7 Problems on the Laplace Transform

14.1 Calculate the Laplace transform of
a) $3\,e^{-4\,t}$ b) $2\,t^2$ c) $4\cos(5\,t)$ d) $\sin(\pi\,t)$ e) $\frac{-3}{\sqrt{t}}$

14.2 Calculate the Laplace transform of
a) $3\,t^4 - 2\,t^{\frac{3}{2}} + 6$ b) $5\sin(2\,t) - 3\cos(2\,t)$
c) $3\sqrt[3]{t} - 4\,e^{2\,t}$ d) $\frac{1}{t^2}$

14.3 Determine the Laplace transform of the time functions

a) $f(t) = \begin{cases} A & 0 \le t \le t_0 \\ A\,e^{-2\,(t-t_0)} & t > t_0 \end{cases}$ b) $f(t) = \begin{cases} 0 & \text{for } t < a \\ A & \text{for } a < t < b \\ 0 & \text{for } t > b \end{cases}$

c) $f(t) = \begin{cases} t & \text{for } 0 \le t \le 3 \\ 3 & \text{for } t > 3 \end{cases}$ d) $f(t) = \begin{cases} \sin t & \text{for } t \le \pi \\ 0 & \text{for } t > \pi \end{cases}$

14.4 Calculate

a) $\mathcal{L}^{-1}\left(\dfrac{5}{s+2}\right)$ b) $\mathcal{L}^{-1}\left(\dfrac{4\,s-3}{s^2+4}\right)$ c) $\mathcal{L}^{-1}\left(\dfrac{2\,s-5}{s^2}\right)$

d) $\mathcal{L}^{-1}\left(\dfrac{1}{s^k}\right)_{k>0}$ e) $\mathcal{L}^{-1}\left(\dfrac{4-5\,s}{s^{\frac{3}{2}}}\right)$ f) $\mathcal{L}^{-1}\left(\dfrac{1}{s^2+2\,s}\right)$

14.5 Apply the theorems of the Laplace transform to calculate the inverse
Laplace transform
a) $\mathcal{L}^{-1}\left(\dfrac{2\,s+3}{s^2-2\,s+5}\right)$ b) $\mathcal{L}^{-1}\left(\dfrac{e^{-2\,s}}{s^2}\right)$
c) $\mathcal{L}^{-1}\left(\dfrac{e^{-5\,s}}{s^4}\right)$ d) $\mathcal{L}^{-1}\left(\dfrac{1-e^{-2\,s}}{s^3}\right)$

14.6 Calculate the associated time function by partial fractional decomposition of the frequency function

a) $F(s) = \dfrac{2\,s^2-4}{(s-2)\,(s+1)\,(s-3)}$ b) $F(s) = \dfrac{3\,s+1}{(s-1)\,(s^2+1)}$
c) $F(s) = \dfrac{5\,s^2-15\,s+7}{(s+1)\,(s-2)^2}$ d) $F(s) = \dfrac{3\,s^2-7\,s+6}{(s-1)^3}$

14.7 Solve the differential equation

$$y''(t) + y(t) = S(t)$$

with initial conditions $y(0) = 1$, $y'(0) = 0$ using the Laplace transform.

14.8 Solve the 4th order differential equation

$$y^{(4)}(t) + 2\,y''(t) + y(t) = \sin(t)\,S(t)$$

with $y(0) = 1$, $y'(0) = -2$, $y''(0) = 3$, $y'''(0) = 0$ using the Laplace
transform.

Applications of the Laplace transform

14.9 R, L, U_B are connected in series with a S switch. The switch is initially closed and opened at $t = 0$: $I(t = 0) = I_0$. Solve the differential equation

$$R\,I(t) + L\,\frac{d}{dt}\,I(t) = 0 \quad , \quad I(0) = I_0$$

with the Laplace transform.

14.10 R, L, U_B are connected in series with a S switch. The switch is initially opened $(I(t = 0) = 0)$. How does $I(t)$ behave, if $U_B(t) = U_0 \sin(\omega t)$? Solve the differential equation

$$R\,I(t) + L\,\frac{d}{dt}\,I(t) = U_B(t)$$

with the Laplace transformation.

14.11 A particle moves on the x axis and is attracted to the origin 0 with a force proportional to the distance of 0. If the particle starts from rest at $x = 5\,cm$ and reaches the position $x = 2.5\,cm$ for the first time after 2 seconds, calculate
a) the position at any time t after the start,
b) the magnitude of its velocity at $x = 0$,
c) the acceleration.

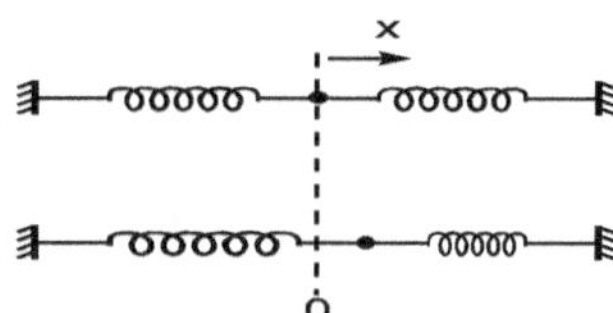

14.12 The position of a particle moving along the x-axis is determined by the equation

$$\frac{d^2}{dt^2}\,x(t) + 4\,\frac{d}{dt}\,x(t) + 8\,x(t) = 20\,\cos(2\,t)$$

If the particle starts from rest at $x = 0$, calculate x as a function of t.

14.13 A $2\,kg$ heavy mass hangs on a spring with spring constants $D = 200\,\frac{N}{m}$ at rest. Calculate the position of the mass at any time t when its damping force is 40 times the current velocity.

14.14 A mass on a vertical spring is subjected to forced vibration. The deflection from the rest position is described by

$$\frac{d^2}{dt^2}\,x(t) + 4\,x(t) = 8\,\sin(\omega t) \qquad (\omega > 0)\ .$$

If $x(0) = 0$ and $\dot{x}(0) = 0$, calculate x as a function of t and the period of the external force for which resonance occurs.

Index of Volume 1

Index

Homepage of the Book: Additional Material

iMath: iMath is an interactive maths application: In this pedagogically appealing app, easy-to-understand exercises from this book are solved in detail. The app can also be used for exam preparation. It can be launched directly from

$$http://www.imathhome.de/iMatheWeb$$

YouTube videos: On the Westermann YouTube channel, many of the topics covered in this book are explained in short videos in German. In the form of summaries, the most important aspects are summarized in a short and easily understandable way. The corresponding links to the videos can be found at

$$https://www.youtube.com/channel/UChzktnND8kk9pmwQmybSx-w$$

Additional material is available on the homepage of the book. All information, MAPLE procedures, MAPLE worksheets and additional chapters can be downloaded free of charge from

$$http://www.imathhome.de/books/mathe/start.htm$$

MAPLE-**Worksheets:** All the MAPLE worksheets for the problems and examples mentioned in the text. In particular, the worksheets for all visualizations can be found here.

Animations: All animations shown or described in the text are available on the homepage as Animated-Gif, so that they can be started directly in the browser.

$$http://www.imathhome.de/animations$$

Solutions to the problems: Complete solutions are given for all problems.